BIOCHEMISTRY PRIMER FOR EXERCISE SCIENCE

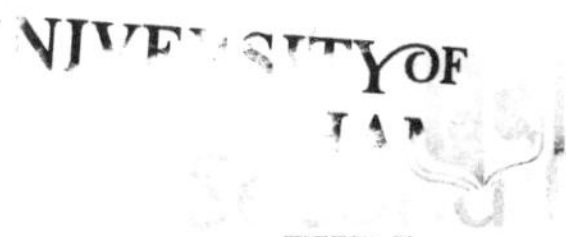

Michael E. Houston, PhD
Virginia Tech
Blacksburg, Virginia

Human Kinetics

Library of Congress Cataloging-in-Publication Data

Houston, Michael E., 1941-
Biochemistry primer for exercise science / Michael E. Houston.--2nd ed.
p. cm.
Includes bibliographical references and index.
ISBN 0-7360-3644-X
1. Biochemistry. 2. Exercise--Physiological aspects. 3. Metabolism. I. Title.

QP514.2 .H68 2001
612'.044--dc21 00-052983

ISBN: 0-7360-3644-X

Acquisitions Editor: Mike Bahrke; **Developmental Editor:** Jennifer Clark; **Assistant Editor:** Laurie Stokoe; **Copyeditor:** Judy Peterson; **Proofreader:** Pamela S. Johnson; **Indexer:** Marie Rizzo; **Permission Manager:** Courtney Astle; **Graphic Designer:** Fred Starbird; **Graphic Artist:** Yvonne Griffith; **Cover Designer:** Nancy Rasmus; **Art Manager:** Craig Newsom; **Illustrator:** Terry N. Hayden; **Printer:** Versa Press

Printed in the United States of America 10 9 8 7 6 5 4 3 2 1

Human Kinetics
Web site: **www.humankinetics.com**

United States: Human Kinetics, P.O. Box 5076, Champaign, IL 61825-5076
800-747-4457
e-mail: humank@hkusa.com

Canada: Human Kinetics, 475 Devonshire Road Unit 100, Windsor, ON N8Y 2L5
800-465-7301 (in Canada only)
e-mail: ray4@mnsi.net

Europe: Human Kinetics, P.O. Box IW14, Leeds LS16 6TR, United Kingdom
+44 (0) 113 278 1708
e-mail: humank@hkeurope.com

Australia: Human Kinetics, 57A Price Avenue, Lo
08 8277 1555
e-mail: liahka@senet.com.au

New Zealand: Human Kinetics, P.O. Box 105-231,
09-523-3462
e-mail: hkp@ihug.co.nz

Contents

Preface

A new understanding of human metabolism during exercise is coming from molecular approaches, and it is becoming increasingly clear that we must adopt such an approach when preparing undergraduate students to understand modern exercise physiology. This can be a daunting task, since many of the biochemistry courses that students of exercise science enroll in focus on a general understanding at the molecular level of plants and single-cell and multicellular organisms. A focus on human biochemistry is rare. Moreover, a modern biochemistry textbook may have more than 1,200 pages. It is expensive and heavy to carry about!

The primary rationale for the first edition of *Biochemistry Primer for Exercise Science* was to present the basic biochemistry that a student of exercise science would need to understand exercise at the molecular level and to enable him or her to read and understand the most recent journal articles. Many reviewers of the book understood this, but many others, including students, wanted an approach that actually integrated biochemistry and exercise physiology. The second edition does this.

The second edition still provides the basic molecular information needed to understand modern exercise science, but it also incorporates many exciting changes. You will see a revised organization, new exercise examples and applications throughout the book, and a more user-friendly approach. Key points are emphasized throughout each chapter. We have highlighted new and important words and concepts. We have added challenging questions to each chapter. We have included a comprehensive reference list. In addition, there is a glossary and a list of abbreviations.

This new edition represents the wisdom gained over 30 years from teaching exercise science to undergraduate and graduate students in Canada, Denmark, and the United States. Your comments and suggestions will be graciously received.

Acknowledgments

As scientists, we must recognize that our inspiration, knowledge and ideas can be traced to notable contacts with teachers, colleagues, collaborators, and students. I am honored to recognize a group of people who have had a major impact on my career in exercise science. To Bengt Saltin, a colleague and friend who inspired and awed me with the comprehensiveness of his insights. To Howie Green, a colleague and friend who taught me tenacity and critical thinking. To Richard Hughson, who has been such a sound mentor and friend to a host of successful graduate students. To Jim Stull, who gave me an opportunity to work with him at the University of Texas Southwestern Medical Center at Dallas. Finally, I owe a tremendous debt to the intelligent young Kinesiology students at Waterloo, the eager Exercise Science students at UBC, and the enthusiastic Hokie grad students at Virginia Tech.

PART I

The Major Players

Everything you do, everything you are, and everything you become depend on the actions of thousands of different proteins. These are large molecules composed of individual units called amino acids. Twenty different amino acids are combined in various sequences to make proteins—as few as 100 amino acids form a small protein. The order in which the amino acids are arranged in a specific protein is critical. Information for assembling a protein in a cell is stored in the nucleus in discrete elements called genes, which are sections of the DNA molecules that make up our chromosomes. Four different bases comprise the information component of DNA. The sequence of these bases in DNA molecules spells out a sequence of amino acids in proteins. In chapter 1 we look at the amino acid building blocks, their properties as acids and bases, how they are assembled to make proteins, and protein structure. Chapter 2 focuses on the body's major protein action molecules, the enzymes. In chapter 3 we consider the structure of nucleotides and how they are assembled to make the master blueprint, DNA, and its working partner, RNA.

CHAPTER 1

Amino Acids, Peptides, and Proteins

Proteins are the molecules responsible for what happens in cells and organisms. We can consider them the body's action molecules: they can be enzymes that catalyze all the chemical reactions in the body; receptor molecules inside cells, in membranes, or on membranes that bind specific substances; contractile proteins involved in the contraction of skeletal, smooth, and heart muscle; or transport molecules that move substances in the blood, within cells, and across membranes. Proteins are also parts of bones, ligaments, and tendons, and in the form of antibodies or receptors on lymphocytes, they help protect us from disease.

Twenty different amino acids are used to make proteins. The genes in our cell nuclei contain the information needed to specify which amino acids are used in making a protein and in what order. The genes in chromosomes are segments of DNA, large molecules containing four different bases. The sequence of bases in a gene spells out the sequence of amino acids for a protein. Because of the huge size of DNA molecules, the base sequence in a gene is copied by making a new molecule, messenger ribonucleic acid (mRNA). The copying process is known as **transcription**. The mRNA information is then used to order the binding of the 100 or more specific amino acids to make a protein. This step, known as **translation**, takes place outside of the nucleus.

Proteins are continually being turned over: Old protein molecules are broken down to their constituent amino acids in a process known as **degradation**, and new protein molecules are synthesized. This continual synthesis and degradation of proteins in cells is known as protein turnover (see figure 1.1). Amino acids released when a protein molecule is broken down can be reutilized. We can express the rate of turnover of an individual protein as the time taken for one half of the protein molecules to be replaced, that is, the **half-life**. Some proteins have a short half-life, measured in minutes or hours, while others have a half-life that may be expressed in weeks. Research examining the role of specific exercise training, nutrition, and

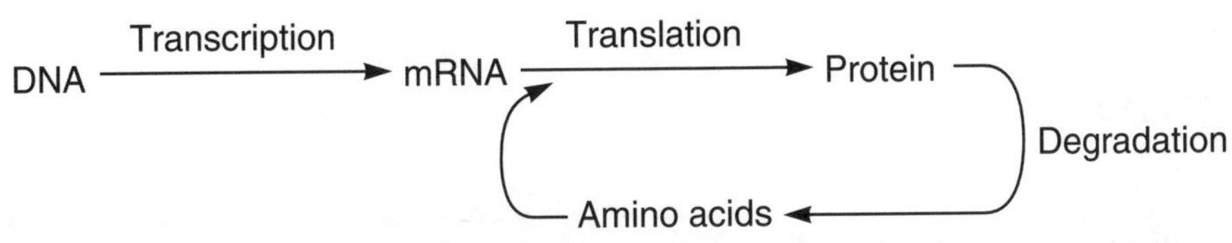

Figure 1.1 A simple outline of the steps involved in the turnover of body proteins.

ergogenic aids on the turnover of skeletal muscle proteins is thriving, both from the perspective of basic science and the practical implications of this for sport and health.

The Nature of Amino Acids

Figure 1.2 illustrates what is typically shown for the general structure of an amino acid. In this figure, each amino acid has a carboxyl group, an amino group, a carbon atom identified as the alpha (α) carbon (also known as carbon 2), and a variable group known as the side chain and indicated by the letter R. In fact, 19 of the 20 common amino acids have this general structure. What makes each amino acid unique is its side chain or R group. Each amino acid has (a) a specific side chain (designated as an R group), (b) a name, (c) a three-letter designation, and (d) a single capital letter to represent it. Differences in the properties of the 20 common amino acids are based on their side chains. The general amino acid structure depicted in figure 1.2 never occurs exactly as shown because of the acid-base properties of the amino and carboxyl groups. Figure 1.3 shows the general structure of an amino acid as it would exist at pH 7. This is often described as the **zwitterion** form. Figure 1.4 shows each of the 20 amino acids, including the name, three-letter short form, single letter identification, and the overall structural formula with the specific side chain. Amino acids are typically organized on the basis of the side chain structure and how this will interact with water. Charged polar and uncharged polar side chains will interact with polar water molecules.

KEY POINT

The 20 amino acids used to make proteins are the same ones used by all animal and plant life. Diversity in the structure of amino acids helps produce the diversity in the nature and function of proteins.

Acid-Base Properties of Amino Acids

The dissociation of the general weak **acid** (HA) is written as

$$\underset{\text{acid}}{HA} \longleftrightarrow \underset{\text{proton}}{H^+} + \underset{\text{conjugate base}}{A^-}$$

Because weak acids do not dissociate completely, we can write an equilibrium expression for the reversible dissociation using square brackets to represent concentration:

$$\frac{[H^+]\,[A^-]}{[HA]} = K_a$$

The stronger the acid (HA), the more it is dissociated, the more the equilibrium is shifted to the right, and the larger the value for K_a, the acid dissociation constant. Therefore, the larger the numerical value for K_a, the stronger the acid.

Suppose we dissolve HA in a solution containing the sodium salt of A^- (i.e., NaA) such that at

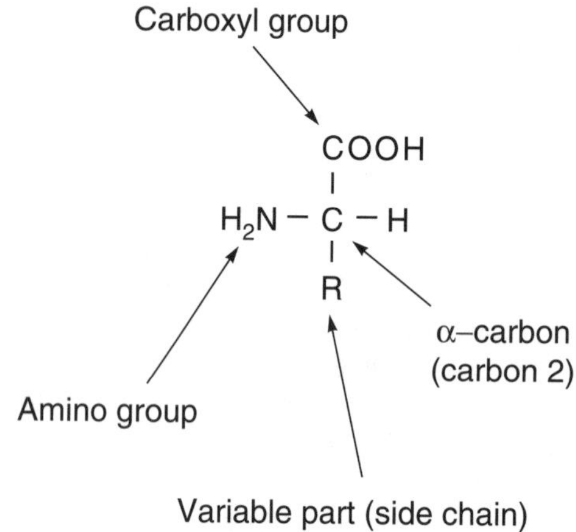

Figure 1.2 The general structure of an amino acid showing the amino group and the carboxyl group attached to the same carbon atom.

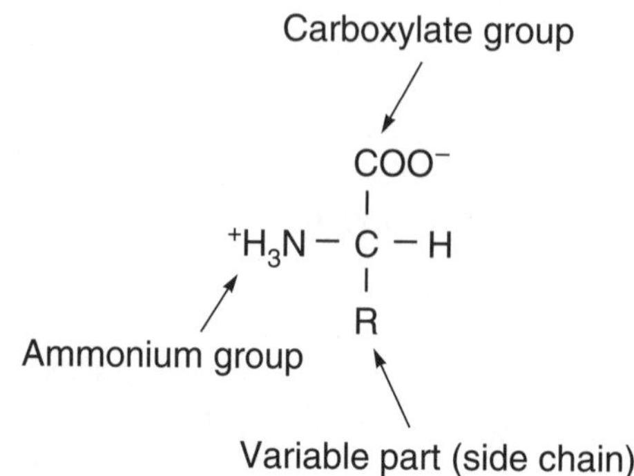

Figure 1.3 The general structure of an amino acid at pH 7.

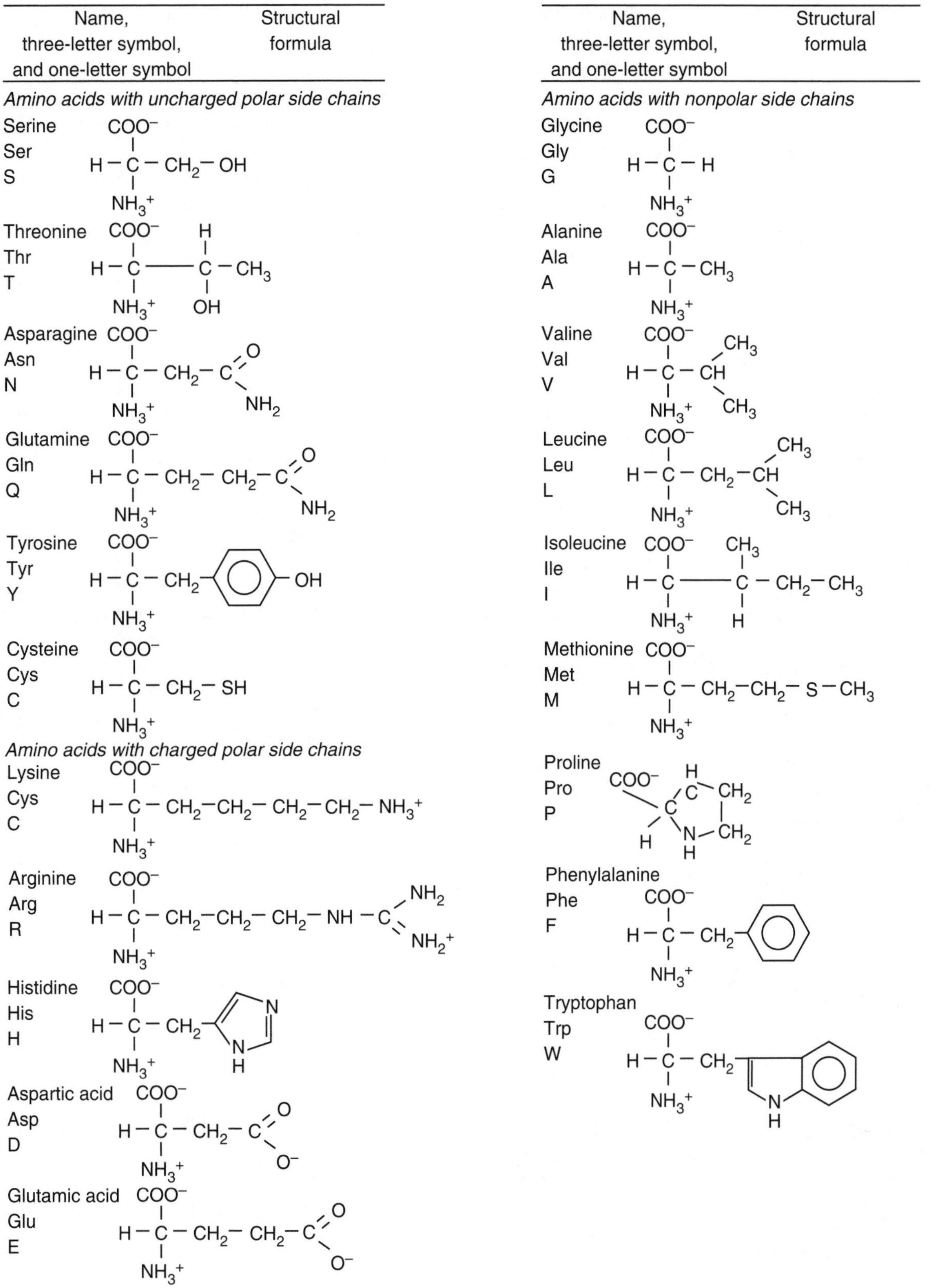

Name, three-letter symbol, and one-letter symbol	Structural formula
Amino acids with uncharged polar side chains	
Serine Ser S	$H-C(COO^-)(NH_3^+)-CH_2-OH$
Threonine Thr T	$H-C(COO^-)(NH_3^+)-C(H)(OH)-CH_3$
Asparagine Asn N	$H-C(COO^-)(NH_3^+)-CH_2-C(=O)NH_2$
Glutamine Gln Q	$H-C(COO^-)(NH_3^+)-CH_2-CH_2-C(=O)NH_2$
Tyrosine Tyr Y	$H-C(COO^-)(NH_3^+)-CH_2-$ (ring) $-OH$
Cysteine Cys C	$H-C(COO^-)(NH_3^+)-CH_2-SH$
Amino acids with charged polar side chains	
Lysine Cys C	$H-C(COO^-)(NH_3^+)-CH_2-CH_2-CH_2-CH_2-NH_3^+$
Arginine Arg R	$H-C(COO^-)(NH_3^+)-CH_2-CH_2-CH_2-NH-C(NH_2)=NH_2^+$
Histidine His H	$H-C(COO^-)(NH_3^+)-CH_2-$ (ring: N, N–H)
Aspartic acid Asp D	$H-C(COO^-)(NH_3^+)-CH_2-C(=O)O^-$
Glutamic acid Glu E	$H-C(COO^-)(NH_3^+)-CH_2-CH_2-C(=O)O^-$

Name, three-letter symbol, and one-letter symbol	Structural formula
Amino acids with nonpolar side chains	
Glycine Gly G	$H-C(COO^-)(NH_3^+)-H$
Alanine Ala A	$H-C(COO^-)(NH_3^+)-CH_3$
Valine Val V	$H-C(COO^-)(NH_3^+)-CH(CH_3)CH_3$
Leucine Leu L	$H-C(COO^-)(NH_3^+)-CH_2-CH(CH_3)CH_3$
Isoleucine Ile I	$H-C(COO^-)(NH_3^+)-C(CH_3)(H)-CH_2-CH_3$
Methionine Met M	$H-C(COO^-)(NH_3^+)-CH_2-CH_2-S-CH_3$
Proline Pro P	COO^-, H, C, CH_2, C, H, N–H, CH_2 (ring)
Phenylalanine Phe F	$H-C(COO^-)(NH_3^+)-CH_2-$ (ring)
Tryptophan Trp W	$H-C(COO^-)(NH_3^+)-CH_2-$ (ring: N–H)

Figure 1.4 Structures of the 20 amino acids used to make proteins, with their names, three-letter abbreviations, and single-letter identifiers.

equilibrium, the concentration of undissociated HA is equal to the concentration of A^-. Now the above equilibrium equation is simplified to

$$[H^+] = K_a$$

because the other terms ([HA] and [A^-]) cancel out since they have the same value.

Now we will take the negative logarithm of both sides of this equation; that is,

$$-\log [H^+] = -\log K_a,$$

or

$$pH = pK_a.$$

By definition, pH is the negative logarithm of [H^+]. Thus pK_a is the negative logarithm of K_a, the acid dissociation constant.

The pK_a for an acid is the pH of a solution where the acid is one-half dissociated, that is, 50% of the molecules are dissociated and 50% are not dissociated. Therefore, the smaller the value of pK_a for an acid, the stronger the acid.

All amino acids have at least one acid group (proton donor) and one basic group (proton acceptor). Amino acids may have both acid and base properties; that is, they are **amphoteric**.

In amino acids, the major acid groups are the carboxyl (–COOH) and the ammonium ($-NH_3^+$) groups. These protonated forms can each give up a proton. The major base groups are the carboxylate ($-COO^-$) and amino ($-NH_2$) groups, which are unprotonated and can accept a proton.

Now let us look at equations describing the ionization of the protonated (acid) forms of the groups:

$$-COOH \longleftrightarrow -COO^- + H^+$$

will have a K_a value and hence a pK_a.

$$-NH_3^+ \longleftrightarrow -NH_2 + H^+$$

will also have a K_a and thus a pK_a.

The carboxyl group (–COOH) is a stronger acid than ammonium ($-NH_3^+$) and will have a lower pK_a value (or its K_a value will be larger). Conversely, the amino group will be a better proton acceptor (base) than the $-COO^-$ and will have a larger pK_a.

The Henderson-Hasselbach equation describes the relationship between the pH of a solution and the pK_a of the weak acid group, and the concentration ratio of the acid (HA) and its conjugate base (A^-).

$$pH = pK_a + \log \frac{[\text{conjugate base}]}{[\text{acid}]}$$

This equation predicts the main form of an ionizable group in a solution at a known pH if we know the pK_a for the ionizable group.

Most of the amino acids have two ionizable groups, each with a pK_a, usually identified as pK_1 (the lower value) and pK_2 (the higher value). Figure 1.5 shows the structure of the amino acid alanine when it is a zwitterion. In this form, alanine has no net charge, but it has one positive and one negative charge. The pH where a molecule has an equal number of positive and negative charges (and therefore no net charge) is its **isoelectric point**, designated pI. For amino acids like alanine, the pI is one half the sum of pK_1 and pK_2.

Monoaminodicarboxylic amino acids, such as aspartic acid and glutamic acid (see figure 1.4), have two groups that can be carboxyl or carboxylate and only one group that can be amino or ammonium. We call such amino acids acidic amino acids, and there will be three pK_a values for these, identified from lowest to highest as pK_1, pK_2, and pK_3. As shown in figure 1.6, the isoelectric point (pI) of aspartic acid will be the pH where each molecule has one positive and one negative charge, yet no net charge. The actual value for the pI will be one half the value of pK_1 and pK_2, that is, 2.8.

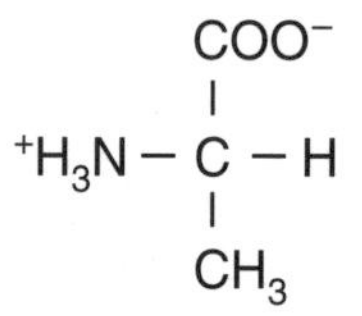

COO^-
|
$^+H_3N - C - H$
|
CH_3

Figure 1.5 The structure of the amino acid alanine at its isoelectric point.

COO^- ← $pK_1 = 1.9$
|
$^+H_3N - C - H$
|
$pK_3 = 9.6$ ↗ CH_2
|
COOH ← $pK_2 = 3.7$

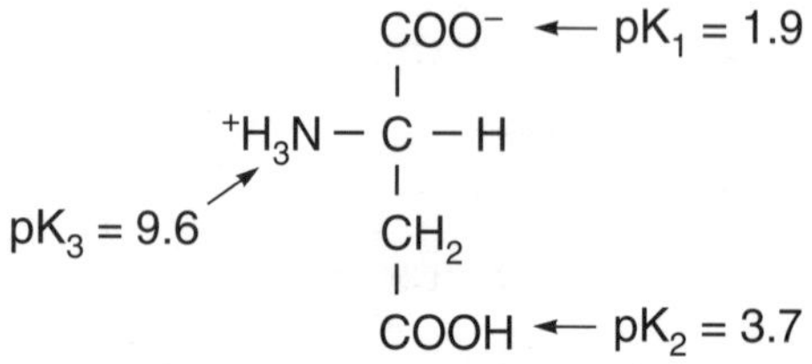

Figure 1.6 The structure of aspartic acid at its isoelectric point. The pK_a values for the three dissociable groups are shown in order of increasing magnitude.

Amino acids such as lysine (see figure 1.4) have two amino groups and only one carboxyl group and are often called basic amino acids. Figure 1.7 shows the structure of lysine at its isoelectric point. Notice that lysine has no net charge due to an equal number of positive and negative groups. The pI value for lysine will be one half the sum of pK_2 and pK_3, that is, 9.8.

In addition to the major ionizable groups discussed so far (i.e., the amino and carboxyl groups attached to the alpha carbon and side chain), others can be quite important (see figure 1.8). Histidine is often found at the active site of many enzymes, for it influences their catalytic ability.

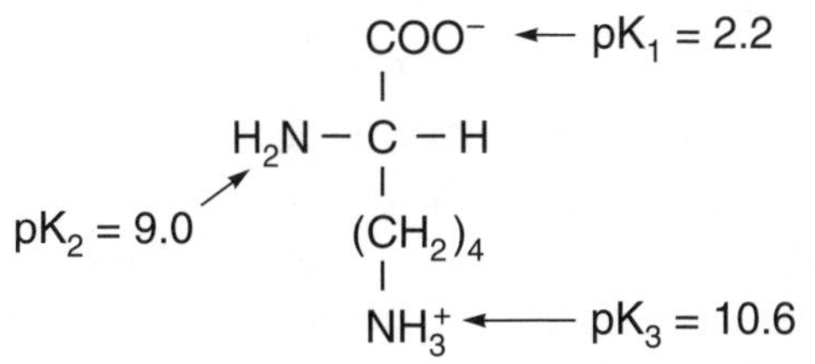

COO^- ← $pK_1 = 2.2$

$H_2N - C - H$

$pK_2 = 9.0$

$(CH_2)_4$

NH_3^+ ← $pK_3 = 10.6$

Figure 1.7 The structure of lysine at its isoelectric point. The pK_a values for the dissociable groups are shown.

KEY POINT

The pH of intracellular and extracellular fluids must be carefully regulated since even a small change in pH can add or remove a proton from an amino acid. Such a change may alter the structure and thus the function of proteins, altering our metabolism.

Stereoisomerism of Amino Acids

As shown in figure 1.4, all amino acids except glycine have four different groups attached to the alpha (α) carbon. Because of this, the α carbon is a chiral center, and the amino acid is chiral or asymmetric, with two different ways of arranging these groups; that is, the amino acid has two different configurations. Figure 1.9 shows the groups around the alpha carbon to be three-dimensional; the dashed lines mean that the bonds are going into the paper, and the wedges mean that the bonds are coming out of the plane of the paper toward you. When the carboxylate group is at the top and going into the paper, the L and D refer to the position of the ammonium group, that is, on the left side (L or levo) or right side (D or dextro). These **stereoisomers** (or space

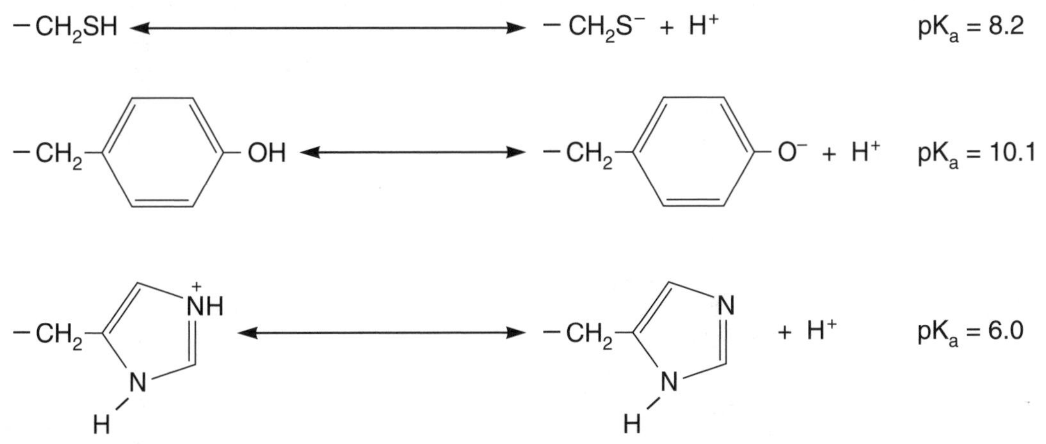

$-CH_2SH \longleftrightarrow -CH_2S^- + H^+$ $pK_a = 8.2$

$-CH_2-C_6H_4-OH \longleftrightarrow -CH_2-C_6H_4-O^- + H^+$ $pK_a = 10.1$

$-CH_2$–imidazolium (NH$^+$, N–H) $\longleftrightarrow$ $-CH_2$–imidazole (N, N–H) $+ H^+$ $pK_a = 6.0$

Figure 1.8 Equations to show the acid dissociation characteristics for the side chains of the amino acids cysteine (top), tyrosine (middle), and histidine (bottom). The numerical value for the pK_a for each dissociation is shown.

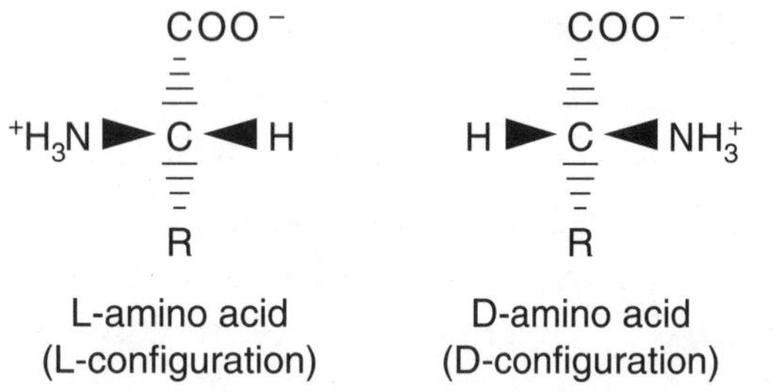

Figure 1.9 The two stereoisomeric forms for the general structure for an amino acid.

isomers) are also enantiomers—pairs of molecules that are nonsuperimposable mirror images of each other. They may be compared to a right and left hand. When you hold a left hand to a mirror, a right hand is the image. Similarly, holding a D-amino acid to a mirror gives an L-amino acid as the image. All naturally occurring amino acids are in the L-configuration.

Characteristics of Peptides

Peptides are formed when amino acids join together via their ammonium and carboxylate groups in a specialized form of the amide bond known as a **peptide bond**. Because the body's amino acids are in an environment with a pH around 7, the amino group is protonated and is an ammonium group, whereas the carboxyl group is unprotonated and is a carboxylate group. Figure 1.10 shows how a peptide is formed from two amino acids. Combination of the ammonium and carboxylate groups yields the peptide bond, and a water molecule is eliminated. The peptide bond is rigid and planar. Peptides are drawn by convention starting with the free ammonium group and ending with the free carboxylate group. We describe these as the N-terminus and the C-terminus, respectively.

The amino acids in a peptide are known as amino acid residues. The prefixes used to describe the number of amino acid residues in a peptide are di—two, tri—three, tetra—four, penta—five, hexa—six, hepta—seven, octa—eight, nona—nine, and deca—ten. Thus a nonapeptide consists of nine amino acid residues. An oligopeptide is one that contains roughly 10 to 20 amino acid residues. The term **polypeptide** refers to a large peptide that contains more than 20 amino acids; indeed some polypeptides have more than 4,000 amino acid residues. A protein and a polypeptide can be the same thing if there is only one polypeptide in the molecule. However, many proteins contain more than one polypeptide chain. In this case the protein is not a polypeptide, but a molecule containing two or more specific polypeptides.

KEY POINT

Many important hormones are peptides, such as insulin, glucagon, and growth hormone. Growth factors have specific, potent effects on tissues, and these are peptides. The hypothalamus, a key regulator of the overall function of our body, releases a number of peptide factors that control the secretions from other glands, particularly the pituitary gland.

Peptides are linear (as opposed to branched) molecules. This is a natural consequence of the linear sequence of bases within a gene in DNA, which is responsible for specifying the linear sequence of amino acids in the polypeptide chain. The **primary structure** of a peptide is the sequence of amino acids starting from the N-terminus.

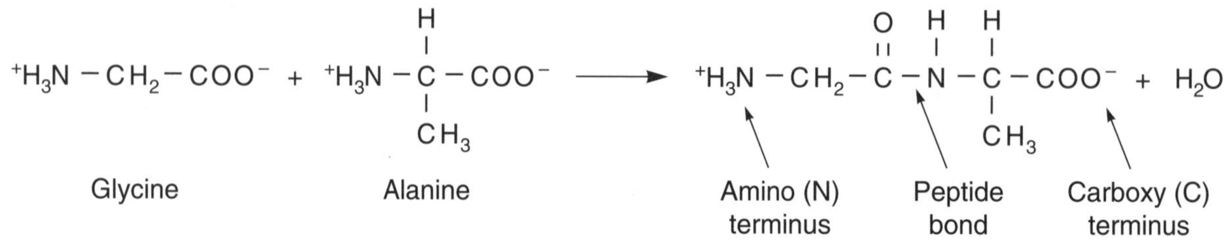

$$^{+}H_3N-CH_2-COO^- + {}^{+}H_3N-\underset{CH_3}{\overset{H}{C}}-COO^- \longrightarrow {}^{+}H_3N-CH_2-\overset{O}{\overset{\|}{C}}-\overset{H}{N}-\underset{CH_3}{\overset{H}{C}}-COO^- + H_2O$$

Glycine | Alanine | Amino (N) terminus | Peptide bond | Carboxy (C) terminus

Figure 1.10 Combining two amino acids, glycine and alanine, to form the dipeptide glycyl-alanine.

Structure of Proteins

Proteins are composed of one or more polypeptide chains. Many also contain other substances such as metal ions (e.g., hemoglobin contains iron), carbohydrates (e.g., glycoproteins contain sugars attached to amino acids; these can be found on cell membranes, pointing into the fluid surrounding cells), and fat or lipid (e.g., blood lipoproteins transport fat such as cholesterol and triglyceride). The precise biological structure of a protein, and hence its function, is determined by the amino acids it contains.

As mentioned, the primary structure of a protein (or peptide) is the sequence of amino acids starting from the N- (amino) terminus. With 20 different amino acids, the different possible sequences for even a small polypeptide are enormous. However, a specific protein in your body must have its proper amino acid sequence. Even replacing one amino acid with another could make a protein totally useless and possibly dangerous. For example, sickle cell anemia, characterized by short-lived erythrocytes of unusual shape, results from the replacement of just one amino acid (valine) with another (glutamic acid) in one chain of the hemoglobin molecule. As shown in figure 1.4, valine has an uncharged hydrophobic side chain, whereas the side chain of glutamic acid carries a negative charge. Even this modest change can have a profound influence on the properties of the hemoglobin molecule.

Bonds and Interactions Responsible for Protein Structure

Proteins have a very specific structure in the cell that is essential for a protein to perform its unique function. As we have seen, the primary structure, or the sequence of amino acids in the polypeptide chain, is maintained by peptide bonds. These are strong, rigid bonds joining the individual amino acids together in the protein. However, the three-dimensional structure of proteins is very important to their function. The three-dimensional structure is maintained by a variety of strong and weak bonds. These are illustrated in figure 1.11.

Disulfide bonds are covalent bonds that join together the side chains of two cysteine residues ($-CH_2SH$) in the same or different polypeptides, resulting in S–S (disulfide) bonds and the loss of

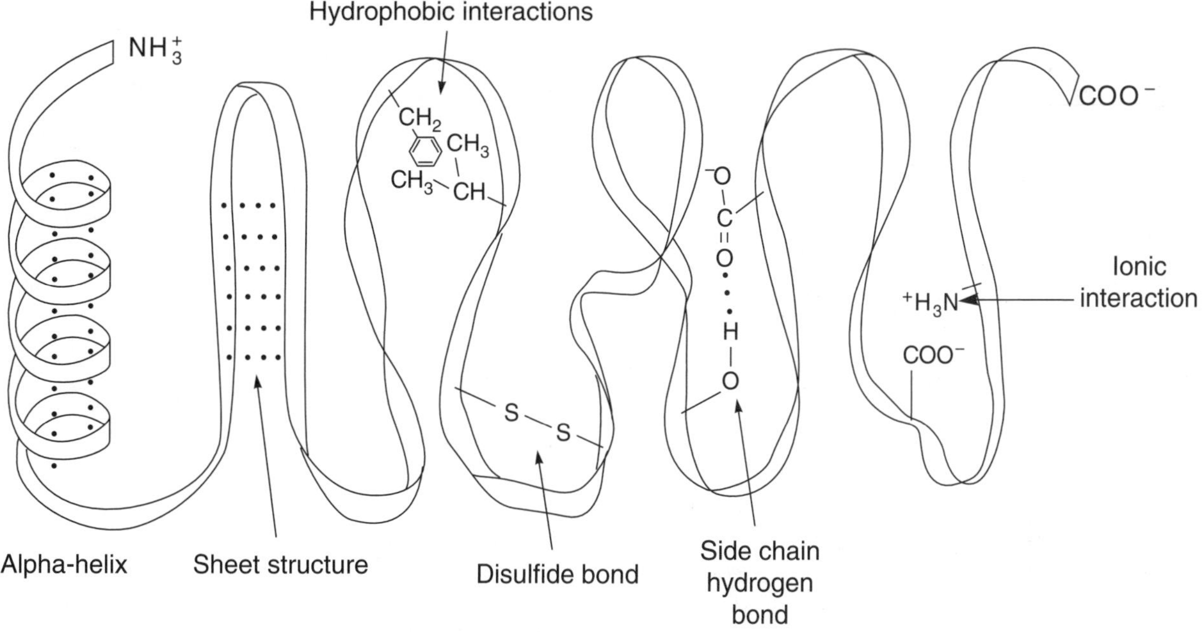

Figure 1.11 The ribbon shows the peptide backbone of a hypothetical protein, indicating structural features, such as the alpha-helix and beta-pleated sheet characteristic of protein secondary structure, and the hydrogen bonding (dotted line), ionic interactions, hydrophobic interactions, and disulfide bonds, which maintain the three-dimensional conformation of proteins.

two hydrogen atoms. Disulfide bonds in the same polypeptide chain are intramolecular bonds, whereas those joining different polypeptide chains are intermolecular bonds. For example, the polypeptide hormone insulin consists of two peptide chains held together by two intermolecular disulfide bonds.

Hydrogen bonds are weak electrostatic attractions between an electronegative element, such as oxygen or nitrogen, and a hydrogen atom that is covalently bonded to another electronegative element. Hydrogen bonds between individual water molecules, for example, are responsible for some of the unique properties of water. Figure 1.11 shows the hydrogen bond between the components of a peptide bond, which are important in maintaining overall protein structure. Hydrogen bonds also form between polar parts of some amino acid side chains and those in the protein molecule and water (see polar side chains of amino acids in figure 1.4). Although hydrogen bonds are weak, they are important because so many of them are involved in the structure of a protein. Figure 1.11 shows ionic interactions between oppositely charged groups, as found in proteins, such as the N- and C-termini and the charged carboxylate or ammonium groups in the side chains.

Amino acids are primarily found in a polar environment because they are in an aqueous medium with charged ions and polar molecules. Proteins spanning membranes are an exception because the interior of a membrane is hydrophobic. Water is a polar molecule and interacts well with polar and charged groups on amino acids. This means that inside cells and in the blood, polar and charged amino acids are most likely to be exposed to their aqueous environment. However, 9 of the 20 common amino acids are **hydrophobic** or nonpolar because their side chains are composed of carbon and hydrogen atoms and thus have no affinity for water (see figure 1.4). In fact, they can disrupt the relatively organized structure of liquid water. These side chains tend to cluster in the interior of the protein molecule, not because they have an affinity for each other but because they cannot interact with water. Fat globules also form tiny spheres in water to keep their exposed surface as small as possible. We call the clustering of hydrophobic groups on amino acids hydrophobic interactions (see figure 1.11).

Secondary Structure of Proteins

So far, a one-dimensional protein structure has been described, that is, the sequence of amino acids. As mentioned, proteins actually have a three-dimensional structure. Imagine a peptide backbone with the side chains of each amino acid radiating out. The backbone consists of three elements—the alpha carbon, the nitrogen group involved in the peptide bond, and the carbonyl group. The **secondary structure** is the spatial path taken by these three elements of the peptide backbone. Side chains (or R groups) of individual amino acids are not considered in secondary structure. Two common, recognizable structural features of the polypeptide backbone, the **alpha-helix** and the **beta-sheet**, are found in many proteins. These result from the way the polypeptide backbone organizes itself. They are stabilized by hydrogen bonding between neighboring elements of the peptide bond, as shown in figure 1.11 with dotted lines.

Tertiary Structure of Proteins

The tertiary structure is the overall three-dimensional arrangement of a polypeptide chain, including the secondary structure and any nonordered interactions involving amino acid residues that are far apart. Most proteins are found in aqueous mediums where they assume a compact globular structure maintained by the forces already described. Examples include proteins in the blood and enzymes within cells. The hydrophobic side chains are buried in the interior of the protein, away from the aqueous medium, with hydrophilic side chains and the N- and C-termini exposed on the surface where they can interact with water or other polar molecules and ions. The tertiary structure also involves the spatial position of ions or organic groups that are part of the makeup of many proteins. The term **amphipathic** is often used to describe a part of a protein that must interact with both a hydrophobic and a hydrophilic region. Amphipathic means that the section of the protein will have a polar region, able to interact, for example, with the cytoplasm of the cell, and a nonpolar region that may interact with the hydrophobic part of the cell membrane.

KEY POINT

The sequence of bases in a gene specifies the amino acid sequence of a polypeptide, which determines secondary and tertiary structure. The biological function of the polypeptide is therefore based on its overall structure.

The secondary and tertiary structure of a polypeptide chain depends on the kind and sequence of its amino acids. In the cell, or wherever a protein is found, its overall structure or conformation must be maintained, because proteins must recognize and interact with other molecules. However, the conformation of a protein in vivo is not fixed but changes in subtle ways as it carries out its particular function. For example, the cross bridges in the contractile protein myosin alter their conformation to generate force during muscle contraction. Membrane transport proteins alter their conformations as they move substances from one side of a membrane to the other.

Quaternary Structure of Proteins

Many proteins consist of more than one polypeptide chain, each containing its own unique primary, secondary, and tertiary structures. We call these chains subunits. **Quaternary structure** refers to the arrangement of the individual subunits with respect to each other. The subunits in an oligomeric protein (i.e., one containing more than one subunit) are held together with noncovalent bonds, that is, hydrogen bonds, electrostatic interactions, and hydrophobic interactions. Oligomeric proteins are common because several subunits allow subtle ways of altering the protein's function. For example, hemoglobin A, the adult form of hemoglobin, is a tetramer consisting of four subunits, two alpha (with 141 amino acids per subunit) and two beta (with 146 amino acids per subunit). Each subunit contains one heme group that binds one oxygen molecule. The quaternary structure of hemoglobin refers to the way the two alpha and two beta subunits interact with each other. The oligomeric nature of hemoglobin enhances its role in loading oxygen in the lung and releasing it at the cell level. In the lungs, where the oxygen concentration is high, binding of one oxygen molecule to one subunit facilitates the binding of oxygen molecules to the other subunits. This helps make hemoglobin saturated with oxygen in the lungs, maximizing its ability to transport oxygen. At the tissue level, where the oxygen concentration is much lower, the subunit interactions facilitate the unbinding of oxygen so that it is available for diffusion into adjacent cells.

Myosin, a contractile protein, is a hexamer (i.e., consists of six subunits). There are two heavy chains, each with a molecular weight of about 200,000 (often expressed as 200,000 daltons or 200 kilodaltons, kD). The two chains wind around each other, forming a single long chain for most of their length. At one end, they each form separate globular regions, known as heads. In each globular head there are two light chain polypeptides each with a molecular weight of about 18 to 25 kD. Two are regulatory light chains and two are alkali light chains, which can be extracted with dilute alkali (see figure 1.12). The globular heads act as the cross bridges that attach to the protein actin to generate force when a muscle contracts. *Skeletal* muscle fiber typing using histochemistry is based on chemical differences in the heavy chains of myosin in different muscle fiber cross sections. Type I fibers have type I myosin heavy chains, type IIA fibers have IIA heavy chains, IIX fibers have the IIX myosin heavy chain, and IIB fibers have IIB myosin heavy chains. There are three major categories of myosin, found in heart muscle (cardiac myosin), smooth muscle, and skeletal muscle. The nature of muscle contraction in skeletal, cardiac, and smooth muscle cells differs quite markedly. Skeletal muscle fibers generally contract rapidly but are seldom continually active, whereas cardiac muscle cells contract continuously over a life

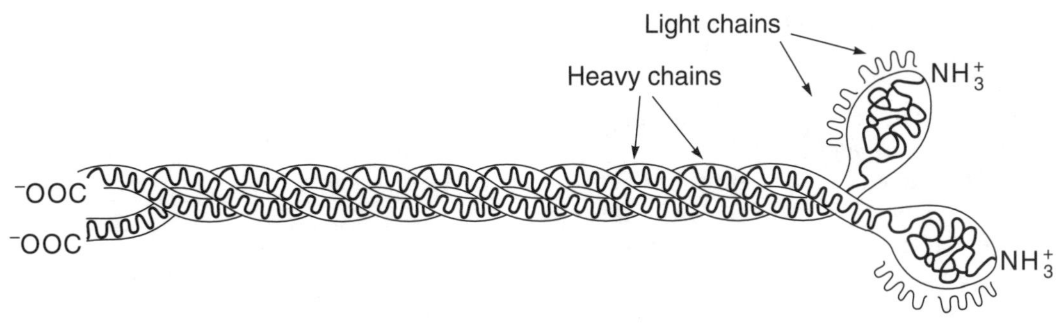

Figure 1.12 Schematic of a myosin molecule, showing the possible position of the two heavy chain and four light chain subunits.

span. Smooth muscle cells contract very slowly and help to regulate our internal environment. Indeed, myosin is found in all cells.

The term **domain** refers to part of a protein molecule, usually globular in shape, that has a specific function, such as binding or catalysis. A single protein may contain a number of domains. For example, each heavy chain in the contractile protein myosin has an actin binding domain and an ATP binding domain. These domains act in concert to utilize the energy of ATP to generate a force and produce movement.

The noncovalent forces that maintain the secondary, tertiary, and quaternary structures of proteins are generally weak. **Denaturation** refers to a disruption in the overall conformation of a protein with loss of biological activity. When denatured, proteins usually become less soluble. Denaturation can be caused by heat, acids, bases, organic solvents, detergents, agitation (e.g., beating egg whites to produce a meringue), or specific denaturing agents such as urea. We seldom encounter such harsh conditions that our body proteins are denatured. However, burning the skin or the reflux of acid from the stomach into the esophagus may cause some protein denaturation. Before proteins are broken down in their normal turnover process, they are denatured by enzyme-catalyzed chemical modification, then broken down to their constituent amino acids. In high concentrations, urea, a normal product of amino acid breakdown in our cells, can be used as a laboratory denaturant. If denaturation is not too severe, some proteins may renature, resume their natural conformation, and regain their biological activity.

Summary

Proteins are the action molecules of life, made from 20 different amino acids. Differences among amino acids are based on their characteristic side chains, known as R groups. Amino acids exist in all cells of the body and in the fluids outside the cells. Because each has at least one group that acts as an acid and one that acts as a base, they are said to be amphoteric. All amino acids have at least one charged group, and at the neutral pH associated with life they have at least two charged groups. When an amino acid has no net charge, it is said to be a zwitterion. The pH where this occurs is known as the isoelectric point.

Peptides are formed when amino acids are joined together by peptide bonds. In general, proteins are large peptides (polypeptides), usually containing 100 or more amino acids. The primary structure of a peptide is the sequence of amino acid residues starting from the end containing the free amino group, known as the N-terminus. Because of their large size, proteins have higher levels of three-dimensional structure, known as secondary and tertiary structures. If a protein consists of more than one polypeptide, the structural relationship of the polypeptide subunits is described as a quaternary structure. The forces such as hydrogen bonds and hydrophobic interactions maintaining the secondary, tertiary, and quaternary structures are generally weak.

Key Terms

acid 4
alpha-helix 10
amphipathic 10
amphoteric 6
beta-sheet 10
degradation 3
denaturation 12
disulfide bonds 9
domain 12
half-life 3
hydrogen bonds 10
hydrophobic 10
isoelectric point 6
myosin 11
peptide bond 8
polypeptide 8
primary structure 8
quaternary structure 11
secondary structure 10
stereoisomers 7
transcription 3
translation 3
zwitterion 4

Review Questions

Using the information found in figure 1.4, answer the following questions about the peptide shown in shorthand: Gly-Arg-Cys-Glu-Asp-Lys-Phe-Val-Tyr-Cys-Leu.

1. How many amino acid residues are there?
2. Identify the N- and C-terminal amino acids.
3. What would be the net charge on this peptide at pH 7.0?
4. Identify the amino acids that have a polar, but uncharged, side chain.
5. Identify the nonpolar amino acids.
6. What are the hydrophobic amino acids?
7. Is it possible for this amino acid to have an intramolecular disulfide bond? Explain.

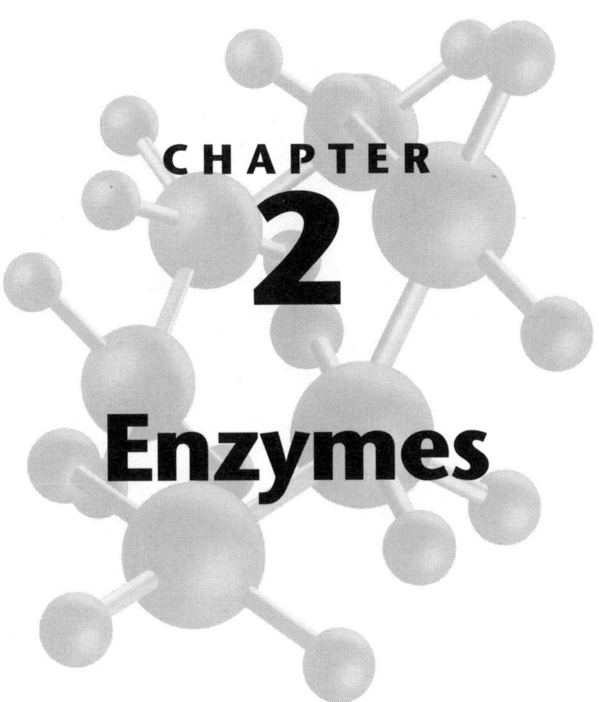

CHAPTER 2

Enzymes

Enzymes are proteins that catalyze the thousands of different chemical reactions that constitute our metabolism. As proteins, they are large molecules, made from the 20 different amino acids. As catalysts, enzymes speed up chemical reactions without being destroyed in the process.

Enzymes as Catalysts

Enzymes speed up chemical reactions by lowering the energy barrier to a reaction—called the energy of activation—so that it takes place at the low temperature of an organism (37 °C for a human). In the laboratory, chemical reactions may take place because we heat the reacting substances, overcoming the energy barrier that prevents the reaction from occurring at a lower temperature. Because our bodies function at a relatively constant, but low, temperature (37 °C), we need enzymes to reduce the energy barrier or energy of activation. In this way, the thousands of chemical reactions that reflect our metabolism can speedily take place. In fact, enzymes are so effective that they can increase the speed of reactions by far more than a billion times the rate of an uncatalyzed reaction.

The molecule or molecules that an enzyme acts on is known as its **substrate**, which is then converted into a product or products. Enzymes are highly specific, catalyzing a single reaction or type of reaction. Without an enzyme present, many reactions would eventually reach an equilibrium in which there is no net change in substrate or product concentrations. The enzyme allows the reaction to reach equilibrium faster than if the enzyme were absent, but the position of equilibrium remains the same.

A part of the large enzyme molecule will reversibly bind to the substrate (or substrates), and then a specific part or parts of the enzyme will catalyze the detailed change necessary to convert the substrate into a product. The enzyme has amino acid residues that bind the substrate, and those that carry out the actual catalysis. There is thus a binding domain and a catalytic domain, although the term **active site** often represents both the binding and catalytic domains of the enzyme protein.

We can write a general equation to describe a simple enzyme-catalyzed reaction in which a single substrate S is converted into a single product P. The reaction could be irreversible, with all substrate molecules converted into product molecules and indicated by an arrow with only one head pointing toward the product. Irreversible reactions are typically described as nonequilibrium reactions. Alternatively, the reaction could be reversible, such that, given time, it establishes an equilibrium, with the ratio of product concentration to substrate concentration a constant described by the **equilibrium constant** (K_{eq}). Reversible reactions are commonly described as equilibrium reactions, and the two descriptions mean the same thing. An equilibrium reaction will be shown with a double-headed arrow, meaning that the product is also a substrate for the reverse reaction. Figure 2.1 illustrates these two types of reactions and outlines their properties.

Most enzyme-catalyzed reactions are considered to take place in discrete steps (see figure 2.2). Enzyme E first combines reversibly with substrate S to form an initial enzyme-substrate complex ES. This complex is then converted to an enzyme-intermediate complex EI, which then changes to an enzyme-product complex EP. EP then dissociates into free product, and the enzyme is released unchanged.

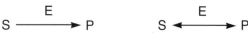

• Irreversible reaction	• Reversible reaction
• Large energy change	• Small energy change
• Nonequilibrium reaction	• Equilibrium reaction
• Less common	• Product is also a substrate for the reverse reaction

Figure 2.1 Irreversible (left) and reversible (right) reactions catalyzed by an enzyme E in which a substrate S is converted into a product P.

Figure 2.2 Possible steps in the conversion of substrate S to product P under the influence of enzyme E. Four steps are identified: An initial enzyme-substrate complex ES is formed, then it is converted to an enzyme-intermediate complex EI and then to the enzyme-product complex EP. EP breaks down to product and free enzyme.

Rates of Enzymatic Reactions

Consider the case of an enzyme-catalyzed reaction in which substrate S is converted to product P, catalyzed by enzyme E. This reaction is reversible, so that P is a substrate for the reverse reaction. Our goal is to measure the rate of this reaction in the direction S⟷P. If we begin with only S plus the enzyme, P will be formed. As the concentration of P increases and that of S decreases the reverse reaction will take place, becoming more important as the concentration of P increases. Accordingly, the rate of the forward reaction must be measured quickly before any appreciable amount of P is formed. We call this quickly measured forward reaction rate the initial velocity.

Effect of Substrate Concentration

Let us carry out an experiment in which we set up 10 test tubes, each containing a solution at 25 °C, pH 7.0, and a fixed concentration of enzyme E. We add a specific amount of substrate S to each test tube, mix it, and quickly measure the rate of the reaction, either by measuring the rate of disappearance of S or the rate of appearance of P. Let us assume that the concentration of S in test tube 1 is 2 micromolar, with higher and higher concentrations in the other tubes such that the concentration of S in tube 10 is 500 micromolar. After getting our initial velocities, expressed in units of micromoles of S disappearing per minute (or micromoles of P appearing per minute), we plot the initial velocity as a function of substrate concentration. The graph should look like the one shown in figure 2.3.

Figure 2.3 shows that initial velocity is higher as substrate concentration is increased. Note that the relationship is not linear but hyperbolic. That is, at low substrate concentration the velocity increases linearly with increasing substrate concentration, but at a higher substrate concentration the velocity flattens out, approaching a **maximum velocity** called $\mathbf{V_{max}}$. When this is reached, increas-

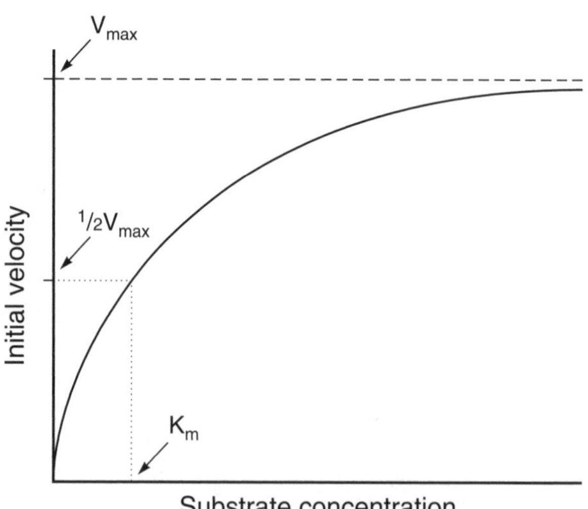

Figure 2.3 The relationship between the initial velocity of an enzyme-catalyzed reaction and the substrate concentration. V_{max} is the maximum velocity at which the amount of enzyme present can catalyze the reaction. K_m is the substrate concentration required to produce one-half V_{max}.

ing substrate concentration will not produce any increase in the rate of the reaction. If we determine the value of the maximum velocity, divide this in half, then determine what substrate concentration will produce one half of V_{max}, we get a concentration known as the **Michaelis constant** and represented by the abbreviation **K_m**. The K_m is defined as the substrate concentration needed to produce one-half the maximal velocity of an enzyme-catalyzed reaction. The K_m has units of concentration. V_{max} is the limiting rate for the velocity of an enzyme-catalyzed reaction at fixed enzyme concentration. It occurs when enzyme active sites are so totally saturated with substrate that as a substrate molecule is converted to product and leaves the enzyme, another substrate molecule immediately binds.

V_{max} and K_m are known as **kinetic parameters** for an enzyme-catalyzed reaction. K_m is said to reflect the affinity of an enzyme for its substrate; the smaller the value of K_m, the greater the affinity an enzyme has for its substrate. The reverse reaction will have its own kinetic parameters, because the product becomes the substrate for the backward reaction. The kinetic parameters for the reverse reaction are unlikely to be identical to those for the forward reaction. Enzymes that exhibit hyperbolic kinetics, as shown in figure 2.3, are said to exhibit Michaelis-Menten kinetics. Not all enzymes behave this way, as we shall discover later in this chapter.

In the cell, the substrate concentration is generally equal to or less than the value for its K_m. This offers two advantages: (a) a substantial fraction of enzyme catalytic ability is being used, and (b) the substrate concentration is low enough that the enzyme can still respond to changes in substrate concentration because it is on the steep part of the curve (see figure 2.3). If the substrate concentration is much greater than K_m we would get efficient use of the enzyme, but it would respond less effectively to changes in substrate levels because it is on the flat part of the curve (see figure 2.3). K_m values for substrates thus reflect in general the concentrations of these substrates in the cell.

KEY POINT

Tissues can adjust the amount of enzyme molecules by altering the rate of expression of the genes for enzymes. The liver can vary the amount of enzymes involved in metabolizing carbohydrates and fats by adjusting the rate of expression of genes for key enzymes in response to changes in our dietary intake of carbohydrates and fats. In this way, the amount of enzyme reflects in general the amount of its substrate.

There are enzymes that catalyze the same reaction in different tissues, but the enzymes have different kinetic parameters for the substrate(s). Often products of different genes, they are known as **isozymes** or isoenzymes. When glucose enters a cell, the first thing that happens is that a phosphate group is attached to it. This reaction is catalyzed by four isozymes known as hexokinase I, II, III, and IV. Three of these isozymes have low K_m values for glucose (20–120 μmolar); the fourth (hexokinase IV, also known as glucokinase and found in the liver) has a high K_m for glucose (5–10 mM). The low-K_m isozymes can phosphorylate glucose even when the blood concentration is low. This is especially important to the brain, which depends solely on glucose as its fuel under normal circumstances. The high-K_m isozyme is found in the liver, where glucose is stored when blood concentration is elevated, for example, following a meal. The liver isozyme thus readily responds to glucose as a substrate only when blood glucose is elevated. Figure 2.4 shows the Michaelis-Menten kinetics for a low-K_m hexokinase and for glucokinase.

KEY POINT

Isozymes of myosin with different heavy chains account for the different muscle fiber types mentioned in the previous chapter. Slow myosin heavy chains play a major role in the slower contraction speed of type I fibers. This is due to a lower V_{max} compared to the faster contracting IIA and IIX myosin heavy chains.

Determining reliable values for the kinetic parameters K_m and V_{max} from the velocity-substrate concentration curve shown in figure 2.3 is difficult. A wide range of substrate concentrations must be tested to ensure a reasonable value of V_{max} because we also need this value to determine K_m. An easier way, shown in figure 2.5, is a

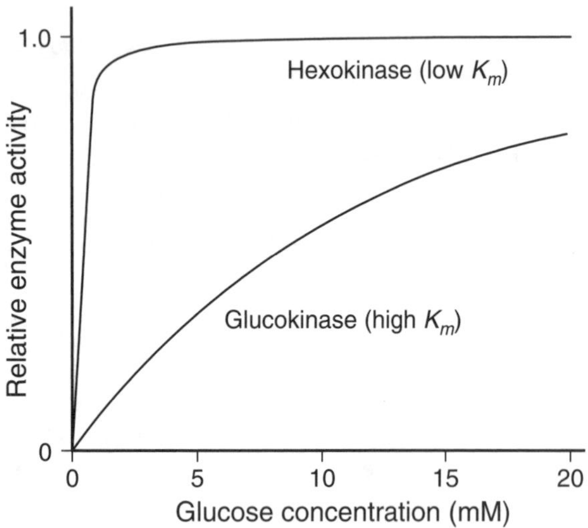

Figure 2.4 The relative velocities of a low-K_m and a high-K_m isozyme of hexokinase. The isozyme with the high K_m is commonly known as glucokinase.

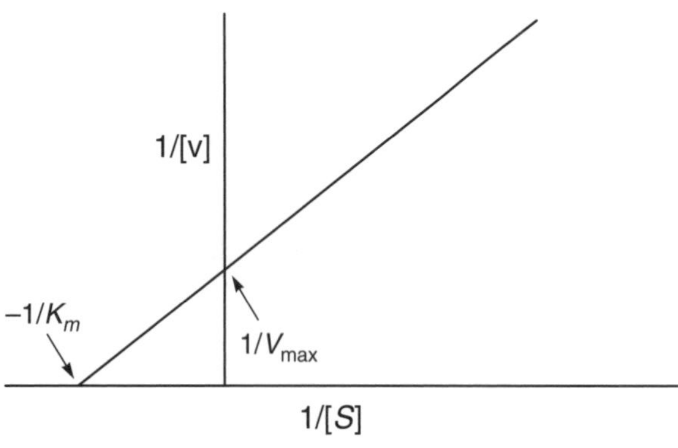

Figure 2.5 A Lineweaver-Burk plot showing the determination of reliable values for K_m and V_{max}.

Lineweaver-Burk plot. Here, the reciprocals of velocity and substrate concentration are plotted, which allows one to determine accurately the kinetic parameters for an enzyme-catalyzed reaction because these values are determined from intercepts on the horizontal and vertical axes. One can thus obtain more accurate values for the kinetic parameters from fewer data points.

Protein Transporters

The cell membrane and internal membranes (such as the mitochondrial membrane) act as barriers to the movement of molecules and ions. Since cells can only function if there is a two-way traffic of ions and molecules across their membranes, specific transport proteins, embedded in or traversing the membrane, are needed to do this. The membrane transport proteins may go by such names as translocases, porters, carriers, or transporters. The substances carried may be small ions (Na^+, Mg^{2+}, Ca^{2+}, etc.) or nutrient molecules such as glucose, amino acids, or fats.

The membrane protein may function as a simple conduit that allows an ion or molecule to flow down its concentration gradient, either into or out of the cell or organelle (mitochondrion). We call this **facilitated diffusion**. It is diffusion because the diffusing substance is moving from a higher to a lower concentration. The membrane transporter facilitates the process. Alternatively, the transport process requires the diffusing substance to move against its concentration gradient. Moving a molecule or ion from a lower to a higher concentration is similar to a ball rolling up a hill. This can only take place if it is driven by a release of energy. The energy release may come from the breakdown of ATP or by the simultaneous movement of another molecule down a steeper concentration gradient. We call this energy-requiring membrane transport process **active transport**.

Although not technically enzymes, catalyzing reactions in which a substrate is converted into a product, membrane transporters function in a way that is consistent with the kinetics of enzymes. They are similar to enzymes because the transporters recognize only specific transport substances and move them in a particular direction across the membrane. Their kinetics are similar to those of many enzymes in that we can describe the transport kinetics with V_{max} and K_m. The transporters may also be subject to inhibition by other substances just as enzymes can be inhibited.

KEY POINT

Transport of substances across cell membranes and intracellular membranes is critical to our metabolism. Transport proteins are selective in terms of the molecule or molecules they recognize and the direction of transport of these molecules. In all cases, transport across a membrane either by diffusion or active transport occurs with the release of energy in much the way energy is released when a ball rolls down a hill.

Turnover Number (k_{cat})

Turnover number (k_{cat}) is defined as the maximum number of molecules of substrate converted to product per enzyme active site per unit of time (usually one second). Since this is considered a measure of the maximum catalytic activity for an enzyme, it must obviously be saturated with substrate for this to occur. Some enzymes are extremely efficient catalysts. For example, catalase, an antioxidant enzyme that breaks down hydrogen peroxide, has a k_{cat} of approximately 40 million. Carbonic anhydrase, which combines carbon dioxide with water to make carbonic acid, has a k_{cat} of approximately one million. However, most enzymes do not operate under conditions in which they are constantly saturated with substrate. A better way to express their catalytic efficiency in vivo is to use the expression k_{cat}/K_m, which gives a truer picture of enzyme function under physiological conditions. The k_{cat} directly reflects kinetic efficiency, whereas K_m is inversely related to the affinity of the enzyme for its substrate. Thus the ratio k_{cat}/K_m provides a measure for the catalytic efficiency of the enzyme. While there may be large differences in values for k_{cat} among enzymes, the ratio k_{cat}/K_m provides far less diversity since most enzymes with large k_{cat} values generally have large K_m values.

Effect of Enzyme Concentration

If an enzyme is saturated with substrate, adding more enzyme increases the reaction velocity. V_{max} is thus proportional to enzyme concentration. However, changing enzyme concentration only increases V_{max} and has no influence on K_m. We can use the relationship between enzyme concentration and maximum velocity to determine how much of a particular enzyme is present in a tissue or fluid (e.g., blood). When we undertake an exercise training program, the amount of some enzymes in the trained muscles may increase by a factor of two. This adaptation is in response to the increased demand placed on certain enzymes during training. The muscle cell responds by stimulating the synthesis of more enzyme proteins to reduce the stress of the training.

Effect of Temperature on Enzyme Reactions

Like all chemical reactions, enzyme-catalyzed reactions increase in rate if the temperature is increased. However, since the forces holding the three-dimensional conformation of an enzyme are generally weak, heating too much disrupts the conformation and decreases enzyme activity. Thus, if we plot the velocity of an enzyme-catalyzed reaction as a function of temperature, the curve rises with increasing temperature until about 50 °C, at which point most enzymes start to denature, and the velocity drops quickly. Biochemists describe the relationship between temperature and reaction rate by the quotient Q_{10}, which describes the fold increase in reaction rate for a 10 °C rise in temperature. For many biological processes, the rate of a reaction approximately doubles for each 10 °C increase in temperature. This means the Q_{10} is about two. Can you see how warming up muscles prior to competition can increase the rates at which the energy-yielding reactions take place by increasing the activities of the enzymes?

Effect of pH

Most enzymes have a pH optimum, that is, a particular pH or narrow pH range where enzyme activity is maximal. Values of pH on either side of optimum produce lower reaction rates. The reason for this is that some of the forces holding an enzyme in its native conformation depend on charged groups. A change in pH can alter these charges, thus reducing enzyme function. The change may directly influence the active site or some other part of the enzyme that indirectly affects the active site. A change in pH may also alter the substrate for an enzyme, which could also influence rate. For most enzymes, their pH optimum reflects where they are active in the body. For example, the stomach enzyme pepsin has a pH optimum around 2 because the stomach is acidic. Severe muscle fatigue is due in part to the acidification of the muscle. This can diminish the activity of certain enzymes, thus compromising muscle function.

Enzyme Inhibition

Enzymes can be inhibited by a variety of substances. We describe these inhibitors according to how they influence the enzyme. **Competitive inhibitors** resemble the normal substrate for an enzyme, for they bind to the active site, but cannot be changed by the enzyme. In this sense they act as mimics, but because of subtle differences between them and the normal substrate they are

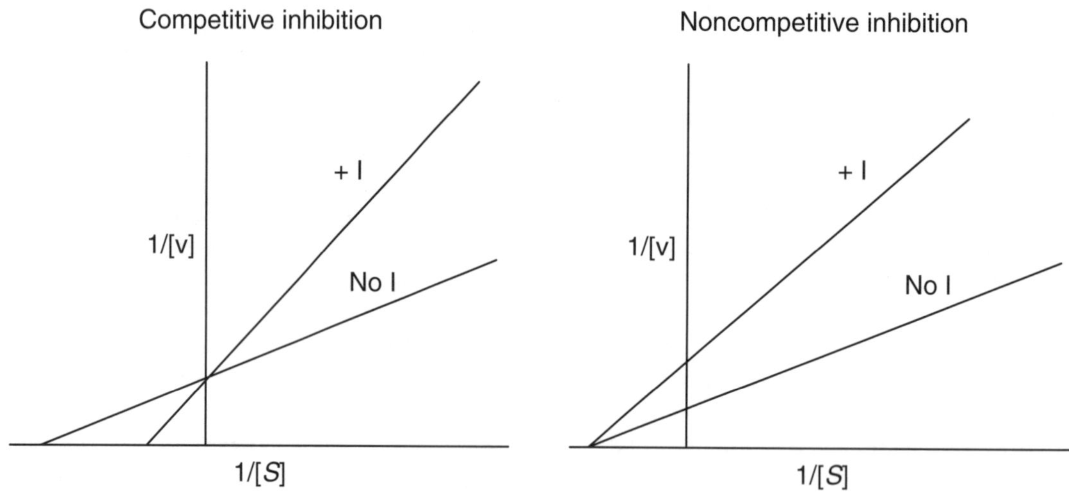

Figure 2.6 Two Lineweaver-Burk plots showing the effect of enzyme velocity on substrate concentration when a competitive inhibitor (left) or noncompetitive inhibitor (right) is present. Note that +I refers to the presence and No I refers to the absence of the inhibitor.

not chemically altered by the enzyme. The inhibitor simply occupies the active site, binding, leaving, binding, and leaving, in a reversible fashion. Competitive inhibitors thus compete with the normal substrate for a place on the active site of the enzyme. This inhibition can be overcome by adding excess substrate. Accordingly, a competitive inhibitor will not affect the V_{max} of the enzyme but will increase the K_m. The Lineweaver-Burk plot in figure 2.6 shows the effect of a competitive inhibitor.

A **noncompetitive inhibitor** does not resemble the normal enzyme substrate and does not bind to the active site. However, when bound to the enzyme, it interferes with its function. Hence, noncompetitive inhibitors lower V_{max} but do not alter K_m. Examples of noncompetitive inhibitors are heavy metal ions (e.g., Hg^{2+}, Pb^{2+}). Figure 2.6 also shows how a noncompetitive inhibitor influences the Lineweaver-Burk plot. Many drugs are developed based on the principles of enzyme inhibition, where specific reactions in bacteria, viruses, or tumors are targeted by the drug. Alternatively, some drugs are designed to decrease the rate of a specific cellular reaction that is too active and thus causing internal problems.

Regulation of Enzyme Activity

Each chemical reaction in a cell is catalyzed by a specific enzyme. As discussed previously, the rate of an enzyme-catalyzed reaction depends on the concentration of substrate. Reaction rates can also be controlled not only by the amount of enzyme protein, but also by where the enzyme is located within the cell. In subsequent chapters, we will encounter a variety of examples in which the activity of existing enzymes in the cell must be modified so that cellular metabolism is appropriate. Simple enzymes, obeying the Michaelis-Menten kinetics shown in figure 2.3, are very common but are rarely involved in controlling metabolism in the cell. For this, enzymes with more complex kinetics or properties are needed. There are two major ways in which the activity of existing enzymes can be controlled, modification by effector molecules and covalent modification.

Allosteric Enzymes

We have already discussed the interaction of enzymes with their substrate molecule(s) at the active site of the enzyme. Some enzymes can also interact with other molecules at locations on the enzyme protein other than the active site. These other sites are known as **allosteric sites** (from the Greek word allo, meaning other). Molecules that bind to large molecules are known as ligands. The ligands may be substrates or products that bind to the active site as well as allosteric effectors that bind to allosteric sites. Therefore, we could classify enzymes as either those whose ligands are only substrates and products that bind to the active site, or those that bind substrates and products as well as one or more ligands that bind at allosteric sites.

In general, **allosteric enzymes** are composed of subunits, that is, they have quaternary structure. The kinetics of allosteric enzymes are more complicated than the simple Michaelis-Menten kinetics shown in figure 2.3. Allosteric enzymes typically demonstrate sigmoidal kinetics, as shown in figure 2.7. Here, the presence of positive or negative allosteric effectors greatly modifies the response of the enzyme to its typical substrate. For example, a much higher concentration of substrate would be needed to achieve one half of V_{max} in the presence of a negative effector, but a much smaller amount of substrate is needed to achieve one half V_{max} with a positive effector bound to its allosteric site. Thus allosteric effectors markedly influence K_m values. Enzymes displaying sigmoid kinetics do so because there is **cooperativity** in substrate binding at the active site in each subunit. This means that the binding of a substrate molecule to the active site of one of the subunits influences subsequent binding of the substrate molecule at the active sites of the other subunits.

Allosteric enzymes play an important role in the regulation of cell metabolism. For example, in a muscle fiber at rest the breakdown of carbohydrate is very low, primarily due to negative effectors binding to the allosteric enzyme phosphofructokinase (PFK). PFK is located near the beginning of the pathway of glycolysis and acts to regulate the breakdown of carbohydrate. When the fiber becomes active in an exercise situation, the rise in the concentration of positive effectors and the decline in the concentration of negative effectors enormously increase the activity of phosphofructokinase.

Covalent Modification of Enzymes

Allosteric regulation is more a "fine-tuning" type of enzyme activity modulation. On the other hand, enzyme activity can be rapidly turned on or off by the covalent modification of specific amino acid residues in the enzyme protein. One example of this is to attach a phosphate group to the hydroxyl part of the side chains of the amino acids serine, threonine, and tyrosine (see figures 1.4 and 2.8). The addition of a phosphate group in place of a single hydrogen atom drastically alters a protein. Not only is the phosphate group far larger, but it contains two negative charges, thus making a major change in protein conformation. Protein phosphorylation is a common and effective mechanism to alter rapidly and reversibly the activity of key enzymes. Addition of the phosphate group may turn on or turn off enzyme activity. Removal of the phosphate group reverses the activity change. The source of the phosphate group is ATP (adenosine triphosphate), a molecule we will encounter throughout all subsequent chapters. Phosphate attachment is

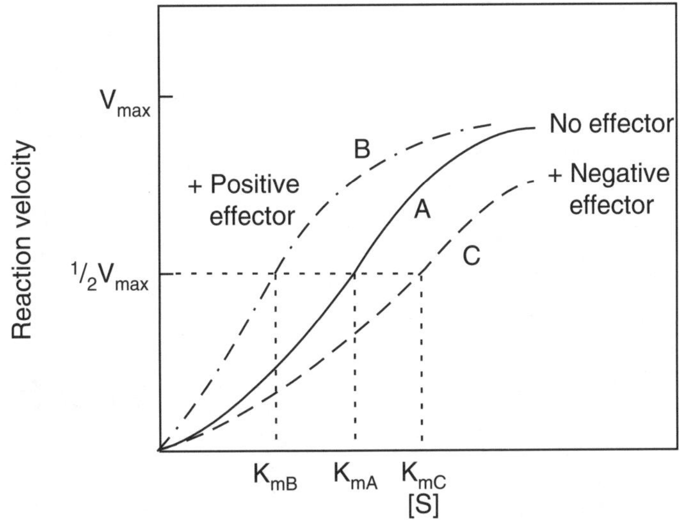

Figure 2.7 The effect of substrate concentration on reaction velocity for an allosteric enzyme: (A) when there is neither positive nor negative allosteric effector, (B) when a positive allosteric effector binds at its allosteric site, and (C) when a negative allosteric effector binds to its allosteric site. In the example shown, V_{max} is the same for the three cases, but the positive allosteric effector reduces K_m (i.e., K_{mB}), while the negative allosteric effector increases the value of K_m (i.e., K_{mC}).

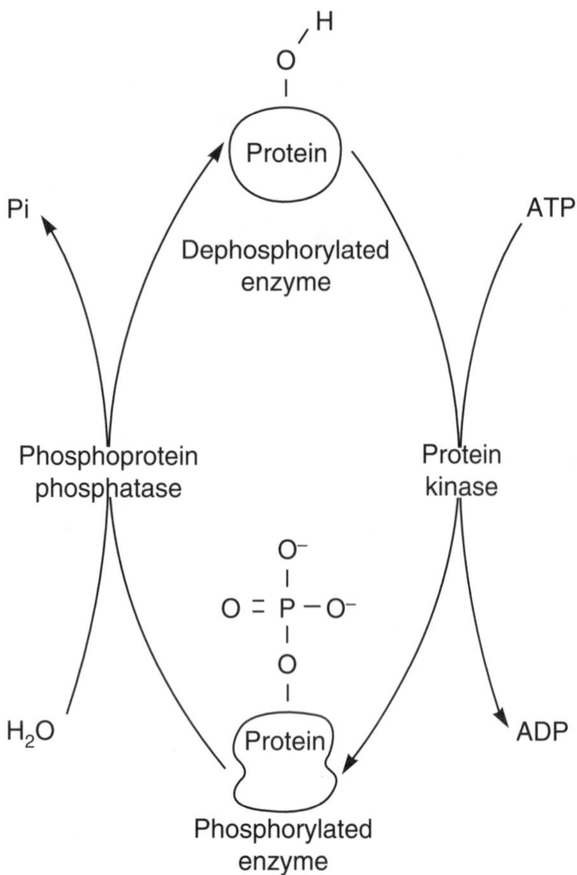

Figure 2.8 Enzyme activity can be turned on or off by the addition of a phosphate group to the OH of a serine, threonine, or tyrosine side chain. A class of enzymes known as protein kinases catalyzes addition of the phosphate group from ATP. Removal of the phosphate group by hydrolysis is catalyzed by phosphoprotein phosphatases.

catalyzed by a class of enzymes known as **protein kinases**, which are specific for serine, threonine, or tyrosine side chains. A class of enzymes known as **phosphoprotein phosphatases** catalyzes removal of the phosphate groups by hydrolysis. A summary of this process is shown in figure 2.8.

KEY POINT

Protein phosphorylation and dephosphorylation works like a light switch, in that it has an all-or-none effect. The enzyme molecule is either active or inactive. Allosteric modification of enzyme activity acts like a dimmer switch, grading the activity of the enzyme.

Control of enzyme activity by phosphorylation plays a critical role in controlling and integrating our metabolism. For example, when we begin to exercise, hormonal signals lead to the phosphorylation and thus activation of hormone-sensitive lipase, the enzyme catalyzing the breakdown of stored fat. In many cases, the protein kinases are activated when external molecules bind to specific sites on the membrane of the cell. There is a class of protein kinases that is activated by a family of extracellular compounds (e.g., peptide growth factors and hormones) that regulate many of the activities of a cell such as cell division, cell death, and cell growth. The external factor binds to its cell membrane receptor, activating one or more protein kinases that phosphorylate proteins, which in turn regulate transcription of genes. Since many of the external factors influence cell growth, they are called mitogens, and the protein kinases activated by mitogen binding are known as mitogen-activated protein kinases, or **MAP kinases**. The MAP kinases phosphorylate serine and threonine side chains in their target tissues. This sequence in which an external molecule such as a growth factor binds to its cell surface receptor and thereby influences events within the cell is known as **signal transduction**. We will encounter many examples of this throughout this book.

KEY POINT

Signal transduction is the process that provides the total network of communication among the various cells, tissues, and organs of the body. Cells cannot survive without communication, but with it, the normal pattern of cell growth, differentiation, and metabolism can be regulated.

Provision of Reactive Groups by Cofactors

Enzymes have reactive groups in the form of side chains of amino acids as well as the N- and C-termini. However, they may need other reactive groups not available on amino acids, called cofactors, in order to carry out their functions as catalysts. Cofactors may be metal ions, such as Zn^{2+}, Mg^{2+}, or Mn^{2+} ions, or they may be organic molecules called **coenzymes**. The polypeptide part of the enzyme is called the apoenzyme; when combined with the cofactor, we have the holoenzyme, illustrated as follows:

apoenzyme (*inactive*) + cofactor = holoenzyme (*active*)

A cofactor tightly bound to the enzyme at all times is called a **prosthetic group**. If the cofactor is not tightly bound but combines with the enzyme and the other substrate during the reaction, we can consider it as a second substrate.

There are eight B vitamins that we need because they form the basic components for coenzymes. Humans and many animals eating a plant and animal diet have lost the ability to synthesize the B vitamins, so they must come from the diet. Table 2.1 lists the eight B vitamins, the coenzymes they form, and their short-form names. The deficiency diseases associated with inadequate intake of specific B vitamins are due to insufficient catalytic power of enzymes because they lack their coenzymes.

KEY POINT

The family of B vitamins performs various functions as coenzymes or critical components of coenzymes. Because they are so important to our overall metabolism and health, and because they are water soluble, we need B vitamins on a daily basis.

People often stress the importance of vitamins but neglect the minerals. We need many more minerals in our diet than vitamins. Many minerals affect the function of enzymes. For example, zinc is an essential component of more than 300 different proteins, including enzymes for synthesizing RNA and DNA, pancreatic digestive enzymes, and enzymes involved in carbohydrate, fat, protein, and alcohol metabolism. Magnesium is associated with the energy-rich ATP molecule, so it is involved in virtually all aspects of our metabolism.

KEY POINT

Minerals not only regulate the activity of many enzymes in a cofactor role, other minerals, such as calcium, act in signal transduction, and some, such as sodium and potassium, are critical to the chemical composition of intracellular and extracellular fluids.

Oxidations and Reductions

Oxidation-reduction, or **redox, reactions** are extremely important for all organisms. In these reactions, something gets oxidized, and something gets reduced. You may have encountered in earlier chemistry courses the term **oxidation**, which means something loses electrons, and **reduction**, which means something gains electrons. The familiar memorization tool that Leo (oxidation loses electrons) says Ger (reduction gains electrons) has been used by students for many years. Redox reactions are absolutely critical to life and underlie all aspects of metabolism. In all cases the electron gain (reduction) is

Table 2.1 The B Vitamins, the Coenzymes They Form, and Their Common Abbreviations

B vitamin	Coenzyme	Abbreviation
Thiamine (B_1)	Thiamine pyrophosphate	TPP
Riboflavin (B_2)	Flavin adenine dinucleotide	FAD
Niacin	Nicotinamide adenine dinucleotide	NAD
Vitamin B_6	Pyridoxal phosphate	PLP
Pantothenic acid	Coenzyme A	CoA
Folate (folacin)	Tetrahydrofolic acid	THFA
Biotin	Biotin	–
B_{12}	Cobalamin coenzymes	–

directly connected with electron loss (oxidation), as seen in the reaction

$$Fe^{3+} + Cu^{+} \longrightarrow Fe^{2+} + Cu^{2+}$$

In this example, the ferric iron (Fe^{3+}) is reduced because it gains an electron from the copper to form ferrous iron (Fe^{2+}), whereas the copper (Cu^{+}) is oxidized by losing an electron to the iron, thereby becoming Cu^{2+} or cupric copper ion.

In the cell, there are redox reactions where ions are oxidized and reduced as the above equation shows. However, in many important redox reactions it is not easy to see that electrons are lost from one molecule and gained by another. For example, figure 2.9 shows two dehydrogenation reactions in which two hydrogens are lost from each of the two partial structures. The important thing is that when the hydrogens leave, they exit with electrons. In the first example, two hydrogen atoms are lost, and a carbon-to-carbon double bond is generated. In the second reaction, two hydrogens are lost from a secondary alcohol, generating a ketone group. In this latter case, one hydrogen comes off as a hydride ion and the other as a proton. The hydride ion is simply a hydrogen atom (with one proton and one electron) plus an additional electron so that it has a negative charge. In summary, dehydrogenation reactions are oxidation reactions in which electrons are lost as part of hydrogen atoms or hydride ions.

KEY POINT

Most of the energy needed by our bodies to grow and survive arises when electrons on fuel molecules are transferred to the oxygen we breathe. Energy released during this process is captured by forming ATP, which is subsequently used to carry out the energy-requiring reactions and processes needed for us to live.

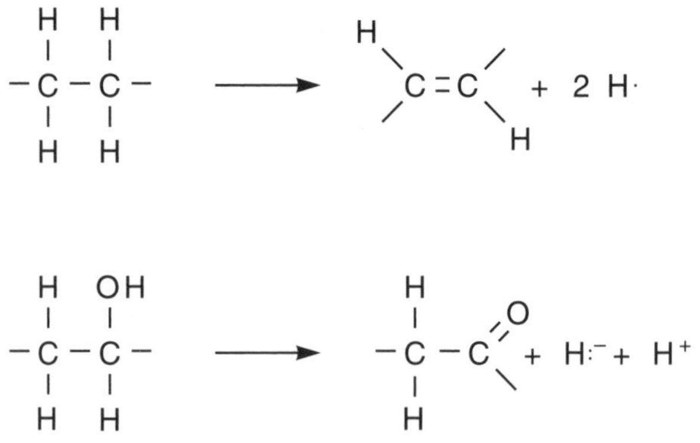

Figure 2.9 Dehydrogenation reactions involve electron loss as hydrogen atoms (top reaction) or a hydride ion (bottom reaction).

Figure 2.10 shows dehydrogenation, but these reactions are also accompanied by hydrogenation reactions, which means a molecule is reduced because it accepts two hydrogen atoms (each with an electron) or a hydride ion (with two electrons). Two major coenzymes are involved in most cell redox reactions. The oxidized form of the coenzyme **flavin adenine dinucleotide** (FAD, see table 2.1) can accept two hydrogen atoms to become $FADH_2$ (the reduced form of FAD). The FAD coenzyme is involved in redox reactions where two hydrogen atoms are removed from chemical structures, such as the top reaction in figure 2.9. Those shown in the lower part of figure 2.9 involve the **nicotinamide adenine dinucleotide** coenzyme (NAD^{+}), which accepts a hydride ion, becoming NADH. NAD^{+} is the oxidized form, whereas NADH is the reduced form of this coenzyme.

Figure 2.10 shows two examples of redox reactions that use the FAD and NAD^{+} coenzymes. In the first reaction, part of the TCA (citric acid or Krebs) cycle, succinate is oxidized to fumarate by losing two hydrogen atoms, forming the carbon-to-carbon double bond. The two hydrogen atoms, each carrying an electron, are picked up by the coenzyme FAD, forming $FADH_2$. Thus, succinate is oxidized to fumarate, and FAD is simultaneously reduced to $FADH_2$ in this redox reaction. The coenzyme (FAD or $FADH_2$) is a tightly bound coenzyme for the enzyme succinate dehydrogenase. In the lower example, lactate loses a hydride ion and a proton when oxidized to pyruvate. At the same time, NAD^{+} accepts the hydride ion, becoming the reduced form of the coenzyme NADH. In this latter reaction, catalyzed by the enzyme lactate dehydrogenase, the coenzyme (NAD^{+} or NADH) binds to the enzyme only at the time of the reaction. Therefore, as the reaction proceeds from left to right as shown in figure 2.10, NAD^{+} is a substrate and NADH is a product. Most redox reactions are reversible, so pyruvate can be reduced to lactate or fumarate reduced to succinate.

The net direction of redox reactions depends on the relative concentrations of the oxidized and reduced forms of the substrates and coenzymes. During exercise, when the rate of pyruvate formation increases in muscle, the enzyme lactate de-

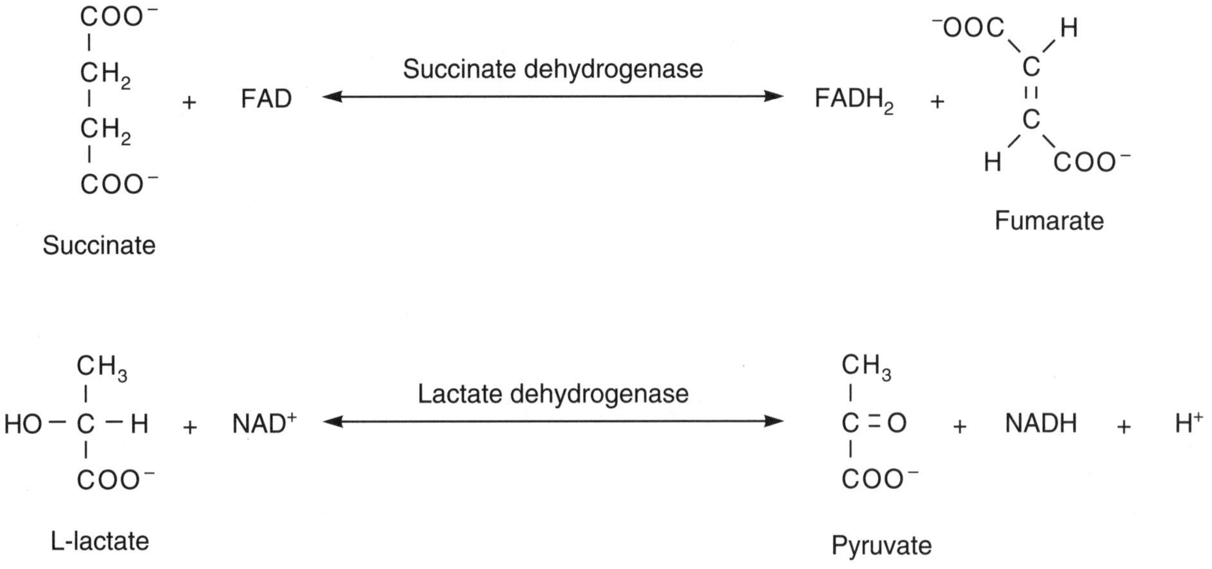

$$\underset{\text{Succinate}}{^{-}OOC-CH_2-CH_2-COO^-} + FAD \xrightleftharpoons{\text{Succinate dehydrogenase}} FADH_2 + \underset{\text{Fumarate}}{^{-}OOC-CH=CH-COO^-}$$

$$\underset{\text{L-lactate}}{CH_3-CH(OH)-COO^-} + NAD^+ \xrightleftharpoons{\text{Lactate dehydrogenase}} \underset{\text{Pyruvate}}{CH_3-C(=O)-COO^-} + NADH + H^+$$

Figure 2.10 Two biologically important redox reactions showing substrates, products, and coenzymes.

hydrogenase produces lactate, attempting to maintain equilibrium. The lactate can travel from the muscle cell to the blood, where its concentration is often used as an indicator of the exercise intensity. Notice in figure 2.10 that we write L-lactate because the middle carbon is chiral, and the L refers to the absolute configuration of the lactate. Lactate comes from lactic acid, a modestly strong acid (pK_a about 3.1). Thus, at the neutral pH where this reaction occurs, lactic acid exists as the ion lactate because it has lost its proton. The same can be said for pyruvate, from pyruvic acid.

Redox reactions represent the molecular basis for energy generation in the cell. Electrons from substrates, generated from the foods we eat, are transferred to coenzymes NAD^+ and FAD generating NADH and $FADH_2$. Subsequently, these electrons are transferred from the reduced coenzymes through a series of carriers embedded in the inner membrane of mitochondria until they are passed to oxygen (O_2). Reduction of oxygen by accepting these electrons and H^+ to form water molecules is the final step (see the following equation). During the passage of electrons from food-derived substrates to oxygen, a tremendous amount of energy is released and captured by forming ATP. The ATP is then used to drive energy-requiring processes in the body. This is the basis of the process called **oxidative phosphorylation** that we will cover in detail in chapter 5.

$$O_2 + 4\,e^- + 4\,H^+ \longrightarrow 2\,H_2O$$

KEY POINT

The harder we exercise, the more ATP we need. This means we must accelerate the rate of transfer of electrons from fuel molecules (dehydrogenation reactions) to oxygen. If we continue to increase the intensity of our exercise, a weak link in the electron transport process is reached. This limitation may be availability of oxygen to accept electrons, or the rate at which we can remove electrons from the fuels, or the actual kind and amount of fuel available.

Measuring Enzyme Activity

We sometimes need to know how many functional enzyme molecules exist in a fluid (e.g., blood) or tissue. Because the number of molecules of functional enzyme is proportional to the V_{max}, measurement results are in units of reaction velocity (i.e., change in substrate or product concentration per minute) per unit weight of tissue, per unit amount of protein, or per volume of fluid. Examples are micromoles of product formed per minute per gram of tissue, millimoles of substrate disappearing per minute per milligram of protein, or micromoles of product formed per minute per milliliter of blood, respectively. The expression of the V_{max} or, as it is commonly called, the activity of the enzyme is important in a variety of physiological and clinical conditions. For example, the activity of mitochondrial enzymes can almost double given the appropriate exercise training stimulus. This means that the maximal flux or traffic of substrate through the enzyme is twice what it was before training. We may want to determine if a particular exercise training program alters the metabolism of a muscle by measuring the activity of selected enzymes. We can also tell if a particular tissue is damaged by measuring the activity of isozymes specific for that tissue that are released due to cell membrane damage.

KEY POINT

If we double the activity of enzymes involved in dehydrogenating fuel molecules, we can potentially double the ability to transfer electrons to oxygen and form the energy molecule ATP.

When measuring the activity of an enzyme certain principles must be established and rigidly followed. First, we need to make the measurements at a substrate (or substrates) concentration high enough to generate a true V_{max}. The pH of the reaction and the temperature at which it is measured should also be standardized so that meaningful comparisons can be made. Finally, we need to have a simple method for measuring the disappearance of substrate or appearance of product.

KEY POINT

When making comparisons between activities of specific enzymes, it is important to know the temperature at which the enzyme reaction velocity (activity) is measured since temperature can profoundly influence the rate of chemical reactions.

This determination can be done if the substrate or product is colored or can be made to generate a colored complex. For example, phosphate appearance can be readily measured because it forms a colored complex with a number of agents. One useful technique takes advantage of two properties of the coenzyme NADH. First, NADH absorbs light at a wavelength of 340 nanometers, so we can measure its rate of formation or disappearance with a spectrophotometer. The relationship is as follows: A 0.1 millimolar solution of NADH has an absorbance of 0.627. Second, NADH fluoresces when bombarded with light of a specific wavelength; thus we can measure its appearance or disappearance with a fluorometer. If the specific reaction does not actually involve NADH, it can be connected to a reaction that does. The rate of the reaction in question then dictates the rate of a connection reaction in which NADH is formed or lost.

Biochemists use the term **International Unit** to express the activity of an enzyme. One International Unit of enzyme activity is the amount of enzyme that converts one micromole of substrate to product in one minute. Thus if an enzyme has an activity of 15 International Units per gram, 15 micromoles of product form per minute per gram of tissue. Since enzyme activity is so sensitive to temperature, the composition of the medium, and pH, one must specify these conditions when describing enzyme activity in International Units.

Summary

Enzymes are biological catalysts—specialized proteins that enormously speed up reactions in cells. Highly specific, they catalyze reactions involving single substrates or a closely related group of substrates. Enzymes have a Michaelis constant, K_m, which is the substrate concentration needed to produce one-half the maximal velocity (V_{max}) of the enzyme reaction. The K_m, a characteristic constant for an enzyme-substrate pair, reflects inversely the affinity of the enzyme for its substrate. The maximal velocity of an enzyme-catalyzed reaction, or V_{max}, is proportional to the amount of enzyme present and can only be determined when the enzyme is saturated with its substrate. Measurement of the V_{max} thus determines the amount of enzyme present. An International Unit is defined as the amount of enzyme needed to convert one micromole of substrate to product in one minute. The action of enzymes can be hindered by the presence of inhibitors—specific substances that resemble the normal substrate and compete with it (competitive inhibitors) or that irreversibly alter the structure of the enzyme (noncompetitive inhibitors). In the cell, regulation and integration of the thousands of chemical reactions can be affected by modulating the activity of a subset of key enzymes. The activity of allosteric enzymes can be changed by the binding of ligands, other than substrates or products, to specific sites known as allosteric sites. Binding of these effector molecules may increase or decrease enzyme activity by changing the enzyme's response to a particular substrate concentration. A more profound change in enzyme activity accompanies the covalent attachment of a phosphate group to an enzyme, catalyzed by a class of enzymes known as protein kinases. Dephosphorylation by phosphoprotein phosphatases reverses the change in activity accompanying phosphorylation.

Membrane transport is carried out by a class of protein molecules with properties similar to those of enzymes. Membrane transport can be described by kinetic constants, V_{max} and K_m, and can be subject to competitive and noncompetitive inhibition. Many enzymes require the presence of nonprotein substances to function. These cofactors can be organic molecules, that is, coenzymes, or they can be metal ions. Most coenzymes are derived from the B vitamins in our diet, while our need for many specific mineral nutrients relates to their role as enzyme cofactors. Isoenzymes (isozymes) are closely related, but different, enzyme molecules that catalyze the same reaction but differ in certain properties, such as K_m. Some of the most important enzymes are the dehydrogenases, which add or remove electrons from their substrates. They play a major role in producing energy in the process of oxidative phosphorylation.

Key Terms

active site 15
active transport 18
allosteric enzymes 21
allosteric sites 20
coenzymes 22
competitive inhibitors 19
cooperativity 21
equilibrium constant 15
facilitated diffusion 18
flavin adenine dinucleotide 24
International Units 26
isozymes 17
kinetic parameters 17
MAP kinases 22
maximum velocity (V_{max}) 16
Michaelis constant (K_m) 17
nicotinamide adenine dinucleotide 24
noncompetitive inhibitor 20
oxidation 23
oxidative phosphorylation 25
phosphoprotein phosphatases 22
prosthetic group 23
protein kinases 22
redox reactions 23
reduction 23
signal transduction 22
substrate 15

Review Questions

1. Exercise training may double the ability of mitochondrial enzymes to transfer electrons from fuel molecules to oxygen. If the overall system is matched, what must happen to balance the ability to deliver the oxygen to the mitochondria?
2. For most people, using pure fat as a fuel as opposed to pure carbohydrate limits the maximum exercise rate by about 50%. Where is the weak link with fat as the fuel?
3. One lab reports that the V_{max} for the mitochondrial enzyme citrate synthase in the quadriceps of a group of elite cyclists averages 40 micromoles of substrate per gram wet weight of tissue per minute. Your lab, on the other hand, measures the same enzyme in a similar group of trained subjects, yet your average enzyme activity is double that of the other

lab. What could contribute to this discrepancy between two apparently similar groups of subjects?

4. Change the units for the activity of citrate synthase from 20 micromoles per gram per minute to (a) millimoles per kilogram per minute, (b) International Units per gram.

5. Some labs determine the activity of muscle enzymes on freeze-dried samples of muscle. If the water content of a muscle sample is 75%, what would be the activity of the citrate synthase in micromoles per gram dry weight per minute if the activity is 20 International Units per gram wet weight?

6. Using the following data generated by measuring the rates of an enzyme-catalyzed reaction at constant enzyme concentration, determine the K_m and V_{max}.

Substrate concentration	**Reaction rate**
1 mM	2 mmol/ml/min
2 mM	4 mmol/ml/min
4 mM	7 mmol/ml/min
10 mM	15 mmol/ml/min
50 mM	25 mmol/ml/min

CHAPTER 3

Nucleotides, DNA, and RNA

As mentioned in chapter 1, DNA is a large molecule containing genetic information in the form of genes. RNA is a working copy of a gene in DNA, used to order the sequence of amino acids in a polypeptide chain. Both DNA (deoxyribonucleic acid) and RNA (ribonucleic acid) are polynucleotides—large molecules (polymers) composed of individual nucleotides (monomers) joined end to end. The nucleotide monomers themselves are made up of three main components: bases, sugars, and phosphoric acid. In this chapter we look at the structures of both DNA and RNA to learn how they carry information. First, we examine the structural composition of the precursor nucleotides.

Nucleotides

Five main bases are found in nucleotides; they are further subdivided into the **purines** and the **pyrimidines**. The two purines are adenine, represented by A, and guanine (G). The pyrimidines are cytosine (C), **thymine** (T), and **uracil** (U). Thymine is only found in DNA and uracil only in RNA. Figure 3.1 shows the chemical structures for the pyrimidines and purines.

Two different sugars exist in nucleotides: deoxyribose, found only in DNA, and ribose, found in RNA. Figure 3.1 also shows these chemical structures. Note that the numbering system for these sugars has a prime (') to differentiate the sugar numbers from the numbers (not shown) associated with the carbon and nitrogen atoms in the ring portion of the bases. Phosphoric acid is found in nucleotides and polynucleotides, but because the pH of the cell is near 7, it is ionized and exists as phosphate (the ionized form of phosphoric acid).

Nucleosides are formed when a base joins to a sugar (see figure 3.2). Thus we have deoxynucleosides (if the sugar is deoxyribose) or simply nucleosides (if the sugar is just ribose). When a specific base joins to a sugar, the name of the resulting nucleoside (or deoxynucleoside) reflects the name of the base. For example, adenine plus ribose or deoxyribose gives **adenosine** or deoxyadenosine, respectively. Thymine plus deoxyribose gives deoxythymidine—sometimes called thymidine because thymine only joins to deoxyribose. The other nucleosides are guanosine, cytidine, and uridine.

Nucleotides are formed when a nucleoside is joined to a phosphoric acid (actually a negatively charged phosphate group). The phosphate is normally attached to one of the hydroxy (OH) groups on the sugar. An acid attached to an OH (or alcoholic group) is an ester; if the acid is derived from phosphoric acid (i.e., phosphate), the ester is a **phosphate ester**. The location of the phosphate is designated by the number of the sugar carbon it is attached to via the ester bond. Figure 3.2 shows an example of a nucleotide. It is very common to

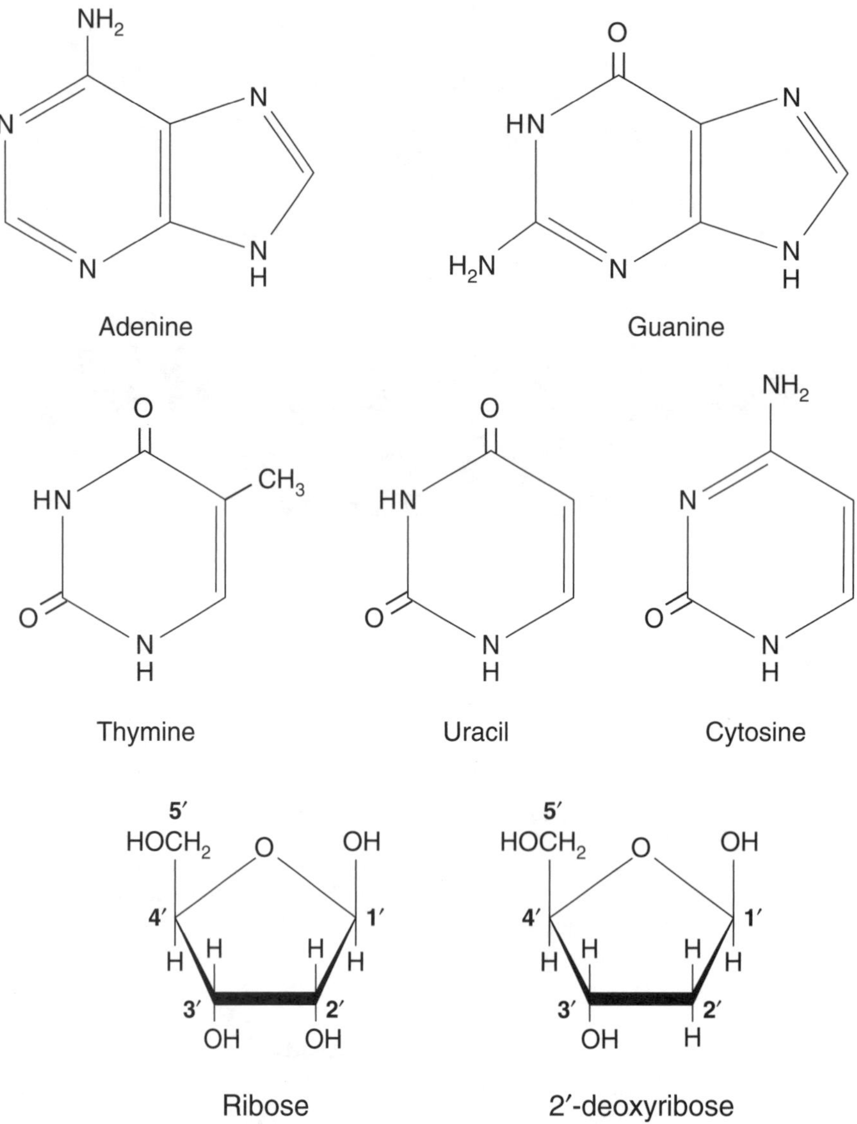

Figure 3.1 The purine and pyrimidine bases and ribose and deoxyribose sugars.

have two or three phosphate groups attached together and then to the sugar. Thus we can have nucleotides with one phosphate (i.e., a nucleoside monophosphate, NMP), with two phosphates (i.e., a nucleoside diphosphate, NDP), or with three phosphates (i.e., a **nucleoside triphosphate**, NTP). A nucleotide made with a deoxyribose sugar yields a deoxynucleoside mono- (dNMP), di- (dNDP), or tri- (dNTP) phosphate.

So far, we have described nucleotides in general terms. If we know the base that is used, we can be more specific. Suppose we have a nucleotide composed of the base adenine, the sugar ribose, and a phosphate group attached to the 5' (five prime) carbon of ribose. It would be adenosine 5'-monophosphate, abbreviated as 5'-AMP (sometimes simply AMP). If two phosphates are attached together and then attached to the 5' position on ribose, it would be adenosine 5'-diphosphate or 5'-ADP (or simply ADP). Three phosphate groups attached to the 5' position of guanosine (the nucleoside) would be known as guanosine 5'-triphosphate or 5'-GTP (or simply GTP).

If two phosphates are joined together, as in 5'-ADP or 5'-ATP, we have an **anhydride bond**—a bond joining two acid groups together. Such bonds

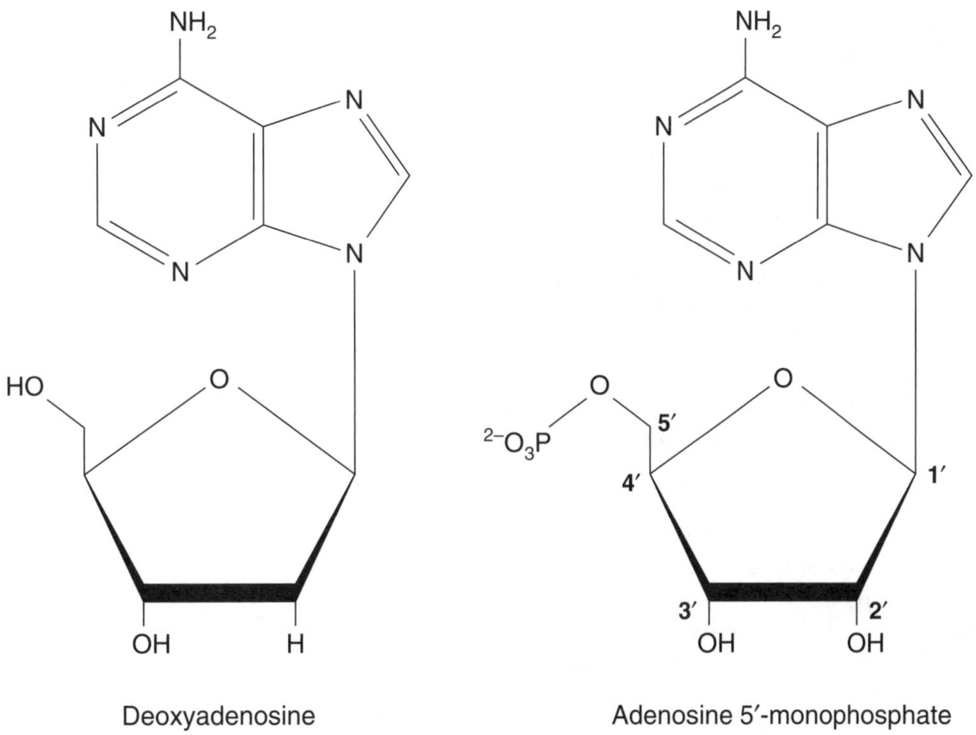

Figure 3.2 Chemical structures of a nucleoside (left) and nucleotide (right).

are energy-rich because their hydrolysis (i.e., splitting) releases much energy. We will discuss hydrolysis later.

> **KEY POINT**
>
> ATP is the energy currency of the cell. When broken down by losing one or two phosphate groups, a great deal of energy is released. Similar hydrolysis of the other nucleoside triphosphates also yields a great deal of energy.

Most of our cells can synthesize purine and pyrimidine bases de novo, which means from the beginning. In addition, there is a purine salvage pathway, used to recover purines when the nucleotides and nucleosides are broken down. Ribose is synthesized in cells from glucose breakdown using a unique pathway known as the **pentose phosphate pathway**. Deoxyribose is formed from ribose by the enzymatic removal of the 2' OH group of ribose. Finally, phosphoric acid in the form of phosphate must be obtained in our diets and is considered an essential mineral.

DNA

When a human egg cell containing 23 chromosomes is fertilized by a sperm cell containing 23 chromosomes, the resulting fertilized egg, containing 46 chromosomes, undergoes repeated cycles of cell division (mitosis). Cells undergo differentiation into specialized cells that further increase in number through cell division, eventually generating a human baby. By adulthood, there are about 10^{13} nucleated cells of many different kinds formed into specialized tissues and about 2×10^{13} nonnucleated red blood cells. The nucleus in each cell (some cells, e.g., muscle cells, contain more than one nucleus) contains 46 chromosomes: two copies of chromosomes 1 to 22 plus two X chromosomes (female) or two copies of chromosomes 1 to 22 plus one X and one Y chromosome (male). Egg and sperm cells contain only 23 chromosomes. Egg cells contain a single copy of chromosomes 1 to 22 plus an X chromosome; sperm cells contain a single copy of chromosomes 1 to 22 plus either an X or Y chromosome. Thus the sperm that fertilizes the egg determines the sex of the offspring.

Each **chromosome** contains one very large DNA molecule, so in a somatic cell nucleus (nonsex cell) there are 46 DNA molecules, but there are only 23 DNA molecules in a sperm or egg nucleus. Figure 3.3 shows part of the structure of a DNA molecule. On the left is part of a single DNA strand indicating the joining of the deoxyribose sugars via phosphate groups. Note that the sugars are linked together by phosphate groups attached to the 3' carbon of one sugar and then to the 5' carbon of the next. The base is attached to the 1' carbon of each deoxyribose. On the right side of figure 3.3 is double-stranded DNA with the two strands (or chains) held together by hydrogen bonding between the bases. In all cases, an adenine base (A) in one chain is linked to a thymine (T) base in the other with two hydrogen bonds, and a cytosine (C) base in one strand is joined to a guanine (G) base in the other with three hydrogen bonds. The two chains are thus **complementary**: if we know the base sequence in one chain, we know the base sequence in the other because A hydrogen bonds to T, and C hydrogen bonds to G. The two chains are also **antiparallel**. This means that one chain runs 5' to 3' from top to bottom, whereas the other runs 3' to 5' from top to bottom. Finally, the two chains are coiled around a common axis forming the familiar double helix, with a full turn of the helix every 3.4 nanometers (see figure 3.4). Each full turn of the helix has 10 stacked pairs of bases. This is the most common form of DNA in the cell and it is known as B-DNA.

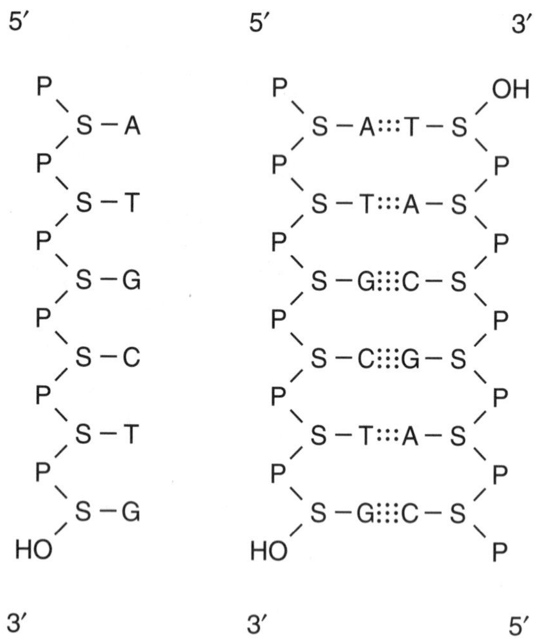

Figure 3.3 A shorthand way of illustrating single-stranded DNA (left) and double-stranded DNA (right). S refers to deoxyribose sugar; P represents the phosphate groups. The sugar phosphate backbone runs from the 5' to 3' end. The complementary strand of DNA on the right is antiparallel. The double-stranded DNA is joined by hydrogen bonds (dotted lines): two join the bases adenine (A) and thymine (T) and three join the bases cytosine (C) and guanine (G).

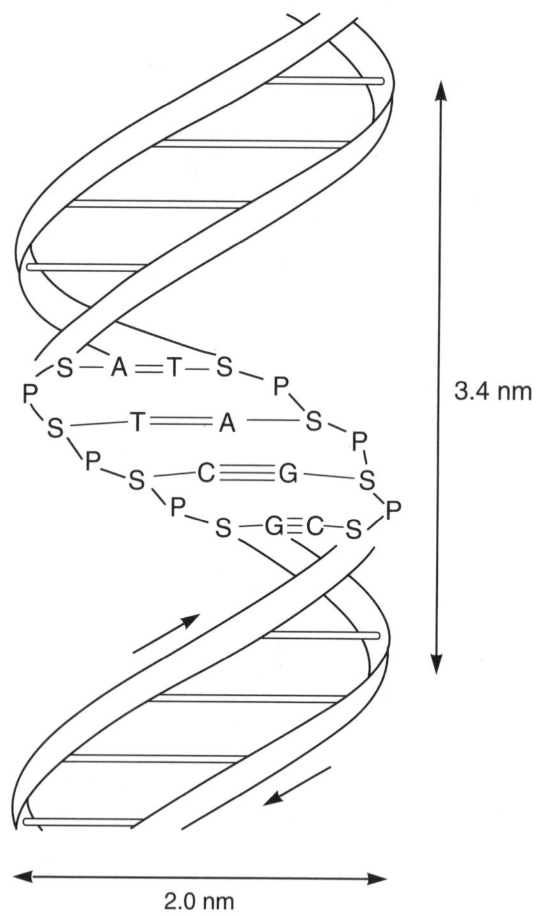

Figure 3.4 The double helical structure of DNA showing the two strands held together by hydrogen bonding. The vertical line represents the central axis; the small arrows show the polarity of the complementary strands. One complete turn of the helix spans a distance of 3.4 nm, in which are 10 stacked pairs of bases.

KEY POINT

With a common sugar phosphate backbone, the message carried by DNA lies in the order of the four bases. Therefore, we can say that the language of DNA is based on only four letters.

DNA molecules are very large and described according to the number of base pairs. Thus terms like base pairs (bp), thousands of base pairs (kilobase pairs, kbp), or megabase pairs (Mbp) are widely used. For example, the total DNA in the 23 chromosomes of a sperm or egg cell contains more than 3×10^9 bp. We could also say it contains more than 3×10^6 kbp or 3×10^3 Mbp. In the nucleus of a somatic cell there would be twice as much DNA since there are 46 chromosomes. If joined end to end and allowed to extend, this DNA would have a length of several meters. Accordingly, in a cell nucleus too small to be seen by the naked eye, the DNA is tightly coiled. Assisting with the coiling is a class of proteins called **histones**. These form an ionic interaction with the DNA, as the histones are rich in positive charges due to an abundance of lysine and arginine amino acid residues, counterbalancing the negative charges of the phosphate groups on the DNA backbone. The combination of DNA and protein in a cell nucleus is called **chromatin**.

KEY POINT

If stretched out, the DNA in a cell nucleus is nearly two meters long, yet it cannot be seen by the naked eye in the cell nucleus. For this to be true, the DNA and histone proteins must be densely packaged. Consider the implications of this packing in terms of copying a section of a DNA molecule (a gene) during transcription.

In summary, DNA is double stranded. The two strands are held together by hydrogen bonds between bases; they are complementary (A = T and G = C), antiparallel (the two chains run in opposite directions), and twisted into a double helix.

DNA Replication

Before a parent cell divides to produce offspring cells, known as daughter cells, the DNA must first be duplicated. Each daughter cell gets an exact copy of the parent DNA. To duplicate the parent DNA, each strand of the parent DNA acts as a template on which a complementary copy is made. Each daughter cell then receives one strand of the parent DNA and a newly synthesized complementary copy. This form of DNA replication is called **semiconservative**.

DNA replication is catalyzed by DNA polymerase. The precursors needed to make a complementary copy of DNA are **deoxyribonucleotide triphosphates** (dNTP): dATP, dCTP, dGTP, and dTTP (or TTP). Since DNA is so large, replication takes place simultaneously at multiple sites on the same molecule. DNA polymerase unwinds the strands where replication is to begin, and a complementary copy is made of each strand by adding nucleotides one at a time, using the dNTP precursors. The incoming dNTP attaches to the 3' OH group of the preceding nucleotide via a phosphate ester bond, and a pyrophosphate molecule (PPi) is released. The next dNTP comes in, attaches to the 3' OH of the preceding nucleotide and another PPi is released. In all cases, the correct dNTP comes in because of complementary base pairing. Figure 3.5 shows this

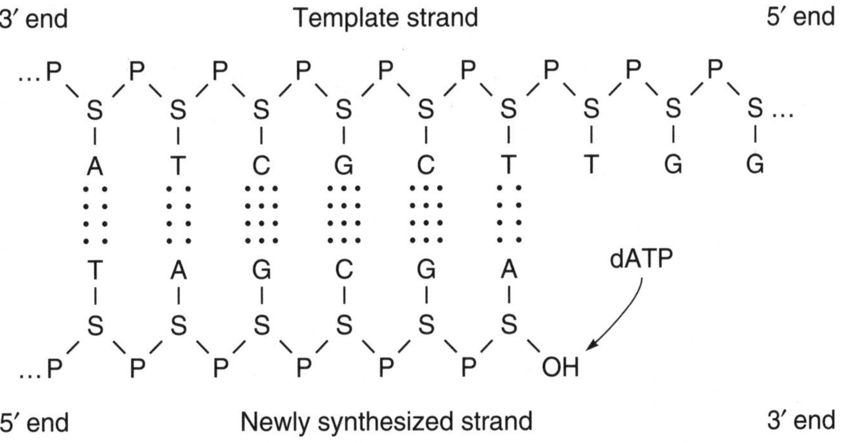

Figure 3.5 A schematic of DNA replication showing the template strand, though both strands are copied. Dotted lines represent hydrogen bonds between complementary bases on the template and newly synthesized strand. The next deoxynucleotide to be added contains the base adenine (A) because the template has the base thymine (T). The 3' hydroxyl group of the last nucleotide will bind to the incoming dATP, adding a new nucleotide to the growing new strand.

schematically. Although both DNA strands are duplicated, this figure shows only one of the two parent strands being copied. As you can see, DNA polymerase reads the template strand in the 3' to 5' direction, whereas the newly synthesized strand grows in the 5' to 3' direction. Magnesium ions (Mg^{2+}) are needed to help the DNA polymerase enzyme function properly.

DNA replication occurs with high fidelity, or very few errors, for an error in base sequence can be very harmful. Between cycles of cell division, the DNA in our cells is constantly damaged due to spontaneous breaks, radiation, harmful chemicals, and oxidation. This leads to breaks in the DNA molecule and loss of nucleotides. If unrepaired, the damaged DNA would be repeated and gradually degraded with each cycle of DNA replication, resulting in mutations and cancers. **Mutations** may be silent, that is, not expressed, or visible, that is, expressed. Visible mutations can be harmful. Mutations may be classified as base substitutions, in which a base is altered or substituted for by a closely related base, or frame shifts, in which a base is added or deleted. Either totally alters the DNA message in that region, if expressed in the cell where it occurs. Fortunately, DNA repair enzymes are constantly scanning nuclear DNA to slice out damaged sequences or join up broken pieces. This process of excision and repair is carried out by a complex of enzymes in the nucleus.

KEY POINT

Mitochondria contain multiple copies of a single circular DNA molecule, which contains 16,569 base pairs and codes for 37 genes. However, the DNA repair mechanisms are much less extensive in mitochondria. Therefore, mutations in the mitochondrial genome accumulate as we age. Many authorities believe that these mitochondrial DNA mutations are the reason for some of the deleterious conditions associated with aging.

Each DNA molecule in a chromosome is linear and therefore has two ends, known as telomeres. There are no genes at the telomeres, but there are repeats of short base sequences up to thousands of times. In humans the repeat is six bases: CCCTAA. The sequence in the complementary strand would be GGGATT. Telomeres act to stabilize DNA. During DNA replication, parts of the telomere are lost. The enzyme telomerase adds individual nucleotides to the end to maintain the telomeres.

RNA

RNA or ribonucleic acid is a polynucleotide that differs from DNA in the following characteristics: (a) It is single stranded; (b) it contains the sugar ribose instead of deoxyribose; (c) instead of the base thymine, it contains the base uracil (U); and (d) RNA molecules are much smaller than DNA molecules. However, RNA molecules can still be fairly large, and we often discuss their size by reference to the number of bases. For example, an RNA molecule that is 2,500 bases long would be described as 2.5 kb. RNA size is also commonly described in terms of the number of nucleotides (Nt); that is, saying 2,000 bases is equivalent to saying 2,000 Nt. DNA constitutes the master file of genetic information, unfailingly reproduced over successive generations of cell division. RNA molecules are working copies of parts of DNA and thus are tools used for making proteins.

KEY POINT

There is a four letter alphabet in DNA: A,G, T, and C. For RNA, which is copied from DNA, there is also a four letter alphabet, but U in RNA replaces T in DNA.

We have seen that single strands of DNA can be joined together by hydrogen bonds if the bases in the two strands are complementary. We call this process **hybridization**, and it is also possible to get RNA-DNA hybridization and RNA-RNA hybridization. This property of hybridization of single-stranded polynucleotide chains is extremely important. For example, one can seek out a particular base sequence in single-stranded DNA or RNA if one has a short length of radioactively labeled polynucleotide of complementary base sequence, that is, a probe. Hybridization utilizing specific probes underlies the techniques of Southern blotting (searching for specific DNA base sequences) and Northern blotting (seeking RNA base sequences).

Some viruses, known as **retroviruses**, contain single-stranded RNA instead of double-stranded DNA as their genetic material. When these viruses enter a cell, they bring with them an enzyme known as **reverse transcriptase**. Inside the host cell, reverse transcriptase catalyzes the formation of a strand of DNA complementary to the RNA molecule of the virus. Accordingly, reverse transcriptase is also known as RNA-directed DNA polymerase. The resulting RNA-DNA hybrid is then acted on by an enzyme, known as a **ribonuclease** (or RNA-degrading enzyme), which breaks down the RNA strand. The action of ribonuclease leaves a single strand of DNA, complementary to the virus RNA. DNA polymerase is then used to make a complementary copy of the single DNA strand, producing double-stranded DNA, which is then incorporated into the chromosomal DNA of the host. The AIDS virus is an example of a retrovirus. Reverse transcriptase is also an important laboratory enzyme, used to make DNA copies of RNA molecules, known as **complementary DNA** or cDNA.

Summary

Deoxyribonucleic acid, or DNA, is the molecule of heredity. Each human somatic cell nucleus contains 46 DNA molecules, which are associated with protein, making up our 46 chromosomes. DNA consists of two polynucleotide chains. Each nucleotide in a DNA polynucleotide chain consists of an organic base attached to a sugar known as deoxyribose, containing a covalently bound phosphate group. Adjacent nucleotides in each chain are held together by phosphate ester bonds joining sugar molecules and phosphate groups. The bases in DNA are adenine (A), guanine (G), cytosine (C), and thymine (T). The two polynucleotide chains in DNA are held together by hydrogen bonds between complementary bases, A bonding with T and G bonding with C. The two chains are antiparallel, with one running in the 5' to 3' direction and the other running 3' to 5'; these directions describe the orientation of the sugar phosphate backbone of the polynucleotide. The complementary polynucleotide chains are wound in a double helix. Information for the sequence of amino acids in polypeptide chains is provided by the sequence of the four bases in each polynucleotide chain.

Before cells divide to form new cells, the DNA is duplicated in a reaction catalyzed by DNA polymerase. Each chain in the parent cell acts as a template on which a complementary strand of new DNA is made. Subsequently, offspring cells each receive a strand of parent DNA and a new complementary copy, such that daughter DNA and parent DNA are identical. This is known as semiconservative DNA replication. Ribonucleic acid or RNA is a single-stranded polynucleotide, a copy of part of a DNA molecule known as a gene. RNA differs from DNA in that the sugar ribose replaces deoxyribose and the base uracil (U) replaces thymine (T). RNA information in the form of the base sequence in the single-stranded molecule is used to order the sequence of amino acids in a protein. This topic is covered in part III.

Key Terms

adenosine 29
anhydride bond 30
antiparallel 32
chromatin 33
chromosome 31
complementary 32
complementary DNA 35
deoxyribonucleotide triphosphates 33
histones 33
hybridization 34
mutations 34
nucleoside triphosphate 30
nucleosides 29
nucleotides 29
pentose phosphate pathway 31
phosphate ester 29
purines 29
pyrimidines 29
retroviruses 34
reverse transcriptase 35
ribonuclease 35
semiconservative 33
thymine 29
uracil 29

Review Questions

1. Estimate the length of DNA in a somatic cell nucleus if the 46 molecules of DNA are laid end to end. Assume that 10 each base pairs of DNA have a length of 3.4 nanometers and that there are 3×10^9 base pairs in the genome.
2. If the sequence for the first 10 bases of a DNA molecule is TTAGCCTAAA, what would be the corresponding sequence of bases in the complementary strand?

PART II

Metabolism

Imagine yourself standing waist-deep in water at the beginning of the Olympic triathlon. You have a swim of 1.5 km, a bike ride of 40 km and a run of 10 km facing you. Competing against you will be the best triathletes in the world. At this particular time you are unlikely to consider the metabolic aspects of your event, but an exercise scientist could tell you that over the next almost two hours, you will

- expend approximately 2,400 kilocalories (10,000 kilojoules),
- oxidize approximately 460 grams of carbohydrates,
- oxidize more than 50 grams of fat,
- utilize about 480 liters of oxygen, and
- break down and re-form about 123 moles of ATP, consisting of 7.4×10^{25} molecules of ATP, and weighing 50 kilograms.

You will use the great majority of this ATP to drive the contractions of the muscles used in the swim-ming, the cycling, and the running. This is material that you have learned about in your physiology of exercise courses. How you will re-form the ATP so that its concentration will remain relatively constant in these muscles throughout the competition represents most of our discussion of metabolism.

The section on metabolism focuses on the detailed processes for the breakdown of fuel molecules (fats, carbohydrates, and amino acids) to make ATP. We also look at the formation and storage of fuels. Throughout these chapters, we emphasize not only the details of the processes, but also the mechanisms that regulate them, which includes the roles played by diet and exercise. We begin with an overview of metabolism, including the pivotal role of energy.

CHAPTER 4

Bioenergetics

It is relatively easy for students to memorize metabolic pathways in which fuel molecules are broken down and the energy released is captured in the synthesis of ATP. Few, however, really understand what makes this work and the critical role of free energy release and capture. By the end of this chapter, you should be able to appreciate energy transduction in the cell.

Metabolism, which represents all the chemical reactions or processes that occur in a living organism, operates in much the same way as a gasoline engine. For the engine to do useful work, it needs to oxidize fuel. For our bodies to grow and function (useful endeavors), we need to oxidize fuels. Our bodies and the gas engine are also open systems that exchange matter and energy with the surroundings. The engine takes in fuel and oxygen and gives off heat and exhaust gases and generates useful work. We take in food, water, and oxygen and give off heat and waste gases, liquids, and solids and, it is hoped, generate some useful work. A gas engine is highly organized with each part in a precise location; this allows the engine to carry out its function. Our cells are also highly organized with proteins arranged in a particular way rather than haphazardly located. This too is essential to cell function. We look first at the energetic aspects of metabolism before focusing on the details of the various pathways and processes.

The Concept of Free Energy

Before we begin, however, let us review a few simple concepts:

- All chemical reactions involve energy changes. If there is no energy change, nothing happens.
- Chemical reactions in living organisms are catalyzed by enzymes.
- Enzymes attempt to drive the reaction they catalyze toward equilibrium.
- As enzyme-catalyzed reactions proceed toward equilibrium, they release energy.
- The farther a reaction is from equilibrium, the more energy it can release.
- Some energy released in a chemical reaction can be used to do work; the remainder is unavailable.

In thermodynamics, we focus only on differences between initial and final states; in this sense, we talk about changes. The following thermodynamic terms are defined as applied to biological processes:

- **Enthalpy change** (ΔH): In any chemical reaction or sequence of reactions, the change in energy of the reactants when they are turned into products is called the enthalpy change. This can be measured as the total heat energy change, described by the abbreviation ΔH. When heat energy is given off or released in a reaction or process, we have a negative value for ΔH; such a reaction is said to be **exothermic**. Reactions or processes with a positive value for ΔH are said to be endothermic.

- **Entropy** change (ΔS): Entropy is a measure of randomness or disorder. If a complex sugar molecule reacts with oxygen so that it is broken down to water and carbon dioxide, the products are more disorganized than the reactants. When a reaction or process becomes more disordered, it has a positive value for ΔS. When a reaction or process has a negative value for ΔS, it means something more organized than the reactants has been created, as when a protein is synthesized from amino acids.
- **Free energy change** (ΔG): Of the total energy released in a reaction or process, not all is available to do something useful. The free energy change, or ΔG, is the maximum energy available from a reaction or process that can be harnessed to do something useful. From a biological perspective, something useful would be a muscle contracting and lifting a load, moving ions across a membrane against their concentration gradient, or synthesizing a protein from amino acids. When free energy is released the ΔG is negative, and we call the reaction **exergonic**. A reaction with a positive value for ΔG cannot occur by itself—as a ball spontaneously rolling up a hill—and is said to be **endergonic**.

The relationship between enthalpy, entropy, and free energy for a reaction or process is given by the expression

$$\Delta G = \Delta H - T\Delta S \qquad (4.1)$$

Exergonic Reactions Drive Endergonic Reactions

Equation 4.1 describes the relationship between the free energy change, the enthalpy change, and the entropy change for a reaction or process that occurs at the temperature T, given in degrees Kelvin (°K = °C + 273).

ATP breakdown (properly termed ATP hydrolysis), the oxidation of fuels, and glycolysis are examples of biological exergonic reactions, whereas protein synthesis, muscle contraction, creating ion gradients across membranes, and the storing of fuels are examples of endergonic processes. Since endergonic reactions cannot occur by themselves, they must be driven by a simultaneous exergonic reaction such that the combined endergonic plus exergonic reaction is exergonic. For example, a muscle cannot contract by itself; it must be driven by the simultaneous hydrolysis of ATP. Moving sodium ions out of the cell and simultaneously moving potassium ions back in requires ATP hydrolysis. In these two examples the free energy released by the hydrolysis of ATP is necessary. There are other processes that require specific nucleoside triphosphates (NTP) other than ATP. For example, joining one amino acid to another during the synthesis of a peptide is an endergonic reaction and is driven by GTP hydrolysis. Glucose molecules cannot join together to make glycogen in muscle or liver unless UTP is simultaneously hydrolyzed. In summary, exergonic reactions can drive endergonic reactions provided the sum of the two is exergonic. In most cases, the exergonic reaction is NTP hydrolysis.

As shown in figure 4.1, processes such as glycolysis or the oxidation of fuels are used to combine Pi and NDP (nucleoside diphosphate, e.g., ADP and

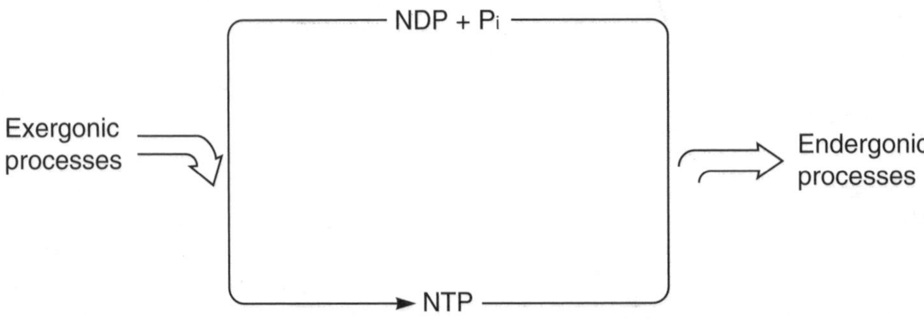

Figure 4.1 Exergonic processes, such as glycolysis and the oxidation of fuels, provide the free energy to combine a nucleoside diphosphate (NDP) with a phosphate (Pi) group to make a nucleoside triphosphate (NTP). Subsequent hydrolysis of the NTP then provides the energy to drive endergonic processes such as muscle contraction, creating ion gradients, and synthesizing larger molecules from smaller precursors.

GDP) to make NTP (i.e., ATP and GTP). Then the NTP drives the endergonic reaction. This process is the essence of life.

Quantitative Values for Free Energy

In thermodynamics, one cannot measure exact values associated with an initial and final state of a reaction. However, we can measure energy changes over the course of a reaction as changes in enthalpy, changes in free energy, and changes in entropy. For the reversible chemical reaction given by the equation

$$A + B \longleftrightarrow C + D$$

the free energy change is given by the expression

$$\Delta G = \Delta G^{o\prime} + RT \ln \frac{[C]\,[D]}{[A]\,[B]} \qquad (4.2)$$

where ΔG is the free energy change for the reaction.

$\Delta G^{o\prime}$ is the standard free energy change for the reaction under what are considered to be standard conditions, when each of the reactants and products is at a concentration of 1.0 M. A prime is added to indicate a pH of 7.0; the traditional thermodynamic standard free energy is defined at a most nonbiological pH of zero. $\Delta G^{o\prime}$ reflects the energy-generating potential of the reaction, and a value for $\Delta G^{o\prime}$ can be determined from the equilibrium constant at a pH of 7.0 (K'_{eq}, discussed later).

R is the gas constant, which has a value of 8.314 joules per mole per °K (1.987 cal per mol per °K). Remember, 1 calorie (cal) is equivalent to 4.18 joules (J) or 1 kilocalorie (kcal) = 4.18 kilojoules (kJ). T is the absolute temperature expressed in °K.

[A] is the concentration of species A, [B] is the concentration of species B, and so forth.

Suppose we start the above reaction by adding A and B together. The reaction will quickly proceed to the right, and as C and D accumulate, the net reaction (toward C and D) will gradually slow down. Eventually the forward (toward C and D) and backward (toward A and B) reactions will occur at the same rate, and we will have reached **equilibrium**. From an energy perspective we know the following:

1. The farther a reaction is from equilibrium, the larger the value of G (as a negative value). In equation 4.2, when the reaction begins, [A] and [B] will be large, whereas [C] and [D] will be small. Accordingly, the natural logarithm (ln) term will be large and negative. Thus at the start of the reaction the ΔG will be large and negative.
2. As the reaction proceeds toward equilibrium, the numerical values for [A] and [B] will decrease, whereas the values for [C] and [D] will increase. Accordingly ΔG will decrease but still be negative.
3. At equilibrium, there is no net reaction and no net energy change, so the value for ΔG is zero. Accordingly, a reaction at equilibrium releases no free energy.

Rewriting equation 4.2 at equilibrium, where $\Delta G = 0$, we get

$$0 = \Delta G^{o\prime} + RT \ln \frac{[C]\,[D]}{[A]\,[B]}$$

However, at equilibrium (and a pH of 7.0) the concentration ratio [C][D]/[A][B] is given by the equilibrium constant K'_{eq}. We use this fact and rearrange the equation:

$$\Delta G^{o\prime} = -RT \ln K'_{eq} \qquad (4.3)$$

This equation allows us to determine the value of the standard free energy change for a reaction at pH 7.0 if we know the temperature and the value of the equilibrium constant. We can say the following about the $\Delta G^{o\prime}$ value for a reaction:

1. It is determined by the value for the equilibrium constant.
2. It reflects the energy-generating potential for the reaction.
3. It does not define the actual energy change for the reaction inside a cell, since this depends on the relative concentrations of reactants and products.

The actual free energy change for a reaction inside a tissue is easy to determine. Take a sample of tissue and quickly freeze it, which immediately stops all chemical reactions. Analyze chemically the amount of reactant(s) and product(s) for the reaction in question. From this, one can determine the **mass action ratio** for the reaction. This is the ratio of product to reactant concentrations as they existed at the moment the tissue was frozen, given by c. The K'_{eq} for the reaction can be determined by allowing it to come to equilibrium in a test tube under conditions of temperature, pH, and ion concentration identical to those of the living tissue, and then analyzing the equilibrium concentrations of reactant(s) and product(s). Using equation 4.4, we could determine the free energy change of our reaction as it existed at the time it was frozen.

$$\Delta G = -RT \ln \frac{K'_{eq}}{c} \quad (4.4)$$

In general, enzyme-catalyzed reactions in the cell are linked to other reactions, often in well-established pathways. The free energy changes for linked reactions are additive. Suppose we want to determine the standard free energy change for ATP hydrolysis.

$$ATP + H_2O \longrightarrow ADP + Pi$$

The reaction proceeds so far to the right that it is almost impossible to determine an equilibrium constant. Thus establishing an actual concentration for ATP at equilibrium would not be possible. However, we can determine this another way. Look at the first two reactions below. In the first, ATP is used to phosphorylate (add a phosphate group to) glucose. In the second, glucose 6-phosphate, the product of this phosphorylation, is dephosphorylated. If we know the standard free energy changes for each of these reactions, and if we add these reactions together algebraically, the net reaction (under the line) is the reaction for ATP hydrolysis. The algebraic sum of the standard free energy changes is then the value for ATP hydrolysis.

(1) ATP + glucose $\longrightarrow$ glucose 6-phosphate + ADP
$\Delta G^{o'} = -23$ kJ/mole

(2) glucose 6-phosphate + $H_2O \longrightarrow$ glucose + Pi
$\Delta G^{o'} = -14$ kJ/mole

(net) ATP + $H_2O \longrightarrow$ ADP + Pi
$\Delta G^{o'} = -37$ kJ/mole

KEY POINT

Many chemical reactions in cells are linked with other reactions to form what we call pathways. We will encounter a variety of these, such as glycolysis and the TCA cycle. While the actual concentrations of intermediates in the pathway may not change, there will be substrate(s) entering the pathway and the final product(s) leaving the pathway. We use the term flux to describe how fast the pathway is converting initial substrate to final product. There will be an overall free energy change for the pathway, but the greater the flux through the pathway the more free energy is released per unit of time.

Energy-Rich Phosphates

Living organisms obtain their energy from the breakdown of fuels such as fat, carbohydrate, and amino acids (from protein). Inside cells, fuel molecules are catabolized to simple products such as CO_2 and H_2O. During this **catabolism**, the free energy released drives the phosphorylation of nucleoside diphosphates to form nucleoside triphosphates. We use the term **phosphorylation** here to mean that we are adding a phosphate group, in this case, to an NDP to make NTP. The major nucleoside diphosphate is ADP, and the product is ATP. The reaction illustrating the formation of ATP is

$$ADP + Pi \longrightarrow ATP + H_2O$$

The hydrolysis of ATP (or GTP or UTP) drives endergonic reactions or processes. Notice that the reactions showing the formation of ATP and the hydrolysis of ATP are exactly reversed, although these are definitely not reversible reactions. Each day, an active person turns over (breaks down and resynthesizes) an amount of ATP approximately equal to his or her body weight.

The structure of ATP, shown in shorthand in figure 4.2, contains three phosphate groups joined by two anhydride bonds. Although the phosphate groups are stabilized by magnesium (Mg^{2+}) in the cell, negative charges are still in close proximity and breaking the bonds between either the α and β or β and γ phosphate groups is energetically favorable. Thus the hydrolysis of anhydride bonds of ATP generates energy to drive endergonic reactions. Extra energy is obtained when the hydrolysis is between the α and β phosphates, because the product PPi can also be hydrolyzed by an ubiquitous enzyme, **inorganic pyrophosphatase**, which results in a further large free energy release (see figure 4.2). In the cell, ATP is always found in close ionic association with Mg^{2+}. Therefore, we

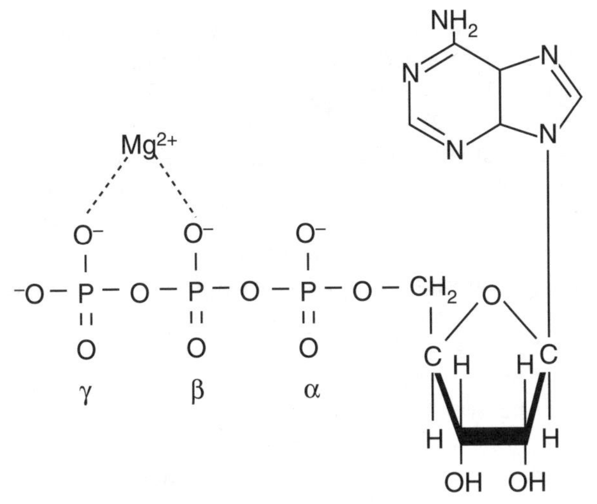

$$(1)\ ATP + H_2O \longrightarrow ADP + Pi$$

$$(2)\ ATP + H_2O \longrightarrow AMP + PPi$$

$$PPi + H_2O \longrightarrow 2\ Pi$$

Figure 4.2 ATP in the cell is found in association with magnesium ions. The three phosphate groups are often identified by Greek letters. Hydrolysis of ATP can take place between the β and γ phosphate groups (equation 1) or between the α and β phosphate groups (equation 2).

should show ATP as MgATP, but for the sake of convenience we won't. In all chemical reactions involving ATP as substrate or product, consider that Mg^{2+} is always present. ADP has an affinity for the Mg^{2+} ion also, but not nearly as strong as ATP. Thus in the cell, all the ATP forms a complex with Mg^{2+}, but only a fraction of the ADP is associated with Mg^{2+}.

KEY POINT

To understand the fact that ATP hydrolysis releases a lot of free energy, consider the analogy of a spring, compressed between your thumb and second finger. It requires force to keep the coils of the spring together. If you quickly release the pressure on the spring, it rapidly pops free, reaching a longer length and releasing free energy.

ATP would be useless as an energy currency in the cell if the concentrations of ATP, ADP, and Pi were equilibrium concentrations. At equilibrium, the concentration of ADP would be more than 10 million times greater than ATP, with no net ATP hydrolysis and no free energy release. For ATP to be the important energy currency it is, the concentration in the cell is kept very far from equilibrium so that the ratio [ATP]/[ADP] is normally greater than 50. Moreover, when ATP is used as an energy source it is replenished at the same rate so that its concentration does not decrease. During very severe exercise, the concentration of ATP does decline but not to less than 50% of what it was at rest. Because large free energy changes accompany the hydrolysis of ATP and the other nucleoside triphosphates, they are known as energy-rich phosphates.

The Pool of Phosphates in the Cell

The adenine nucleotides (i.e., ATP, ADP, AMP) are primarily involved in coupling exergonic and endergonic reactions. ATP is formed from ADP when fuel molecules are broken down, and ATP hydrolysis drives most endergonic processes. GTP drives peptide bond formation during protein synthesis, and one GTP molecule is formed during the tricarboxylic acid (TCA) cycle. Also, UTP is used to make glycogen in liver and muscle. As we will see, GTP, UTP, CTP, and ATP are used to make the various molecules of RNA. Furthermore, when RNA molecules are broken down, the products are NMP molecules, that is, GMP, UMP, CMP, and AMP. Since there is a continual synthesis and breakdown of RNA molecules in the cell, the need to maintain a balanced level of NTP molecules is evident.

This means that there must be specific reactions to interconvert the various components of the nucleoside phosphate pool. For example, nucleoside diphosphates must convert to nucleoside triphosphates and nucleoside monophosphates to nucleoside diphosphates. For the first interconversion, a nonspecific enzyme known as nucleoside diphosphate kinase transfers phosphate groups to and from ATP and ADP, and NTP and NDP, as follows:

$$NDP + ATP \xleftrightarrow{\text{nucleoside diphosphate kinase}} NTP + ADP$$

In this freely reversible reaction, NDP (such as UDP, GDP, and CDP) is phosphorylated to NTP by accepting a phosphate group from ATP. Because the reaction is freely reversible, its net direction depends on the relative concentrations of ATP, ADP, and the other NTPs and NDPs. The overall effect of the nucleoside diphosphate kinase reaction is to maintain a balance in the ratio of NTP/NDP in a cell.

To convert NMP to NDP, nucleotide-specific enzymes transfer a phosphate group from ATP to the nucleoside monophosphate, making ADP and a nucleoside diphosphate. They are said to be specific because one exists for each nucleoside monophosphate. For example, AMP kinase catalyzes the following reaction:

$$\text{AMP} + \text{ATP} \xleftrightarrow{\text{AMP kinase}} \text{ADP} + \text{ADP}$$

UMP kinase catalyzes a similar reaction:

$$\text{UMP} + \text{ATP} \xleftrightarrow{\text{UMP kinase}} \text{ADP} + \text{UDP}$$

These freely reversible reactions also need magnesium ions because these ions are bound to the ATP molecules. The net effect of the nonspecific nucleoside diphosphate kinase and the specific nucleoside monophosphate kinase (e.g., AMP kinase and UMP kinase) is to maintain a balance between the NTP, NDP, and NMP in the cell.

KEY POINT

To keep our metabolism in balance, it is very important to the cell to keep constant ratios between the various NMPs, NDPs, and NTPs. ATP, which is the product of most of our catabolic processes, is the direct source of energy-rich phosphates to maintain this balance.

Phosphagens

The ATP concentration in most tissues is fairly low, about 3 to 8 millimoles per liter of cell water. Because the water content of skeletal muscle is approximately 75% by weight, we can also express the ATP concentration range as about 2 to 6 millimoles of ATP per kilogram of tissue. Since ATP represents the immediate energy source to drive endergonic processes, problems could arise if ATP is needed at a rapid rate and therefore is quickly used up. In cells with a slow acceleration of ATP-consuming reactions, ATP concentration can be easily maintained by a gradual acceleration of ATP-producing reactions, such as fuel oxidation. However, in muscle this could be a big problem because, during the transition from rest to maximal exercise, the rate of energy expenditure in a human muscle can increase by more than 100 times. The rate of energy turnover in a rested muscle is about 1 millimole of ATP per kilogram of muscle per minute. During sprinting, an elite athlete is able to turn over ATP at a rate of about 4 millimoles of ATP per kilogram of muscle per second. For such a highly trained athlete, this is more than 200 times the value in rested muscle. For normally active individuals, achieving 75% of the maximum rate of ATP turnover of an elite athlete is certainly feasible. During peak activity, all of the muscle cell's ATP could be consumed in about 2 seconds, if it were not regenerated.

KEY POINT

Although a muscle's ATP can be turned over at an astonishingly high rate, during most kinds of physical activity its content seldom decreases much below the resting level.

ATP in cells is regenerated from ADP by breaking down fuel molecules using aerobic or anaerobic catabolic processes. However, ATP-regenerating processes cannot produce ATP at the same rate at which it is hydrolyzed to drive muscle contraction during sprinting. Moreover, these processes take time to gear up to maximum speed, whereas at the start of a sprint, the rate at which ATP is hydrolyzed is about maximal. To prevent muscle cells from using up their ATP at the start of maximal or near maximal contractions, an alternate energy-rich molecule, known as a phosphagen, is capable of regenerating ATP at a very high rate. In vertebrate muscle the phosphagen is **phosphocreatine** (abbreviated PCr), also called creatine phosphate (abbreviated CP). In some invertebrate muscles, the phosphagen is arginine phosphate.

Phosphocreatine (or creatine phosphate) has its phosphate group transferred to ADP to yield ATP and creatine, in a reaction catalyzed by an enzyme known as creatine kinase:

$$\text{ADP} + \text{PCr} + \text{H}^+ \xleftrightarrow{\text{creatine kinase}} \text{ATP} + \text{Cr}$$

This reaction is freely reversible. During muscle contraction, the forward direction is favored in order to regenerate ATP. During recovery, the backward reaction is favored to regenerate PCr. A favorable difference in the ΔG for hydrolysis of PCr compared to ATP means that ATP formation (and thus PCr disappearance) is favored during rapid rates of ATP hydrolysis. Note that a proton (H^+) is consumed as a substrate when the creatine kinase reaction proceeds toward ATP formation. This is due to the acid-base characteristics of the phosphate groups. Consumption of a proton is advantageous since it can partially reduce the acidification of muscle during very vigorous exercise. We mention this later in this chapter.

The activity of creatine kinase is very high in muscle in order to match the regeneration of ATP during the most vigorous muscle contractions. The actual concentration of PCr in muscle is about 3 to 4 times that of ATP (about 18–20 mmol/kg of muscle), not that much considering how fast ATP can be used in a vigorously contracting muscle. However, it is enough to act as a temporary ATP buffer until other ATP-regenerating processes reach maximal rates. In general, both chemical analysis and the use of phosphorus nuclear magnetic resonance spectroscopy (NMR) reveal that the content of PCr and ATP is higher in muscle or muscle fibers with the highest rates of ATP hydrolysis. The ATP level in muscle must not be allowed to drop very much, for if it does, a condition known as rigor occurs, which can be damaging to muscle cells. In fact, rigor mortis is caused by the loss of muscle ATP some time after death, as a result of the inability of the muscle cells to regenerate ATP.

In skeletal muscle, the activity of creatine kinase exceeds that of all other enzymes. This means that the creatine kinase reaction is very important in muscle for sprinting or burst-type activities such as bounding. Animals in which the gene for creatine kinase is knocked out demonstrate very poor performance during high intensity exercise. Because there is more creatine kinase in skeletal muscle than any other enzyme, damage to a muscle cell membrane, typically through unaccustomed exercise or hard eccentric exercise, results in creatine kinase in the blood. Damage to the heart muscle can also result in the appearance of the cardiac isozyme of creatine kinase in blood. Determination of the activity and isozyme type can reveal the nature and extent of damage to cardiac and skeletal muscle.

> **KEY POINT**
>
> There is a great deal of creatine kinase in muscle. It is found near the contractile proteins, where ATP is hydrolyzed. It is found with the sarcoplasmic reticulum, at the level of the mitochondrial membranes, and it is found free within the muscle cell cytoplasm. In short, creatine kinase is found at the places where ATP is both consumed and produced.

Muscle Contraction

Skeletal muscle, the largest tissue in the body, consists of a number of elongated cells known as fibers, which contain a number of parallel myofibrils (see figure 4.3). Neurons innervate each fiber, and nerve impulses crossing from the neuron to the muscle fiber membrane (or sarcolemma) activate the fiber to make it contract. As shown in figure 4.4, the steps in this process may be summarized as follows:

1. Acetylcholine released by the neuron at the neuromuscular junction diffuses across the gap between neuron and sarcolemma, binds to its receptor, and causes depolarization of the sarcolemma.

2. A wave of depolarization passes over the **sarcolemma** and down into the interior of the fiber via surface invaginations known as **T-tubules**. The depolarization is associated with sodium ions (Na^+) moving into the fiber and some potassium ions (K^+) moving out of the fiber.

3. T-tubule depolarization is linked to calcium ion (Ca^{2+}) release from a specialized form of endoplasmic reticulum known as **sarcoplasmic reticulum** or SR.

4. The calcium ion concentration in the fiber interior increases from about 10^{-7} molar to about 10^{-5} molar (a 100-fold increase).

5. The calcium ions bind to the protein **troponin** in the thin filament in the sarcomere.

6. This binding allows **cross bridges**—protein projections on myosin molecules in the thick filament—to bind to the globular protein, **actin**, in the thin filament.

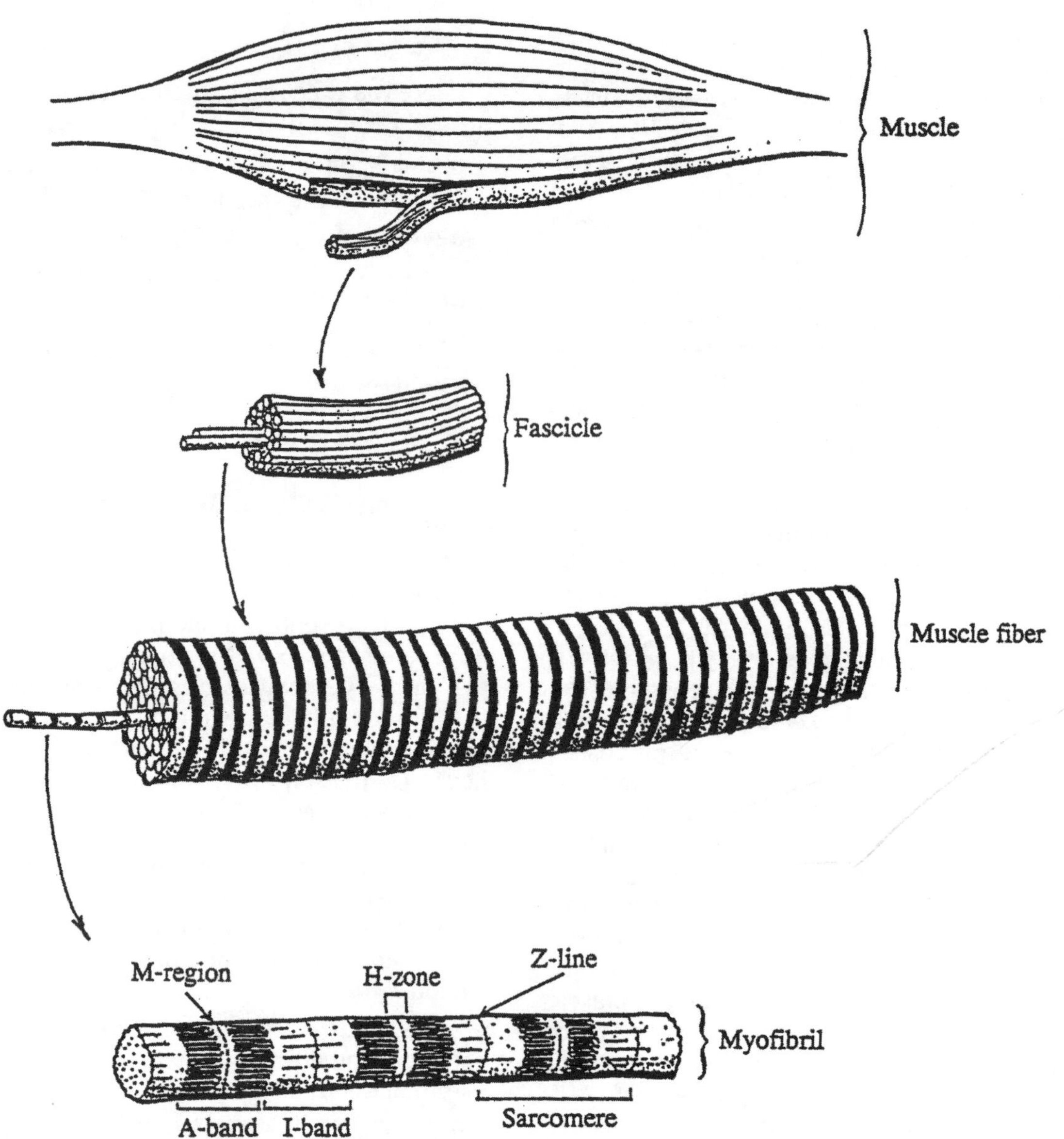

Figure 4.3 Skeletal muscle is composed of elongated cells known as fibers. These are typically arranged in bundles known as fascicles. Within a muscle fiber are myofibrils, aligned in parallel with the fiber long axis. Striations in the fiber and myofibril are due to the arrangement of contractile proteins, creating bands in the sarcomere.

Reprinted, by permission, from Alan J. McComas, 1996, *Skeletal Muscle: Form and Function* (Champaign, IL: Human Kinetics), 5.

7. Using the free energy of hydrolysis of ATP, the bound myosin cross bridges exert force on the actin in the thin filament.

8. The force exerted by millions of cross bridges in many fibers causes the overall muscle to shorten and do work.

9. The Ca^{2+} is pumped back into the SR in a process requiring the hydrolysis of ATP.

Figure 4.5 shows the cross bridge cycle and ATP hydrolysis. The free energy released when ATP is hydrolyzed is used to create a relative force between the thick and thin filaments. This results in relative sliding between the two filaments, generating muscle shortening. As shown in figure 4.5, this process takes place in multiple steps. (1) In the relaxed muscle, each myosin cross bridge contains one bound ATP molecule. (2) ATP is hydro-

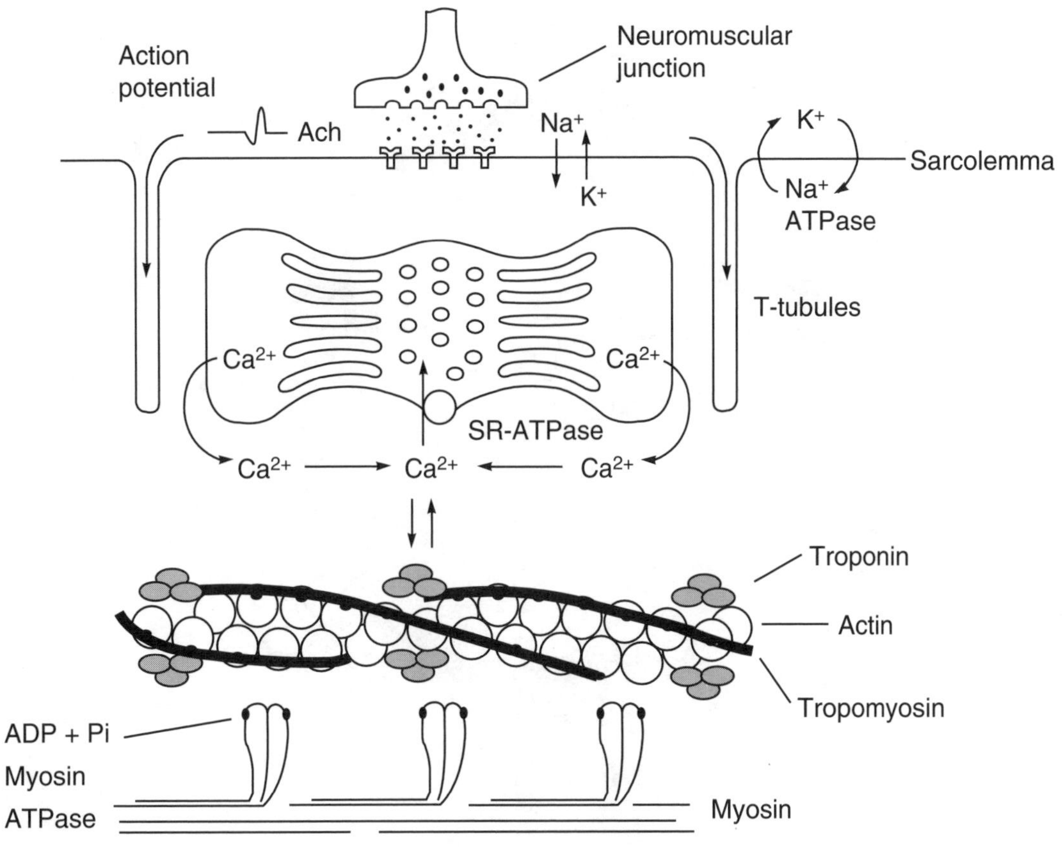

Figure 4.4 An overview of the steps in muscle contraction. The nerve impulse causes depolarization of the sarcolemma when acetylcholine (Ach) binds to its receptor. The resulting depolarization spreads over the surface of the muscle fiber and into the interior through the T-tubules, resulting in the release of Ca^{2+} ions from the sarcoplasmic reticulum (SR). Ca^{2+} binds to troponin on the thin filament, allowing myosin cross bridges to bind to actin on the thin filament creating force with ATP hydrolysis. When the nerve impulse ceases, Ca^{2+} unbinds from troponin, and is transported into the SR by the SR-ATPase at the expense of ATP hydrolysis. The Na^+-K^+ ATPase restores the balance of Na^+ and K^+ ions across the sarcolemma.

lyzed on the myosin molecule, and a strong bond is formed between a cross bridge and actin. (3) The Pi is released, and the myosin cross bridge assumes a different orientation, exerting a strong force on actin. (4) The ADP is released; a new ATP binds, and the binding between myosin and actin is relaxed.

Myosin has two major properties: (a) the ability to bind actin and (b) the ability to hydrolyze ATP as an enzyme. The ATP-hydrolyzing activity of myosin is known as **ATPase** activity. Actin also has two important properties: (a) myosin binding and (b) an ability to activate the ATPase activity of myosin. When a myosin cross bridge binds to actin its ATPase activity is increased at least 100-fold. We call this actin-activated myosin ATPase activity or actomyosin ATPase activity, illustrated as follows:

$$\text{ATP} + H_2O \xrightarrow{\text{actomyosin ATPase}} \text{ADP} + \text{Pi}$$

As we mentioned earlier in this chapter, MgATP is the real substrate for all reactions involving ATP. For convenience, it is traditional to just describe the ATP.

ATPases hydrolyze ATP so that work can be done. In muscle contraction, ATP hydrolysis by the actomyosin ATPase allows the muscle to shorten. Two other ATPases are also involved in muscle contraction. One is the sodium-potassium ATPase found in muscle sarcolemma. Its function is to pump sodium ions out of the cell and potassium ions back in after the muscle membrane depolarization ends, using energy derived from the hydrolysis of the ATP. The other ATPase, in the SR, hydrolyzes ATP and uses the free energy

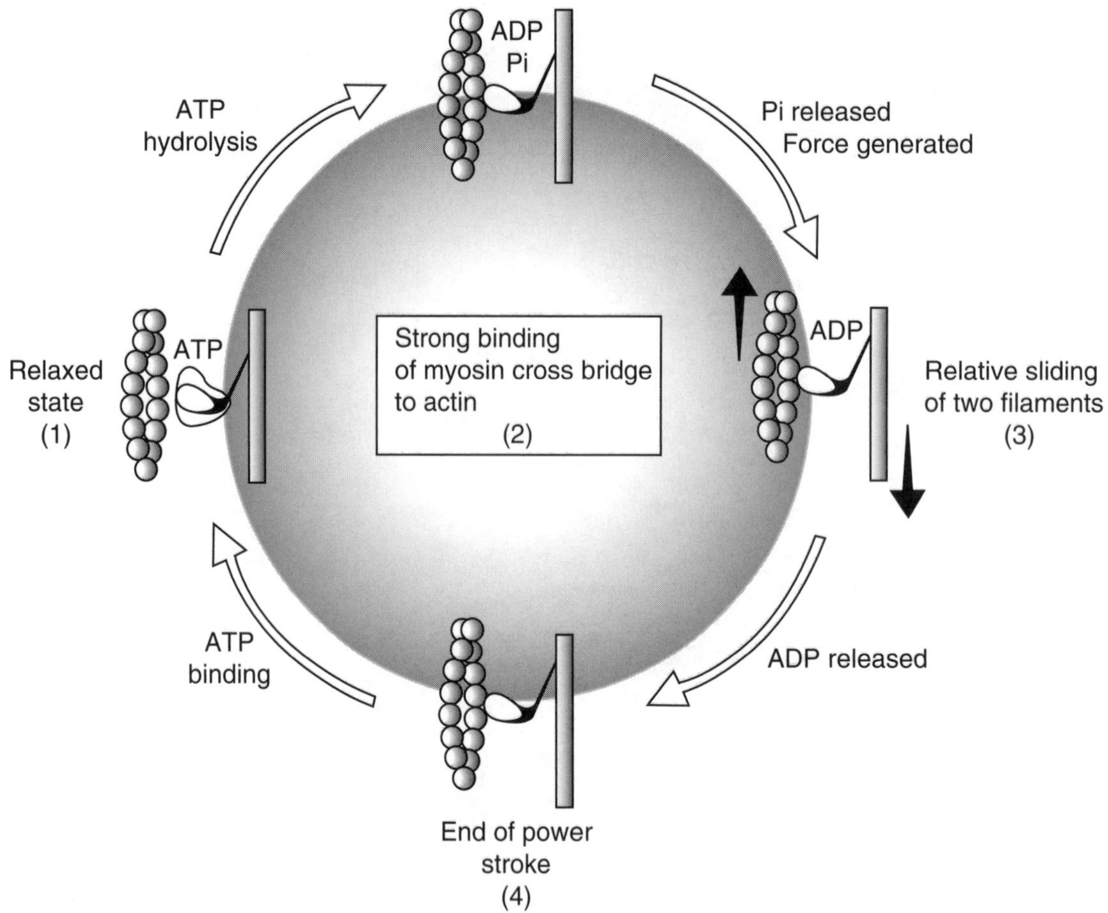

Figure 4.5 Cross bridge cycle. (1) Starting from the relaxed state with bound ATP, the ATP is hydrolyzed and the products ADP and Pi remain. (2) When Pi is released, a strong binding state is produced between the myosin cross bridge and actin. (3) Tilting of the cross bridge produces a relative sliding of the thick and thin filaments. ADP is released at the end of the power stroke, and (4) myosin and actin remain strongly bound to each other. Binding of a new ATP molecule produces the relaxed state.

released to pump calcium ions back into the SR when the muscle is to relax; it is known as the SR-Ca^{2+} ATPase (see figure 4.4).

KEY POINT

The importance of having an abundance of creatine kinase located at sites of ATPase activity should be clear, since it must regenerate ATP from the ADP produced by the various ATPases at a very rapid rate so that ATP concentration does not decrease much.

Three Energy Systems

The contraction of muscle involves ATP hydrolysis at three locations. Of course, this ATP must be regenerated quickly because, as previously stated, little ATP is stored in muscle. The ATP concentration in muscle is effectively maintained by three major processes, which are illustrated in figure 4.6. The ATPases represent the actomyosin ATPase, the ATPase associated with the sarcolemma to pump sodium and potassium ions, and the ATPase that pumps calcium ions back into the SR. ATP can be regenerated by the oxidation of fuels (oxidative phosphorylation), the creatine kinase reaction, and glycolysis.

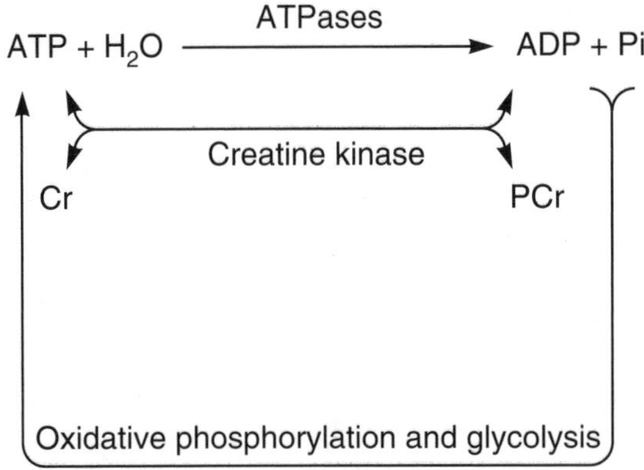

Figure 4.6 In muscle, three ATP-hydrolyzing enzymes (ATPases) use ATP at a rapid rate. The ATP concentration diminishes only slightly due to the rapid response of the creatine kinase reaction and the aerobic and anaerobic (glycolysis) breakdown of fuel molecules. Although the creatine kinase reaction is freely reversible, during muscle contraction the net direction at the site of the ATPase enzymes is to regenerate ATP at the expense of phosphocreatine (PCr).

Oxidative Phosphorylation

Oxidative phosphorylation is the formation of ATP from ADP and Pi in association with the transfer of electrons from fuel molecules to coenzymes to oxygen. Products of oxidative phosphorylation are H_2O and CO_2, and the free energy released is used to drive the phosphorylation of ADP with Pi to make ATP. Oxidative phosphorylation is the name employed primarily by biochemists to describe the major source of ATP to living organisms. Others also describe it as aerobic metabolism, oxidative metabolism, or cellular respiration.

Figure 4.7 outlines the overall scheme of oxidative phosphorylation. In this scheme, SH_2 represents fuel molecules such as carbohydrates, fats, amino acids, or alcohol. Electrons associated with the hydrogen are transferred from SH_2 to coenzymes (represented by NAD^+), and then the electrons on the now reduced coenzyme (NADH) are transferred on to oxygen, forming H_2O. During this process enough free energy is generated to phosphorylate ADP with Pi to make ATP. In chapter 2, electron transfer using dehydrogenase enzymes was illustrated. We devote the next chapter to the steps involved in oxidative phosphorylation. For the present, we can summarize it by the following simple equation, based on figure 4.7:

$$SH_2 + O_2 + ADP + Pi \longrightarrow S + H_2O + ATP$$

This equation does not show the role of the coenzymes NAD^+ or FAD during oxidation and reduction, which we discussed in chapter 2. Moreover, it does not reflect the true stoichiometry of ATP production, nor does it reveal that during oxidative phosphorylation CO_2 is produced from the carbon atoms of fuel molecules, represented by SH_2.

It is fairly easy to measure the rate of oxidative phosphorylation by measuring the disappearance of the substrate oxygen. Indeed, this is the method employed by exercise physiologists when they talk about **oxygen consumption**, measured at the mouth with special breath sampling techniques. The resulting $\dot{V}O_2$ is used as an index of whole body metabolism. During exercise $\dot{V}O_2$ increases in proportion to the rate of whole body energy expenditure. With gradually increasing loads of exercise, for example, treadmill running or pedaling on a cycle ergometer, $\dot{V}O_2$ reaches a maximum or peak that is considered to be the highest rate of oxidative phosphorylation for the individual during that activity. If a large enough muscle mass is activated, this peak rate of oxidative phosphorylation is the maximal rate the individual can achieve. This maximal rate of oxygen consumption or $\dot{V}O_2max$ can be described in absolute units, liters of oxygen consumed per minute, or relative to body weight in milliliters of oxygen per kilogram body weight per minute.

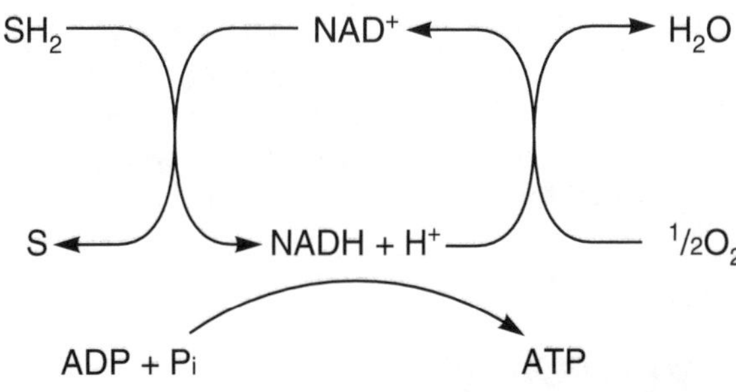

Figure 4.7 A brief summary of oxidative phosphorylation. Electrons in the form of hydrogen are removed from reduced substrates (SH_2) and transferred to the oxidized coenzyme NAD^+. The electrons on the reduced coenzyme NADH are transferred through a series of carriers (not shown) to oxygen, forming water. During the process of electron transfer, free energy is released to phosphorylate ADP, making ATP.

From a simple bioenergetics perspective, we can use the approximate relationship that a $\dot{V}O_2$ of 1.0 liter per minute is equivalent to an energy expenditure of about 5 kilocalories per minute (21 kJ/min). The precise relationship between $\dot{V}O_2$ in liters/minute and energy expenditure in kilocalories/minute depends primarily on the relative proportions of carbohydrate and fat that are being catabolized. The value is slightly higher for pure carbohydrate oxidation compared to pure fat.

KEY POINT

As we will see in the next chapter, oxidation of carbohydrate produces more ATP per unit of oxygen consumed than does the oxidation of fat. That an oxygen consumption of one liter per minute is approximately equal to a rate of energy expenditure of 5 kilocalories per minute helps us understand that these are simply different ways to describe the rate of oxidative phosphorylation.

Oxidative phosphorylation is considered a low power process because it relies on oxygen from the atmosphere as the final acceptor of electrons in fuels. With the steps in oxygen transport from the air, to the lungs, to the blood, to the individual cells, and finally to the mitochondria within cells, oxidative phosphorylation cannot quickly reach a maximal rate. Indeed the time needed to double the rate of oxidative phosphorylation is approximately 15 to 20 seconds. Although its power is limited in most cases by the oxygen transport system, oxidative phosphorylation is considered to have a high capacity (i.e., large fuel tank), because a major fuel for oxidative phosphorylation is fat. Even a very lean person has enough stored fat to provide fuel for days of oxidative phosphorylation.

KEY POINT

We mentioned in chapter 2 that the maximum rate of electron transport from fat to oxygen is approximately one half that when carbohydrate is the sole fuel. This means that work at intensity levels beyond 50% of maximal oxygen consumption can be compromised if body carbohydrate stores are low.

Phosphocreatine

Use of phosphocreatine to regenerate ATP is sometimes referred to by exercise physiologists as the anaerobic alactic system. As the creatine kinase equation below shows, during exercise the concentration of PCr declines, ATP is maintained, and no oxygen (anaerobic) or lactic acid formation (alactic) is involved.

$$\text{ADP} + \text{PCr} + \text{H}^+ \xleftrightarrow{\text{creatine kinase}} \text{ATP} + \text{Cr}$$

The creatine kinase reaction can regenerate ATP rapidly because the activity of creatine kinase is so high. As we have mentioned, creatine kinase is found at locations not only where ATP is hydrolyzed, but also where ATP is regenerated. This means that there are specific locations for creatine kinase—at the level of actin and myosin where actomyosin ATPase hydrolyzes ATP, at the sarcoplasmic reticulum where ATP is broken down to pump Ca^{2+} into the SR, and at the sarcolemma where the sodium-potassium ATPase maintains the proper balance of sodium and potassium ions across the muscle fiber membrane. Creatine kinase is also found on the outside of the inner mitochondrion membrane where it functions to regenerate PCr using the ATP produced by oxidative phosphorylation in the mitochondrion.

Because creatine kinase activity is so high, it can maintain ATP levels remarkably well even during intense exercise. This ATP buffering effect is substantial, and for this reason we can say that creatine kinase has a high power for regenerating ATP. The actual phosphocreatine concentration in resting, human skeletal muscle based on muscle biopsy sample analysis is approximately 18 to 20 millimoles per kilogram wet weight of muscle. Assuming a muscle water content of 76%, the PCr concentration could be described as 76 to 84 millimoles per kilogram dry weight of muscle (or 23–26 mM). No matter how it is expressed, there is a limited supply of PCr, and therefore the creatine phosphate system has a low capacity. This is one of the reasons why ingestion of creatine supplements by athletes is such a common phenomenon. Normally, creatine is synthesized in tissues such as the liver and taken up by muscle cells. One rationale behind creatine supplementation is that by saturating the system in the sarcolemma responsible for taking up creatine from the blood, more creatine will enter the muscle fiber. Since the creatine kinase reaction is considered to be at equilibrium at rest and at least during moderate intensity exercise, theoretically, more creatine in the muscle should mean more PCr and thus improved performance during sprint-type activities. Actual evidence supporting improved high intensity exercise performance with creatine supplementation can be found, but there are also studies showing that there is no effect. Although the body synthesizes creatine, we can also get it in our diets, since it is found in meat and fish. Not surprisingly, vegetarians often show the greatest responses if they take supplemental creatine.

During exercise PCr levels fall in proportion to the relative intensity of the exercise, and the free creatine concentration rises in parallel. For all-out efforts to fatigue, PCr levels can decrease by 90% or more. During recovery or rest periods the reverse of the creatine kinase reaction dominates (see equation on page 50), and phosphate transfer, from ATP produced by oxidative phosphorylation to Cr, creates PCr. The half-time for recovery of PCr during rest periods is approximately 30 seconds. This can vary, since a fit person with a higher capacity for oxidative ATP formation should recover PCr stores at a faster rate. Notice also in the equation that a proton (H^+) is consumed when ATP is formed from PCr. This can be helpful during very severe exercise because such exercise can cause a marked rise in H^+ concentration. Thus this acid buffering role for the creatine kinase is beneficial during high intensity exercise or games.

> **KEY POINT**
>
> Because of the relatively slow half-time for PCr resynthesis, the most commonly observed improvement in performance with creatine supplementation occurs with repeated bouts of very high intensity exercise, when recovery periods are too short to fully restore PCr levels.

Glycolysis

Glycolysis is a sequence of 11 enzyme-catalyzed reactions in which glucose or glycogen stored in the muscle is converted to lactate. At the same time, ATP is regenerated from ADP. The reaction that follows outlines the stoichiometry of glycolysis when glucose is the starting fuel:

$$\text{glucose} + 2\ \text{ADP} + 2\ \text{Pi} \longrightarrow 2\ \text{lactate}^- + 2\ \text{H}^+ + 2\ \text{ATP}$$

In this process, one glucose molecule is converted to two lactate ions and two hydrogen ions, and the free energy released is used to phosphorylate two ADP with two Pi to make two ATP. With no direct involvement of molecular oxygen, **glycolysis** is **anaerobic**. The net production of ATP during glycolysis and when the PCr concentration is decreased is known as **substrate-level phosphorylation**, in contrast to the ATP produced by oxidative phosphorylation.

In the sequence of reactions in which glucose and glycogen are broken down in cells, the second-to-last step produces pyruvate. The last step is the reduction of pyruvate to lactate, as we have shown in chapter 2. However, pyruvate may not be reduced to lactate. For example, pyruvate can enter mitochondria where it is completely broken down. The carbon atoms of pyruvate appear as CO_2, and electrons associated with hydrogen are transferred to oxygen. Some people call the breakdown of glucose or glycogen to pyruvate, and the subsequent complete oxidation of pyruvate to CO_2 and H_2O, aerobic glycolysis. Whatever the name used, the oxidation of pyruvate is very important in ATP formation because, from each glucose molecule, two pyruvate are formed, and the complete oxidation of each pyruvate in mitochondria generates at least 12 ATP.

Glycolysis has a moderate power to generate ATP. This is based on the activities of the enzymes in the cytoplasm that break down glycogen or glucose to lactate. As we will see, the power of glycolysis in fast twitch muscle fibers is larger than in slow twitch muscle fibers because there is a richer concentration of enzymes of glycolysis in the former. The capacity for glycolysis is moderate, much larger than the PCr system, but much less than oxidative phosphorylation. The moderate capacity may be related to the availability of glycogen to feed into the glycolytic process. Since H^+ is a product formed in glycolysis, the moderate capacity may reflect the fact that active glycolysis acidifies muscle, rendering it less able to produce force. Glycolysis can be quickly started at the onset of exercise, and if the exercise is intense enough, the glycolytic pathway can reach a maximum rate in about five seconds. We discuss glycolysis in detail in the chapter on carbohydrate metabolism.

KEY POINT

Since ADP is a substrate, glycolysis will be turned on whenever the rate of ATP hydrolysis by a fiber abruptly increases. For low intensity exercise, little lactate formation will take place, but for higher rates of exercise the concentration of lactate in muscle and blood will increase over time.

AMP Kinase and AMP Deaminase

When we consider the metabolism of muscle during exercise, two other reactions must be considered. The first is one we have seen previously. It is written here in the direction opposite to the way it was shown earlier in this chapter, but this does not matter because the reaction is freely reversible with an equilibrium constant near one.

$$2\ ADP \xleftrightarrow{\text{AMP kinase}} ATP + AMP$$

The enzyme for this reaction is named AMP kinase, but you will also see it described as adenylate kinase or myokinase (*myo* for muscle). The significance of this reaction is as follows: During hard muscle activity the rate of ATP hydrolysis is high. Although muscle efficiently regenerates ATP from ADP by the three energy systems previously described, an increase in ADP concentration occurs. What the AMP kinase (adenylate kinase or myokinase) does is cause two ADP molecules to interact and make one AMP and one ATP. This interaction keeps the ADP concentration from building up to the extent it would without this reaction.

In the second reaction, AMP deaminase (also called adenylate deaminase) irreversibly deaminates (removes the amino group from) the base adenine of AMP, producing ammonia, NH_3. Deaminated AMP becomes IMP (inosine monophosphate).

$$AMP + H_2O \xrightarrow{\text{AMP deaminase}} IMP + NH_3$$

Because NH_3 is a weak base, it takes a proton (H^+) from the medium, becoming ammonium ion (NH_4^+). AMP deaminase is found in higher levels in the fast twitch (type II) muscle fibers. These fibers are more active during intense exercise activity. The activity of AMP deaminase is low at rest but increases due to changes within a muscle fiber under vigorous exercise conditions such as the fall in muscle pH and rise in ADP concentration. The ammonium ion formed during the AMP deaminase reaction can leave the muscle cell and travel in the blood to the liver and kidney for further metabolism.

AMP kinase and AMP deaminase can work in concert during vigorous muscle work. In effect, two ADP molecules are converted into an ATP and an AMP by the AMP kinase reaction. Then in the adenylate deaminase reaction, the AMP is irreversibly converted into IMP. These related reactions play a major role in maintaining optimal energy status in the muscle fiber, especially during intense muscle activity. The irreversible AMP deaminase reaction drives the reversible AMP kinase reaction to the right, to diminish the increase in the concentration of ADP ([ADP]). This reaction also maintains a high [ATP]/[ADP] ratio, which is absolutely important to maintain a large, negative free energy of ATP hydrolysis. If the ATP/ADP concentration ratio decreased too much, less free energy would be available to drive endergonic reactions. If ADP concentration increased too much, it could slow down ATPase reactions by a process known as product inhibition.

The product ammonia, NH_3, in the AMP deaminase reaction is a base (i.e., a proton acceptor) and gets protonated to the ammonium ion, NH_4^+. Formation of ammonium ion from ammonia removes one proton from the muscle cytosol. However, this proton buffering effect is unlikely to be very significant given the much higher rate of proton formation with lactate production in glycolysis. The ammonium ion stimulates the process of glycolysis by activating one of the rate-controlling enzymes. We discuss the significance of this effect in the next chapter.

KEY POINT

Since the AMP deaminase reaction is increased by a decrease in muscle pH, we would expect that this reaction is dominant in fibers with a high capacity for glycolysis and a lower capacity for oxidative phosphorylation. Not surprisingly, type II fibers are the primary sources of blood lactate and ammonia during exercise.

The two reactions involving AMP are very important in muscle. Individuals with a genetic AMP deaminase deficiency have a poor ability to perform intense exercise and attempts at this produce muscle pain, cramping, and early fatigue. The AMP deaminase reaction is important when muscles are working hard. If you walked 800 meters (one half mile), this would not result in a significant displacement of adenine nucleotides from rest levels. However, if you ran 800 meters as fast as possible, changes in muscle adenine nucleotides would be significant due to the combined effects of AMP kinase and AMP deaminase. Compared to rest values, muscles severely fatigued after running as fast as possible for 800 meters would have muscle ATP levels reduced by approximately 20%, a rise in IMP concentration in near proportion to the decline in ATP, and modestly increased AMP and ADP concentrations. In the recovery period after exercise, IMP is slowly converted back to AMP in a sequence of reactions known as the purine nucleotide cycle. We discuss this cycle in chapter 8.

We can summarize as follows the changes that one might measure in muscle during an all-out activity lasting a few minutes (e.g., running 800 meters as fast as possible) or during repeated bouts of high intensity activity (e.g., during interval training):

- A high muscle lactate concentration, related to the fact that glycolysis is an important route for generating the needed ATP.
- A much lower pH. Glycolysis produces one proton for each lactate formed. The acid content of muscle can change by up to a factor of 10 (i.e., a pH decrease of from approximately 7.0 to 6.0) during the most severe exercise. Low muscle pH is an important reason why a fatigued muscle is forced to stop working. Following severe exercise, the pH of the blood may decrease from 7.4 to 7.0 due to transport of lactate and H^+ from muscle to blood.
- A greatly reduced concentration of PCr. During severe exercise PCr is an important source for regenerating ATP.
- An increase in total Pi concentration and, in addition, an increase in the ratio of $[H_2PO_4^-]$ to $[HPO_4^{2-}]$. Since the PCr concentration would decline (leading to an increase in creatine), there would be a corresponding increase in Pi. In addition, because the muscle pH would decline, the dihydrogen form of Pi ($H_2PO_4^-$) would increase due to protonation of the more basic, monohydrogen (HPO_4^{2-}) form.
- A decrease in **total adenine nucleotides** (TAN)—the sum of [ATP] + [ADP] + [AMP]. This is due to the conversion of AMP to IMP.
- An increase in [IMP] and $[NH_4^+]$. The decrease in TAN and the increase in IMP and ammonium ions result from the combined effect of the AMP kinase and AMP deaminase reactions. The increase in IMP concentration is in proportion to the decline in TAN.
- A decrease in muscle glycogen concentration. Glycogen is an important fuel for glycolysis. Its concentration can drop rapidly during intense exercise.

Summary

The thousands of chemical reactions in our bodies involve energy changes. Some reactions release energy while others require an input of energy. The part of the total energy change from chemical reactions that can be harnessed to do useful work is known as free energy change (ΔG). In spontaneous or exergonic reactions, free energy is given off, that is, ΔG is negative. Reactions in the body that require free energy (endergonic reactions) are driven by exergonic reactions. Energy-rich phosphates such as ATP and GTP are used to drive endergonic reactions and processes because, when one or more of the phosphate groups is removed by hydrolysis, much free energy is released. For most endergonic processes, ATP is the energy currency.

In muscle, a tissue with wide variations in energy requirements, the ATP concentration is only sufficient to drive about two seconds of maximal work. Therefore, rapid ways to replenish the ATP must exist. Transfer of a phosphate group from phosphocreatine (PCr) to ADP is a mechanism to rapidly renew ATP, but supplies of PCr are limited. Glycolysis, an anaerobic process in which glycogen or glucose is broken down to lactate, can provide ATP at a fairly rapid rate, but there is a limited capacity for this process. Oxidative phosphorylation, however, provides ATP at a low rate for a prolonged period of time. Exercise at a high rate to fatigue can lead to severe metabolic displacement in muscle as the contracting fibers attempt to generate ATP to match its rate of breakdown by the ATPases. Severe decreases in PCr and pH and large increases in lactate, ammonium, and phosphate ions characterize the fatigued state.

Key Terms

actin 45
anaerobic 51
ATPase 47

Review Questions

1. Using one of the equations in the chapter, show how an increase in muscle temperature can alter the free energy change for a reaction.
2. If the standard free energy change for ATP hydrolysis is –37 kilojoules/mole, what is the concentration of ATP at equilibrium, assuming that the temperature is 27 °C and that ADP and Pi are equal in concentration?
3. If the free concentration of magnesium ions in the cell is 2 millimolar, what is the approximate total concentration of magnesium in the cell?
4. Algebraically sum the ATPase reaction and the creatine kinase reaction to show that the overwhelming source of increased Pi in an actively working muscle fiber is from PCr.

CHAPTER 5

Oxidative Phosphorylation

Humans are open thermodynamic systems. This means we take in matter and energy from our environment and release matter and energy to our environment. On the intake side, we take in oxygen, sometimes a bit of heat, food, and water from the world around us. We release heat, carbon dioxide, urine, and fecal waste back to the environment. Our source of energy to carry out all the actions associated with life comes from the food we eat. In this chapter, we go through the steps in which oxygen is involved in the generation of more than 90% of all the ATP we need. Because this chapter on oxidative phosphorylation represents our first detailed study of a metabolic pathway, we begin with a general overview of cellular energy metabolism.

Overview of Metabolism

The food we eat is broken down in the digestive tract to simplest components, absorbed into the blood, and distributed to various tissues. There, the simple, absorbed molecules may be broken down to yield energy in the form of ATP, stored as energy molecules for future needs, or used to create large molecules (e.g., proteins) specific to our needs.

We have already described briefly the two major energy-generating pathways important to muscle: glycolysis and oxidative phosphorylation. The latter can utilize carbohydrate, fat, and amino acids, whereas glycolysis is restricted to carbohydrate as the fuel. The amino acids that are oxidized are obtained from the normal breakdown of proteins or from excess proteins in the diet that are not immediately converted into fat or glucose. Most fat is stored in specialized cells known as fat cells or adipocytes. Small amounts of fat are also stored in other cells, such as muscle. When metabolism needs to be increased, nerve and hormonal signals cause fat to be broken down, and fatty acids are released to the blood for cells to use as fuel. The body carbohydrate stores are not large; most carbohydrate is stored in liver and muscle as glycogen—a polymer made of individual glucose molecules joined together. Approximately 80% of all the body's carbohydrate is found in muscle as glycogen. Although liver has a higher concentration of glycogen compared to muscle, it is so much smaller in mass compared to the skeletal muscles that it holds only 10 to 15% of total body carbohydrate. The remainder of the carbohydrate in the body is the glucose in blood and extracellular fluids. When needed, glycogen in liver is broken down to maintain a proper glucose concentration in the blood. In muscle, glycogen is broken down and fed into the glycolytic reactions. Glucose is normally the sole fuel for the brain, the tissue that the body treats as most important from an energy perspective.

Figure 5.1 illustrates a section of a muscle fiber, representing the tissue that is or has the potential to be the main energy consumer. For athletes in hard training the consumption of fuel during training may almost double the energy expenditure of the body for the remainder of the day. Figure 5.1 illustrates the major energy pathways in the cell, which are located in the cytosol and mitochondria.

The fuels illustrated for these pathways are glucose from blood, either used directly in glycolysis or temporarily stored as glycogen, and fatty acids (FA), which can be taken up from the blood and used immediately in mitochondria or temporarily stored as intramuscular triglyceride (IMTG). Although our focus is on muscle, these same pathways operate in other cells as well.

In chapter 4 we defined oxidative phosphorylation as the formation of ATP from ADP and Pi in association with the transfer of electrons from fuel molecules to coenzymes to oxygen. Although probably the best term to use, oxidative phosphorylation is also known by other names. The biologist may call it cellular respiration, whereas the exercise physiologist may describe it as oxidative metabolism. As shown in figure 5.1, oxidative phosphorylation takes place in mitochondria. It encompasses the tricarboxylic acid (TCA) cycle (shown by the circle of arrows with oxaloacetate and citrate) and the **electron transport chain** (also known as the respiratory chain), which is shown below the TCA cycle. This cycle starts with acetyl CoA, which can be derived from any kind of fuel molecule. The TCA cycle produces electrons in the form of the reduced coenzymes (mostly NADH). These electrons feed into the electron transport chain and are passed through a series of carriers to the oxygen we breathe. Addition of electrons to oxygen allows it to combine with two protons (H^+) and produces water. Free energy produced in the electron transport chain is used to phosphorylate ADP with Pi to make ATP.

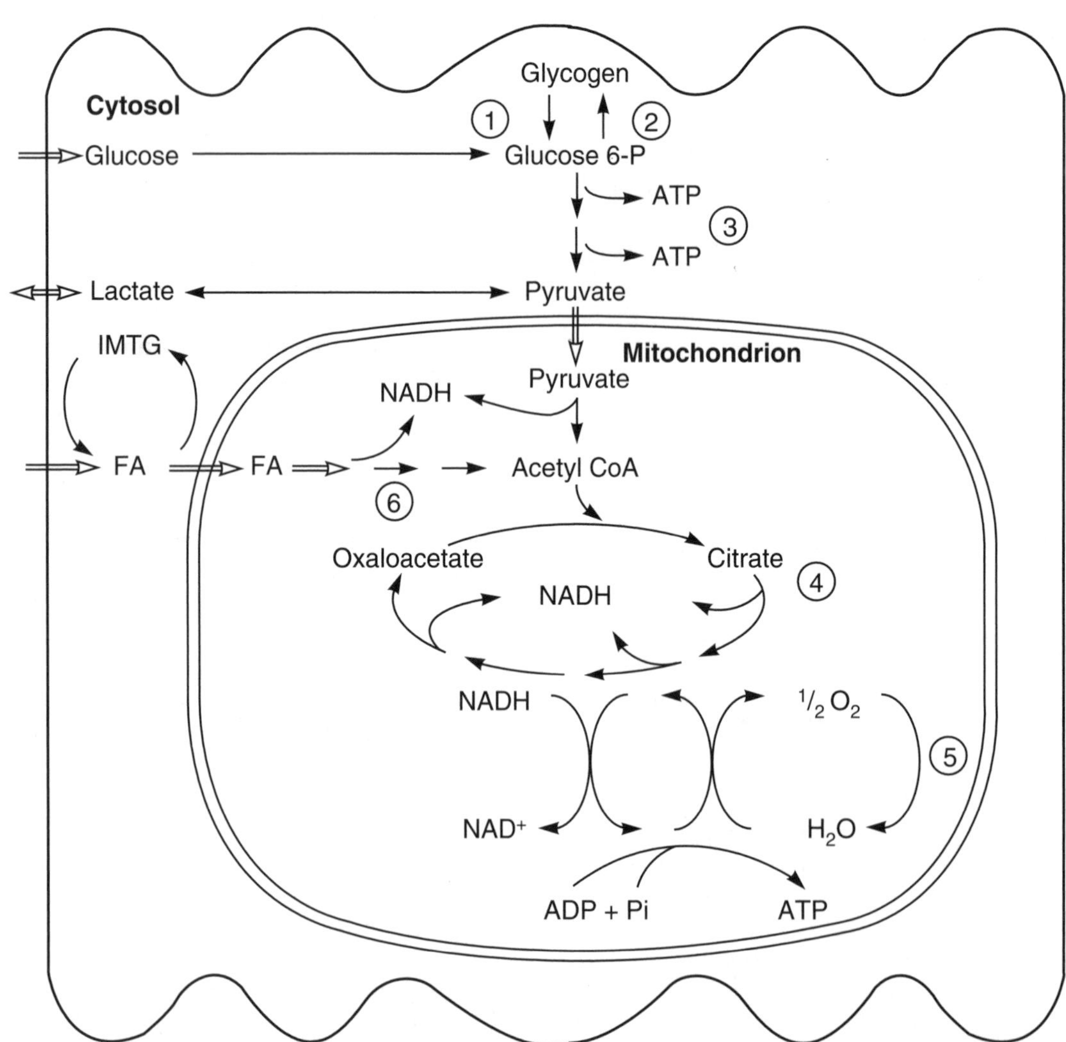

Figure 5.1 Overview of energy metabolism in a cross section of a muscle cell illustrating the cytosol and mitochondrial compartments and the major pathways we discuss: (1) glycogen breakdown, (2) glycogen synthesis, (3) glycolysis, (4) tricarboxylic acid cycle, (5) electron transport chain, and (6) beta-oxidation. A double-headed arrow means a reaction or sequence of reactions is reversible. An open arrow means a substance is crossing a membrane. IMTG represents intramuscular triglyceride and FA represents fatty acids.

As shown in figure 5.1, acetyl CoA is the substrate for the TCA cycle. This can arise from the breakdown of fatty acids using **beta-oxidation** (shown with the number 6 in figure 5.1). Acetyl CoA is also formed in mitochondria from pyruvate, which is produced during glycolysis in the cytosol. In exercise, the proportion of acetyl CoA coming from pyruvate in the breakdown of glycogen and glucose increases in proportion to the intensity of exercise. We discuss this in chapter 6.

> ***KEY POINT***
>
> When we exercise we place demands for ATP use on our muscles. This must be matched by ATP synthesis. For most activities the bulk of the ATP is generated in mitochondria, when electrons on fuel molecules are transferred to coenzymes, through the electron transport chain, to oxygen. Free energy released during this process is captured by forming ATP from ADP and Pi. In this process, fuels and oxygen are consumed; the harder we exercise the greater the requirement for fuels and oxygen.

Mitochondria

As shown in figure 5.1, oxidative phosphorylation takes place in cell mitochondria. Oxygen is delivered to mitochondria by transport in the blood attached to hemoglobin molecules in red blood cells and then by diffusion from capillaries into the cells. Certain cell types, chiefly type I or slow twitch muscle fibers and heart muscle cells, contain a protein known as myoglobin that helps the diffusion of oxygen through the cytoplasm to mitochondria. These myoglobin-containing cells have a high capacity for oxidative phosphorylation.

As might be expected, the density of mitochondria in a cell parallels its capacity for oxidative phosphorylation. Endurance training has been shown to increase mitochondrial density in the muscles involved in the exercise. Figure 5.2 illustrates the essential features of a mitochondrion, with its two membranes. The outer membrane is permeable to most ions and molecules up to a molecular weight of 10,000 daltons (D). The outer membrane is thus leaky because it contains channels or pores composed of proteins known as porins, through which solutes can pass. The outer membrane is rich in protein (60–70%) and may be important to maintain the shape of the mitochondrion. The enzyme hexokinase is typically found attached to the cytosolic side of the outer membrane. The inner membrane is even higher in protein content (nearly 80%) and is impermeable to most ions and polar molecules unless they have specific transporters or carriers. Between the two membranes is the **intermembrane space**.

The inner membrane is formed into bulges called **cristae**, which greatly increase its surface area. The density of cristae in mitochondria is generally much higher in tissues where the rate of oxidative phosphorylation is high, such as the heart. The inner membrane is loaded with enzymes and proteins for transferring electrons to oxygen and the enzyme ATP synthase that converts ADP and Pi to ATP. In figure 5.2, ATP synthase

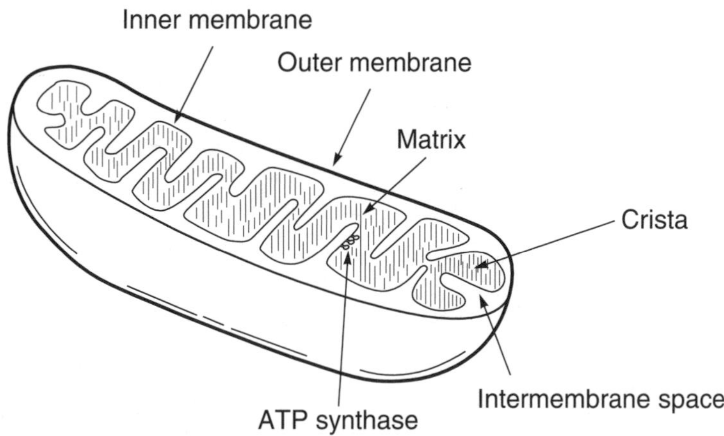

Figure 5.2 The essential components of a mitochondrion. Although only a few are shown, the knobby ATP synthase is found throughout the inner membrane.

is shown as small knobs in the mitochondrial inner membrane.

In the center of a mitochondrion is the **matrix**, a viscous medium containing all the enzymes of the TCA cycle (except succinate dehydrogenase), three enzymes of beta-oxidation of fatty acids, other enzymes, and mitochondrial DNA. This small, circular DNA molecule codes for 13 of the 700 or so mitochondrial proteins (found in the outer and inner membranes, the intermembrane space, and the matrix); the remainder is coded by nuclear genes. Following transcription and translation, proteins are imported into the mitochondrion. Mitochondrial DNA, which in humans has 16,569 base pairs, also codes for two rRNA molecules and 22 tRNA molecules; in sum, a total of 37 genes. In each mitochondrion, there are multiple copies of mitochondrial DNA. Researchers have discovered that mitochondrial DNA is more sensitive to damage by a variety of substances. Lacking the repair enzymes found with nuclear DNA, mitochondrial DNA deteriorates as we age, compromising our ability to generate ATP. Virtually all mitochondrial DNA is maternally contributed—less than 0.1% arises from sperm.

As we shall see, the mitochondrial inner membrane maintains an **electrochemical gradient**. The matrix side has a more negative charge with a higher pH than the side facing the intermembrane space. This gradient allows the mitochondrion to take up Ca^{2+} ions when the concentration of these is elevated in the cytosol. The positively charged calcium ions can flow down both a chemical and electrical gradient into the matrix through a special protein called a **uniporter**. A **sodium-calcium antiport** or proton-calcium antiport removes the calcium from the matrix. Calcium can be a dangerous substance to the cell because it can initiate cell death or apoptosis. Accordingly, its cytosolic concentration is generally kept very low and carefully regulated. On the other hand, calcium has many positive effects for the cell since changes in its concentration can signal alterations in many reactions and processes such as initiating muscle contraction. The ability of the mitochondrion to buffer calcium ions is very important.

KEY POINT

The electrochemical gradient across the inner mitochondrial membrane is critical for the conversion of free energy released during electron transport from fuel molecules, to coenzymes, to oxygen, into ATP formation from ADP and Pi.

General Mechanism of Oxidative Phosphorylation

The chemical equations that follow illustrate the complete oxidation of two types of fuels: glucose, a representative carbohydrate, and palmitic acid, a very common fatty acid.

$$C_6H_{12}O_6 + 6\ O_2 \longrightarrow 6\ CO_2 + 6\ H_2O$$
$$C_{16}H_{32}O_2 + 23\ O_2 \longrightarrow 16\ CO_2 + 16\ H_2O$$

These balanced equations summarize the complete oxidation of these fuels, as would occur in the cell by oxidative phosphorylation. These equations also describe the oxidation of these fuels if they were completely broken down by burning in a very hot flame.

The **respiratory quotient** (RQ) is the molar ratio of CO_2 produced divided by the O_2 consumed during fuel oxidation. Using the previous equations, RQ is 6/6 = 1.0 for glucose and 16/23 = 0.7 for palmitic acid. When the oxygen consumed and carbon dioxide produced are measured at the mouth of an animal or human as $\dot{V}O_2$ and $\dot{V}CO_2$, respectively, we use the term **respiratory exchange ratio** or RER (i.e., $\dot{V}CO_2/\dot{V}O_2$) instead of RQ.

KEY POINT

Measurement of oxygen consumed and carbon dioxide produced helps us understand the nature of the fuels oxidized by all mitochondria-containing cells in the body during the measurement period. As exercise intensity increases, the RER reflects more and more the fuels oxidized by the active muscles.

In the cell, oxidation of glucose and fatty acids is tightly coupled to ADP phosphorylation, that is, phosphorylation of ADP with Pi to make ATP. Let us rewrite the above equations, but now include the phosphorylation part; that is, we will include the number of moles of ATP formed for each mole of fuel oxidized in the cell.

$$C_6H_{12}O_6 + 6\ O_2 + 32\ (ADP + Pi) \longrightarrow 6\ CO_2 + 32\ ATP + 38\ H_2O$$

$$C_{16}H_{32}O_2 + 23\ O_2 + 106\ (ADP + Pi) \longrightarrow 16\ CO_2 + 106\ ATP + 122\ H_2O$$

The number of water molecules generated is increased by 32 for glucose and 106 for palmitic acid compared to the equations that only showed oxidation, because when ATP is formed from ADP and Pi, a water molecule results. In contrast, with ATP hydrolysis, a molecule of water is needed to hydrolyze the ATP.

The **P/O ratio** (sometimes called the ATP/O ratio) is the number of ATP formed for each atom of oxygen consumed. For palmitic acid, P/O is

$$106/(23 \times 2) = 2.3$$

For glucose, P/O is

$$32/(6 \times 2) = 2.7$$

Comparing these two numbers reveals that, for the same amount of oxygen consumed, you get more ATP from glucose than you do from a fatty acid. As we discuss later in this chapter, the P/O ratio for glucose and fatty acids is not universally agreed on. Some believe that it is 3.0 and 2.8, respectively, for glucose and fatty acids. We use a more conservative estimate, the reasons for which are elaborated later in this chapter.

There are two major metabolic pathways involved in oxidative phosphorylation. The TCA cycle breaks down acetyl units derived from fuel molecules and generates the reduced coenzymes **NADH** and $FADH_2$ as well as CO_2. In the second pathway, the electron transport chain, the free energy released when electrons are transferred from reduced coenzymes (NADH and $FADH_2$) to oxygen gets channeled into the phosphorylation of ADP with Pi to make ATP, that is, it drives the reaction.

$$ADP + Pi \longrightarrow ATP + H_2O$$

The mechanism behind oxidative phosphorylation parallels the way electricity is generated by falling water. A dam creates potential energy by raising water up to a high level. When the water falls down through special channels, the kinetic energy rotates turbine blades in a magnetic field, producing an electric current. During electron transfer from NADH and $FADH_2$ to oxygen, free energy is released. This energy is employed to pump protons (H^+) from the matrix side of the inner membrane of the mitochondria to the outside or cytosolic side. As a result, an electrochemical gradient is created in which the cytosolic side of the inner membrane is more positive in charge (the *electro* part of the gradient) and has a higher concentration of H^+ (the *chemical* part of the gradient). When protons (H^+) return down the gradient through a special protein complex, the free energy released is used to make ATP from ADP and Pi. In other words, return of protons down their gradient is harnessed into driving ADP phosphorylation much like the energy of falling water is harnessed to make electricity. Figure 5.3 summarizes this process.

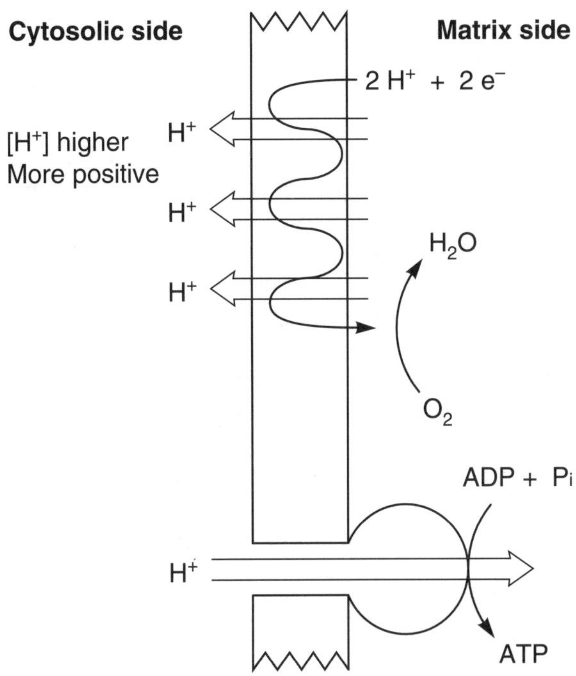

Figure 5.3 A section of the inner mitochondrial membrane showing how electron transfer from substrates through coenzymes to oxygen, forming water, is coupled to ATP formation. During electron transfer, protons (H^+) are pumped across the membrane, creating an electrochemical gradient on the cytosolic side. Their return through ATP synthase drives ATP formation from ADP and inorganic phosphate (Pi).

> **KEY POINT**
>
> We describe the regions on either side of the inner mitochondrial membrane as the matrix and cytosolic sides. Yet as figure 5.2 illustrates, what we call the cytosolic side is in reality the intermembrane space. However, cytosolic side is used in this text because the outer membrane is so porous that the ion composition of the inner membrane space is essentially cytosolic in nature.

Recall that oxidation involves a loss of electrons, and reduction involves a gain of electrons. The two coenzymes we encountered in chapter 2, FAD and NAD^+, can be reduced to $FADH_2$ and NADH, respectively, when they accept electrons from substrate molecules. In their reduced forms they each have a strong tendency to return to their oxidized states. With electron transfer during oxidative phosphorylation, electrons on NADH and $FADH_2$ are transported to oxygen, generating water (after combination with protons). The driving force is called a **redox potential.** In reactions involving energy-rich phosphates, we describe their standard free energy potential by $\Delta G^{\circ\prime}$. Under standard conditions the driving force for electron transfer can be described by the **standard redox potential** or $\Delta E^{\circ\prime}$. Values for $\Delta E^{\circ\prime}$ are measured in volts (V), based on the standard electrode potential for hydrogen gas. The larger the value of $\Delta E^{\circ\prime}$, the greater the tendency for a substance to be reduced (accept electrons). For the reduction of an oxygen atom (that is, one half an oxygen molecule) by two electrons, the $\Delta E^{\circ\prime}$ is 0.82 volts, shown with equation 5.1.

$$1/2\ O_2 + 2\ H^+ + 2\ e^- \longrightarrow H_2O,\ \Delta E^{\circ\prime} = 0.82\ V \quad (5.1)$$

Equation 5.2 illustrates the reduction of NAD^+ to form NADH + H^+. Note the negative sign, indicating a very weak tendency to become reduced.

$$NAD^+ + 2\ e^- + H^+ \longrightarrow NADH,\ \Delta E^{\circ\prime} = -0.32\ V \quad (5.2)$$

KEY POINT

Since oxidation and reduction are two opposite processes, the fact that NAD^+ has a weak tendency to be reduced means that NADH has a strong tendency to become oxidized.

In oxidative phosphorylation, electrons are transferred from NADH to oxygen, reducing the oxygen (which combines with protons to produce water), and at the same time NADH is oxidized to NAD^+. We could summarize this process by combining equations 5.1 and 5.2, but we would need to reverse equation 5.2 to show it as oxidation, which is what actually happens in mitochondria. The two equations and their algebraic sum (below the line) appear as follows:

$$1/2\ O_2 + 2\ H^+ + 2\ e^- \longrightarrow H_2O,\ \Delta E^{o\prime} = 0.82\ V$$
$$NADH \longrightarrow NAD^+ + H^+ + 2\ e^-,\ \Delta E^{o\prime} = 0.32\ V$$

$$NADH + 1/2\ O_2 + H^+ \longrightarrow NAD^+ + H_2O$$
$$\Delta E^{o\prime} = 1.14\ V$$

We can convert $\Delta E^{\circ\prime}$ values in volts to $\Delta G^{\circ\prime}$ values in kilojoules/mole by using the following equation:

$$\Delta G^{\circ\prime} = -nF\Delta E^{\circ\prime}$$

In this equation, n is the number of electrons transferred (2 in our example). F is the **faraday constant** with a value of 96.5 kilojoules per mole per volt. We can now determine the standard free energy resulting from electron transfer from NADH to oxygen by substituting these numbers into the equation.

$$\Delta G^{\circ\prime} = -2 \times 96.5\ kJ/(mol \times volt) \times 1.14\ volts$$

The $\Delta G^{\circ\prime}$ is equal to −220 kilojoules per mole, a number almost six times greater than the standard free energy for ATP hydrolysis, shown previously. The message from this is that a large amount of free energy is released when a pair of electrons on NADH is transferred to oxygen, reducing the oxygen. Electrons are removed from a substrate in pairs when the substrate is oxidized by a dehydrogenase enzyme. In these reactions, coenzymes become reduced to NADH or $FADH_2$. For this reason, the term **reducing equivalents** is often used to describe a pair of electrons.

When the other reduced coenzyme $FADH_2$ is oxidized to FAD, and two electrons are transferred to oxygen as shown in the following equation, slightly less free energy is released.

$$FADH_2 + 1/2\ O_2 \longrightarrow FAD + H_2O,\ \Delta E^{\circ\prime} = 1.04\ V$$

Using the equation to convert to free energy, we get a value for $\Delta G^{\circ\prime}$ of −200 kilojoules per mole. This value is about 10% less than when NADH is oxidized.

Role of the Tricarboxylic Acid Cycle

The tricarboxylic acid cycle, abbreviated TCA cycle, is also known as the citric acid cycle (CAC) or Krebs cycle. You should be aware of all three names, because you will find all of them used. The prime function of the TCA cycle is to completely oxidize acetyl groups in a way that will result in ATP formation. The TCA cycle removes electrons from acetyl groups and attaches them to NAD^+ and FAD, forming NADH and $FADH_2$, respectively. In the electron transport chain, the electrons on the reduced coenzymes NADH and $FADH_2$ will be transferred through a series of carriers to oxygen. In two reactions in the TCA cycle, the carbon atoms in the acetyl group are released as carbon dioxide. Each kind of fuel is converted to acetyl groups, attached to **coenzyme A** (abbreviated CoA) (see figure 5.4).

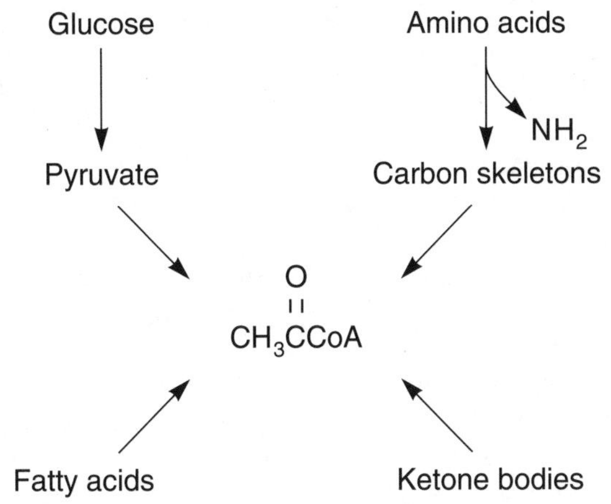

Figure 5.4 To oxidize fuels, glucose, fatty acids, ketone bodies, and amino acids are broken down into two-carbon acetyl groups attached to coenzyme A (CoA). The resulting acetyl CoA feeds into the tricarboxylic acid cycle. The amino groups of amino acids, which cannot be oxidized, are removed and excreted in the urine.

CoA is derived from pantothenic acid, a B vitamin. It acts as a handle to attach to a number of acyl groups, some of which we will see later. CoA has a terminal SH (sulfhydryl) group to which the acetyl group is attached, forming an energy-rich thioester bond. CoA and acetyl CoA are often written as CoASH and $CH_3COSCoA$, respectively. However, for simplicity, we use CoA and acetyl CoA.

Oxidation of acetyl CoA accounts for about two thirds of the ATP formation and oxygen consumption in mammals. Acetyl groups enter the TCA cycle where their two carbon atoms appear as CO_2, while the hydrogens and their associated electrons are removed. Figure 5.5 summarizes and simplifies this process. The :H^- + H^+ represents the hydride ion and proton that are removed from many fuel substrates. As we saw in chapter 2, electrons accompany the hydrogen removed from substrates. The hydride with its two electrons, shown as dots, is attached to NAD^+, forming NADH; the negative charge on the hydride ion balances the positive charge with NAD^+. The two H• represent two hydrogen atoms removed from succinate, although succinate is not identified in the figure. The two hydrogen atoms, each containing one electron, become attached to FAD forming $FADH_2$.

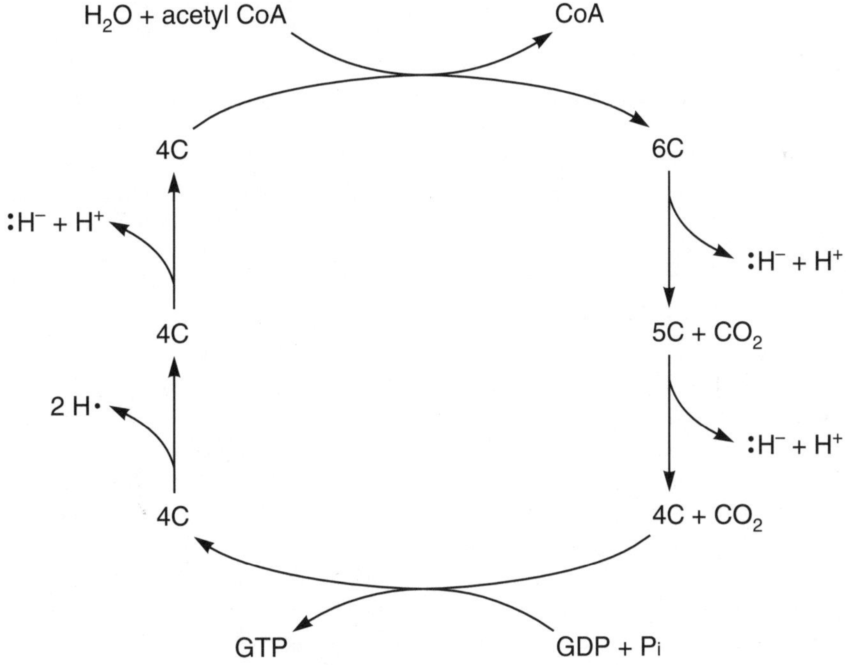

Figure 5.5 An overview of the TCA cycle showing the path of carbon atoms and electrons that are recovered in association with hydrogen. The :H^- is recovered in the form of NADH, while the two H• are recovered in the form of $FADH_2$. The two-carbon acetyl unit feeds into the cycle attached to coenzyme A (CoA). The carbon atoms are recovered as carbon dioxide (CO_2). During one step in the cycle, enough free energy is released to phosphorylate GDP to make GTP.

Figure 5.5 shows that one turn of the TCA cycle consumes one acetyl group and produces four pairs of electrons, one GTP, and two CO_2. Two water molecules are also consumed, although in the simplified diagram in figure 5.5 only one H_2O is shown. The GTP produced is an example of substrate-level phosphorylation, that is, formation of an energy-rich phosphate without using oxidative phosphorylation. The ATP produced in glycolysis also occurs via substrate-level phosphorylation.

KEY POINT

The TCA cycle consumes acetyl groups. One limit to oxidative phosphorylation is the rate at which acetyl groups are provided to the TCA cycle. As mentioned previously, the maximum rate of oxidative phosphorylation is markedly reduced if available carbohydrate stores are depleted. This suggests that the ability to produce acetyl CoA from fat is a limiting factor.

Reactions of the Tricarboxylic Acid Cycle

Figure 5.6 shows the complete tricarboxylic acid cycle including the structures of intermediates. Figure 5.7 summarizes the TCA cycle, without the use of chemical structures. As is the custom for this book, double-headed arrows are used to show reactions that, if isolated, would reach an equilibrium. Arrows showing one direction only define reactions where there is a large free energy change, and if studied in isolation these reactions would use up virtually all substrate. The overall direction of the tricarboxylic acid cycle is to consume acetyl groups.

Hydrolysis of acetyl CoA to acetate and CoA has a $\Delta G^{\circ\prime}$ of –32 kilojoules/mole. Thus the reaction combining acetyl CoA and oxaloacetate to form citrate, catalyzed by citrate synthase, is virtually irreversible. In the next step, the tertiary alcohol group on citrate is converted to a secondary alcohol group by two steps in the reaction catalyzed by the enzyme aconitase. The first step is a dehydration reaction in which the tertiary OH group is removed, producing cis-aconitate, which remains attached to the aconitase enzyme. A hydration reaction follows that results in the formation of isocitrate. Notice that the only difference between citrate and isocitrate is the position and type of OH group—tertiary alcohol on citrate and secondary alcohol on isocitrate.

Isocitrate undergoes oxidative decarboxylation, catalyzed by isocitrate dehydrogenase, to form α-ketoglutarate. First oxidation generates NADH and H^+ and then decarboxylation forms CO_2. The decarboxylation spontaneously follows the oxidation but needs the cofactor Mg^{2+} or Mn^{2+} (not shown). At this point in the cycle, one of the two carbon atoms on the acetyl group is removed as carbon dioxide.

In the next step, an α-ketoglutarate (some call this 2-oxoglutarate) undergoes oxidative decarboxylation to succinyl CoA. The first step is decarboxylation, followed by oxidation, generating NADH and H^+. In this step, the second of the two acetyl carbon atoms is lost as CO_2. The enzyme α-ketoglutarate dehydrogenase (2-oxoglutarate dehydrogenase) catalyzes the same kind of reaction as pyruvate dehydrogenase, a reaction we will encounter soon, and contains the same kinds of subunits and coenzymes. Both enzymes contain three types of polypeptide subunits and five coenzymes. The coenzymes are NAD^+ and CoA, which are loosely bound, and TPP, **lipoic acid**, and FAD, which are tightly bound and not seen in the simple way the reaction is presented in figures 5.6 and 5.7. Recall from table 2.1 that **TPP (thiamine pyrophosphate)** is derived from the B vitamin, thiamine. During the α-ketoglutarate dehydrogenase reaction, enough free energy is released to generate the energy-rich succinyl CoA.

In the next step, succinyl CoA is broken down to succinate and CoA in a reaction catalyzed by succinyl CoA synthetase. When this occurs, the free energy released drives the substrate-level phosphorylation of GDP to make GTP. The succinyl CoA synthetase reaction is freely reversible, and the enzyme's name describes the backward reaction, in keeping with the naming of similar reactions in biochemistry. The GTP produced may be used (a) for peptide bond formation during the process of translation, (b) to phosphorylate ADP to make ATP using the enzyme nucleoside diphosphate kinase, described previously, or (c) in signal transduction.

Once succinate is formed, the remaining three reactions of the TCA cycle regenerate oxaloacetate (one of the starting substances of the TCA cycle, along with acetyl CoA) and generate electrons for the electron transport chain (using the enzymes succinate dehydrogenase and malate

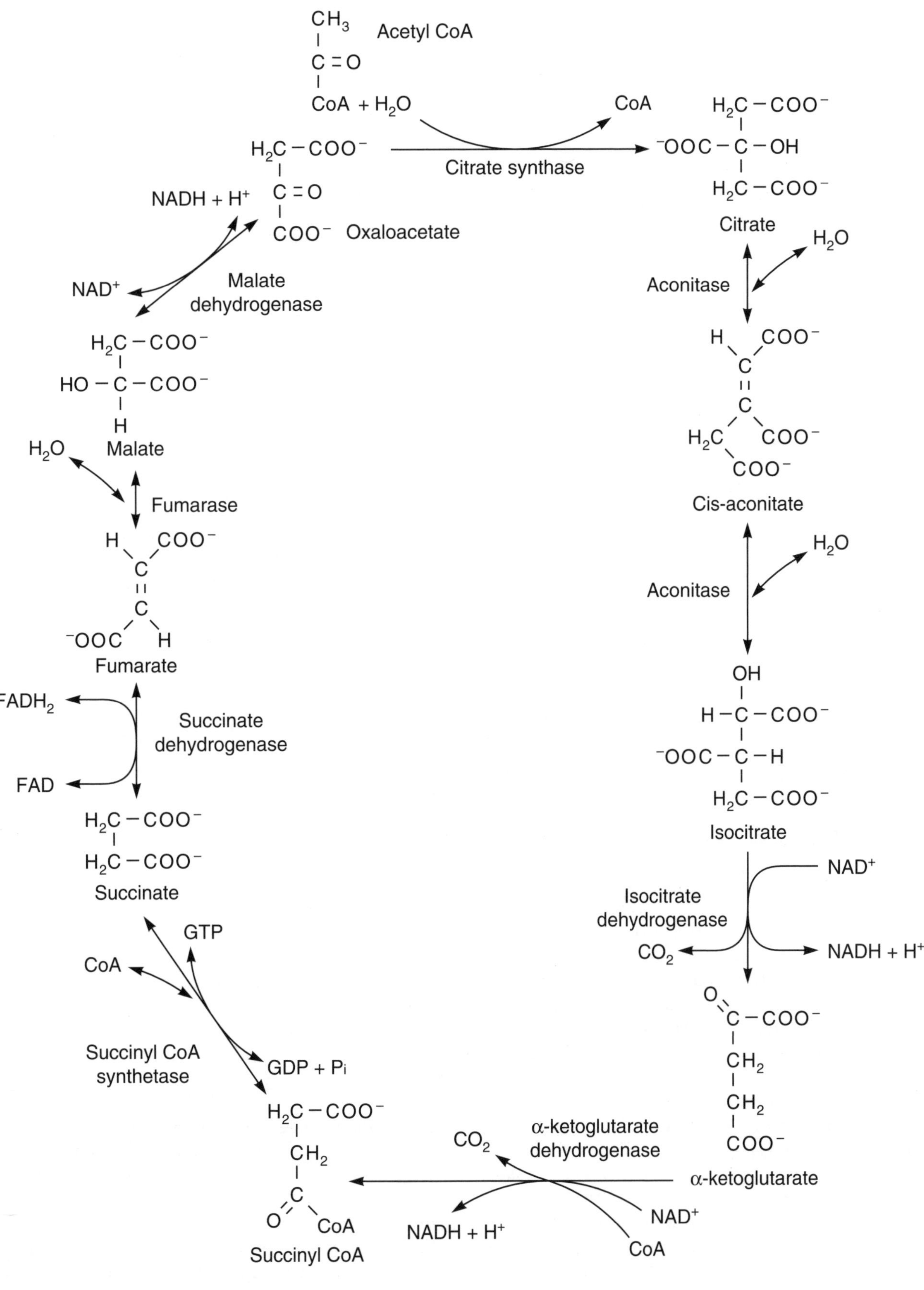

Figure 5.6 The tricarboxylic acid cycle, showing the structures of all intermediates.

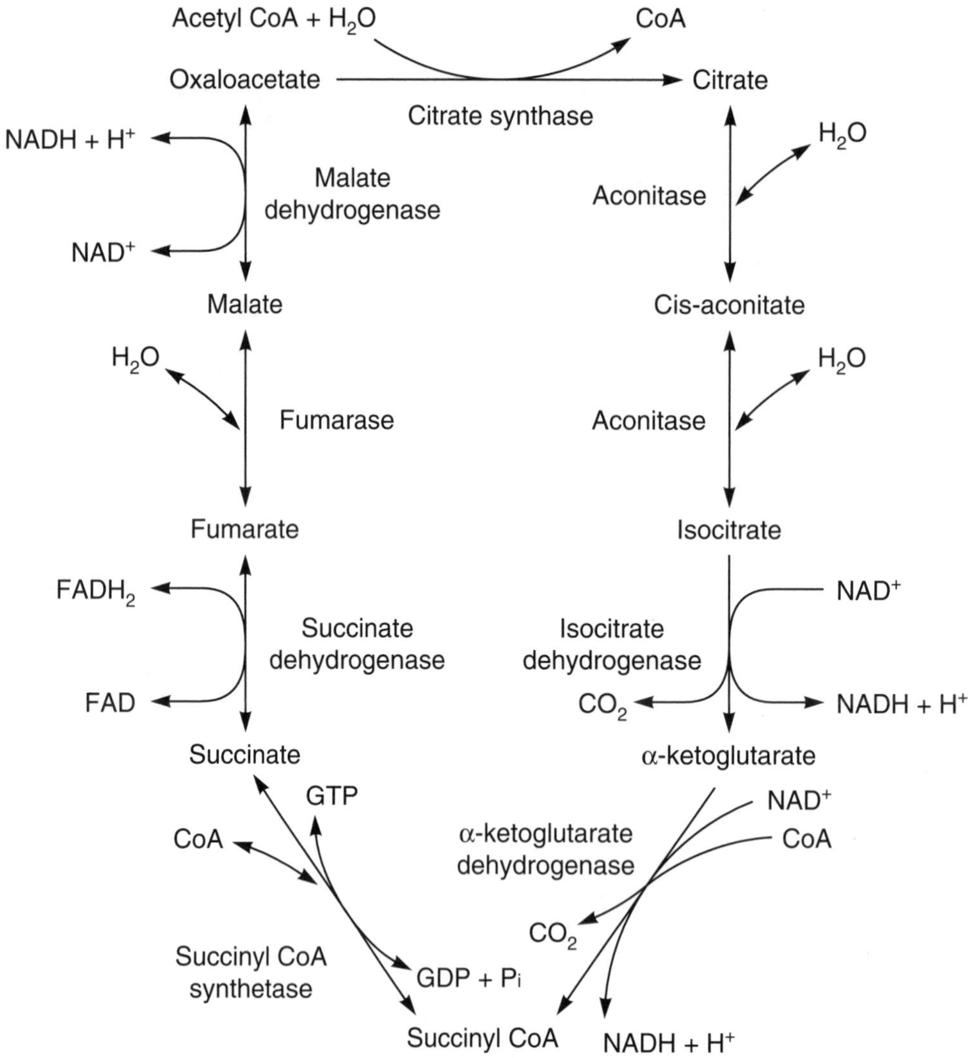

Figure 5.7 The tricarboxylic acid cycle, shown without chemical structures.

dehydrogenase). Succinate dehydrogenase (SDH) contains a tightly bound FAD. Unlike the other enzymes of the TCA cycle that are found in the mitochondrial matrix, SDH is a component of the inner mitochondrial membrane. In the SDH reaction, electrons are transferred from succinate to FAD and then to **coenzyme Q**. We will see later in this chapter that SDH and coenzyme Q are part of the electron transport chain. The product from the SDH reaction, fumarate, is hydrated to malate by the enzyme fumarase. Malate contains a secondary alcohol group that is oxidized in the malate dehydrogenase reaction, generating NADH + H^+ and oxaloacetate, a starting substrate for a new round of the cycle.

If we add algebraically all the reactions of the TCA cycle we get the following:

$$\text{Acetyl CoA} + 3\ NAD^+ + FAD + GDP + Pi + 2\ H_2O \longrightarrow 2\ CO_2 + GTP + 3\ NADH + 3\ H^+ + FADH_2 + CoA$$

The standard free energy change for one turn of the cycle ($\Delta G^{\circ\prime}$) is approximately –40 kilojoules/mole. Thus the TCA cycle is highly exergonic. As shown in the summary equation, the TCA cycle does not involve the net production or consumption of oxaloacetate or any other constituent of the cycle and the only things consumed are an acetyl group and two water molecules.

KEY POINT

For the TCA cycle to operate at a high rate, there must be sufficient oxaloacetate to accept acetyl groups from acetyl CoA in the first reaction of the cycle. If oxaloacetate is used for any other purpose, the maximal power of the TCA cycle may be reduced.

The reduced coenzymes produced in the TCA cycle (NADH and $FADH_2$) are oxidized in the electron transport chain and their electrons transferred to oxygen. We can show this electron transfer as follows:

$$3\ NADH + 3\ H^+ + FADH_2 + 2\ O_2 \longrightarrow 3\ NAD^+ + FAD + 4\ H_2O$$

Associated with the transfer of electrons from the reduced coenzymes to oxygen are the tightly coupled ADP phosphorylation reactions, producing ATP. Based on conservative estimates, the transfer of electrons from three NADH to oxygen will yield 7.5 ATP. In some sources, this is shown as 9 ATP. The transfer of electrons from one $FADH_2$ to O_2 will generate 1.5 ATP; other sources show 2 ATP. In counting the energy-rich phosphates, one must include the GTP formed during the succinyl CoA synthetase reaction. In summary, the complete oxidation of one acetyl group is associated with the formation of 10 ATP. Figure 5.8 illustrates the close coupling of the TCA cycle, the electron transfer chain, and ADP phosphorylation.

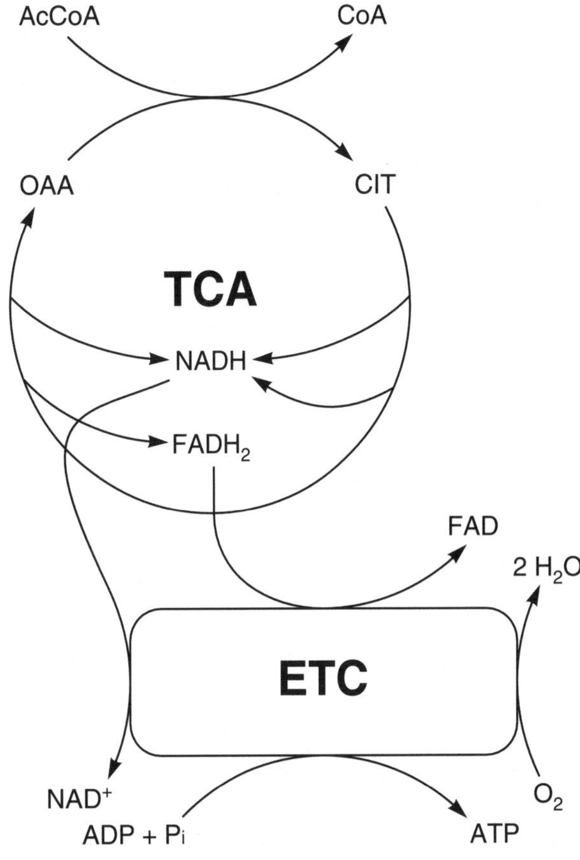

Figure 5.8 The tricarboxylic acid cycle (TCA), electron transfer chain (ETC), and ADP phosphorylation to ATP are tightly coupled in the mitochondrion.

Electron Transfers

From a functional perspective, the electron transport system consists of five protein-lipid complexes located in the inner membrane of the mitochondrion. Four of the complexes make up the electron transport chain, also known as the respiratory chain. The fifth complex is the ATP synthase. In three of the four complexes that make up the electron transport chain, the free energy released is associated with proton pumping from the matrix to the cytosolic side of the inner membrane, as shown in figure 5.9. Two electrons are transferred via NAD^+, requiring dehydrogenases from fuel substrates, through a series of electron carriers to oxygen. The sequences of electron flow are nothing more than a series of oxidation-reduction reactions, each of which can be shown as

$$\text{reduced A} + \text{oxidized B} \longrightarrow \text{oxidized A} + \text{reduced B}$$

There would be a larger value for $\Delta E^{\circ\prime}$ for B than for A, which would favor the reduction of B at the expense of A. We can show this more clearly in terms of electron transfer using hydrogen in the equation:

$$AH_2 + B \longrightarrow A + 2\ e^- + 2\ H^+ + B \longrightarrow A + BH_2$$

Ultimately, the electrons are transferred to molecular oxygen, reducing it and forming water. At stages during the electron transport, enough free energy is released to pump protons across the mitochondrial inner membrane from the matrix to the cytosolic side. Return of these electrons through the ATP synthase is coupled to ADP phosphorylation with Pi to make ATP. Present understanding of this system shows that 2.5 ATP are generated per pair of electrons transferred from NADH to oxygen. This is known as the P/O ratio. In addition, two electrons can be transferred from other fuel substrates to oxygen via FAD-containing dehydrogenases. When this occurs, 1.5 ATP are generated per pair of electrons transferred from $FADH_2$ to oxygen. We use the above nonintegral P/O ratios, but remind ourselves that whether the above numbers are correct or whether they are 3/1 and 2/1 for NADH and $FADH_2$ is a matter of controversy.

Figure 5.10 illustrates the flow of electrons from two representative substrates through their coenzymes and on to oxygen using the electron transfer (respiratory) chain. The four complexes

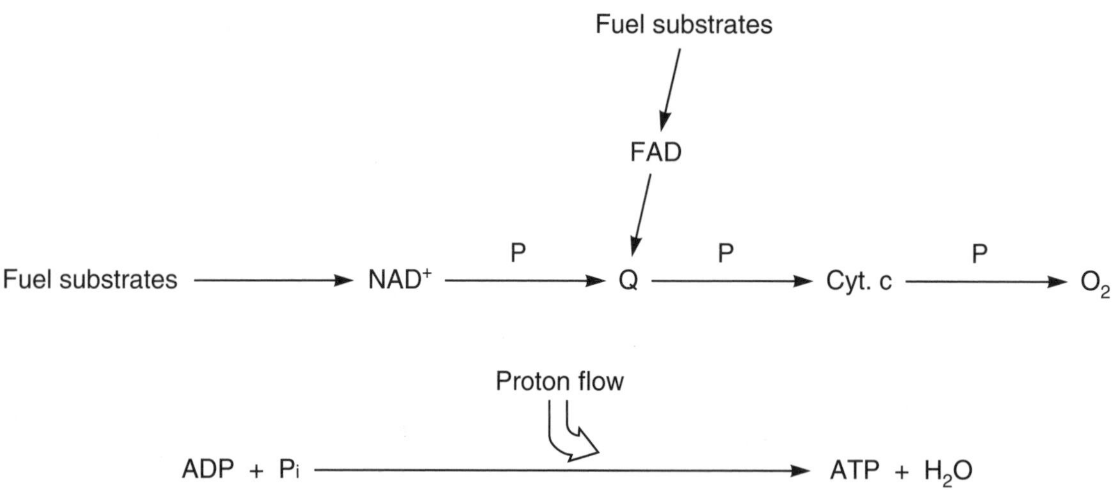

Figure 5.9 Electrons are transferred from fuel substrates to oxygen via NAD^+ dehydrogenases or FAD dehydrogenases. During transfer by NAD^+ dehydrogenases, enough free energy is released to pump protons across the mitochondrial membrane at three sites (P). With FAD dehydrogenases, the lower free energy release pumps protons at only two sites. In the second stage, the proton flow releases free energy, which is harnessed to synthesize ATP from ADP and inorganic phosphate Pi. FAD represents flavin adenine dinucleotide, Q represents coenzyme Q, and Cyt. c represents cytochrome c.

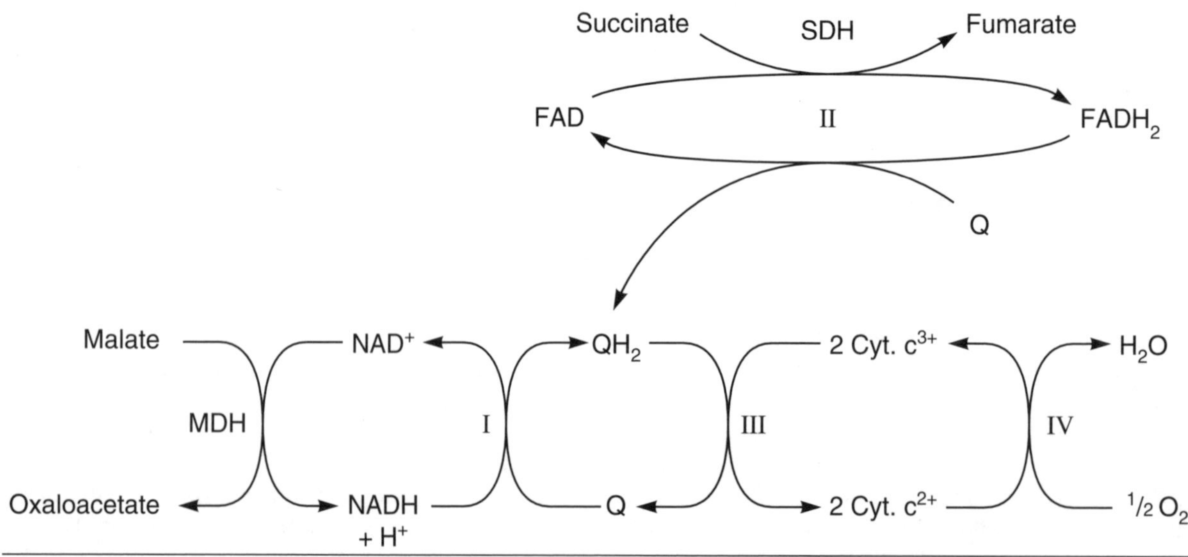

Figure 5.10 Electron transfer from malate and succinate through the four complexes (identified with Roman numerals) of the electron transfer chain to oxygen. Proton pumping across the mitochondrial inner membrane involves complexes I, III, and IV. MDH represents malate dehydrogenase, and SDH represents succinate dehydrogenase, a representative member of complex II. FAD and $FADH_2$ represent the oxidized and reduced forms of flavin adenine dinucleotide. Q and QH_2 represent the oxidized and reduced forms of coenzyme Q. Cyt. c represents cytochrome c.

in the electron transport chain are shown with Roman numerals. Let us start with the substrate malate. Oxidized by the TCA cycle enzyme malate dehydrogenase (MDH) using NAD^+, it yields oxaloacetate and NADH plus H^+. The two electrons, now on NADH, are passed through the complexes I, III, and IV to oxygen. Another substrate, succinate, is oxidized by the enzyme succinate dehydrogenase (SDH, a representative of complex II), using FAD and generating $FADH_2$ (the reduced form) and fumarate. The two electrons on $FADH_2$ are then transferred to oxygen, using complexes III and IV. During electron transfers involving complexes I, III, and IV, protons are pumped across the inner mitochondrial membrane, creating the electrochemical gradient. Proton pumping is a form of active transport, since the protons are being transferred across a membrane from a region of lower concentration (the matrix) to one that is higher in concentration (the cytosolic side of the inner membrane). Electron transfer to oxygen from substrates such as succinate involves only two proton pumping complexes (III and IV), whereas electron transfer from substrates to oxygen in which NAD^+ is the coenzyme involve three proton pumping complexes (I, III, and IV); more ATP is thus generated with substrates such as malate.

> **KEY POINT**
>
> The energy needed to actively transport the protons across the membrane comes from the free energy release during electron transfer at complexes I, III, and IV.

Complex I: NADH-Coenzyme Q Reductase

Complex I, which is also known as NADH dehydrogenase, is a huge complex consisting of at least 42 different subunits. Its role is to transfer a pair of electrons from NADH in the matrix to the oxidized form of coenzyme Q (abbreviated Q; also known as **ubiquinone**) in the inner membrane. Reduction of coenzyme Q produces the reduced form known as **ubiquinol** (QH_2). Figure 5.11 summarizes the electron transfer reaction of complex I and illustrates the structural changes in coenzyme Q that occur with the acceptance of two electrons and two hydrogen ions. The NADH dehydrogenase complex also contains an FMN-protein (another type of flavoprotein) and an iron-sulfide protein as intermediates between NADH and Q. During electron transfer from NADH to coenzyme Q, protons are pumped across the inner mitochondrial membrane.

Complex II: Other Flavoprotein Dehydrogenases

Complex II is represented by a number of flavoprotein dehydrogenase enzymes that transfer electrons from substrates to FAD (flavin adenine dinucleotide) and then to coenzyme Q. An example of an FAD-containing dehydrogenase is succinate dehydrogenase (shown as SDH in figure 5.10), the only tricarboxylic acid cycle enzyme located in the inner membrane and not in the matrix of the mitochondrion. We will encounter two more of these FAD-containing complexes, all of which contain a

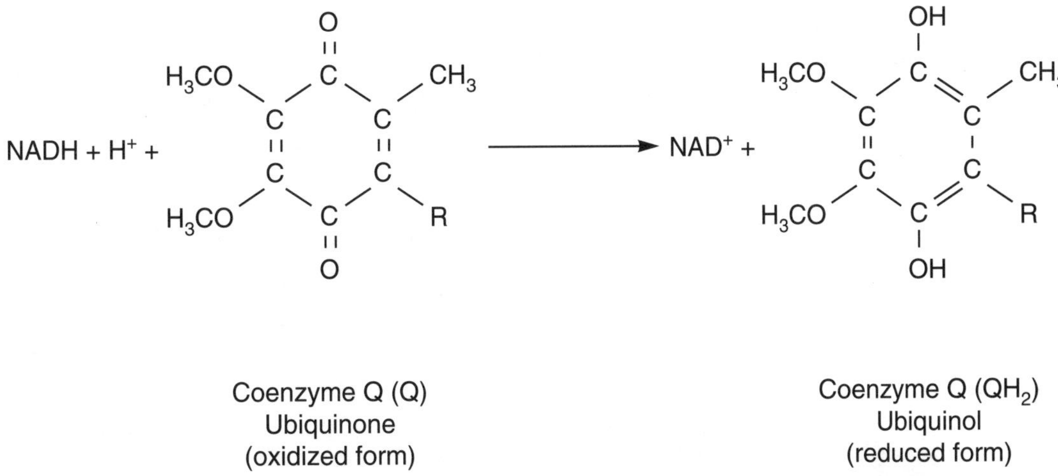

$NADH + H^+ +$ [Coenzyme Q structure] → $NAD^+ +$ [Coenzyme Q, reduced structure]

Coenzyme Q (Q)
Ubiquinone
(oxidized form)

Coenzyme Q (QH_2)
Ubiquinol
(reduced form)

Figure 5.11 Complex I of the electron transfer chain involves reduction of coenzyme Q (ubiquinone) through the transfer of electrons from NADH. The R attached to the ring structure represents a long and repeating isoprene unit.

tightly bound coenzyme known as FAD. In the case of succinate, electrons in the form of two hydrogen atoms are transferred from succinate to FAD to make it $FADH_2$. Then the electrons on $FADH_2$ are passed to the oxidized form of coenzyme Q (ubiquinone) to make QH_2 (ubiquinol). Unlike complexes I, III, and IV, the free energy change associated with complex II is not large; therefore, no protons are pumped across the inner membrane during electron transfer during this type of dehydrogenation. Accordingly, when fuel substrates are oxidized by the FAD-containing dehydrogenases, one less ATP molecule is formed (i.e., 1.5 versus 2.5 ATP per pair of electrons transferred to oxygen).

Complex III: Coenzyme Q-Cytochrome c Reductase

Complex III transfers electrons from reduced coenzyme Q (QH_2) to cytochrome c. The **cytochromes** are a class of heme proteins located in the inner membrane of the mitochondrion. In the center of the heme group is an iron ion that can exist in an oxidized (Fe^{3+}) or reduced (Fe^{2+}) state. In complex III, electrons on reduced coenzyme Q (i.e., QH_2) are transferred to cytochrome c, changing the iron from Fe^{3+} to Fe^{2+}. Since reduction of oxidized iron involves accepting only one electron, two cytochrome c molecules must be reduced to accept the electrons from each QH_2. Although complex III involves electron transfer from coenzyme Q to cytochrome c, there are two other types of cytochromes known as cytochrome b and cytochrome c_1 that are intermediates in the stage from Q to cytochrome c. During electron transfer from QH_2 to cytochrome c, enough free energy is released so that protons are pumped across the inner mitochondrial membrane. In the following summary equation, note that only the electrons are transferred from QH_2 to cytochrome c. As a result, two protons are left over.

$$QH_2 + 2\ \text{cytochrome c-}Fe^{3+} \longrightarrow Q + 2\ H^+ + 2\ \text{cytochrome c-}Fe^{2+}$$

Complex IV: Cytochrome c Oxidase

Cytochrome c oxidase is a complicated protein that contains more than 10 subunits. Two of these are cytochrome a and cytochrome a_3; there are also two copper ions. This complex accepts electrons from reduced cytochrome c and passes them to an oxygen molecule, reducing it, after combination with four protons, to two water molecules. The sequence is as follows. A single electron on reduced cytochrome c is transferred to a protein-bound Cu^{2+} reducing it to Cu^+. The electron is then transferred to cytochrome a, then to another protein-bound copper ion, to cytochrome a_3, and finally to oxygen. Because the oxygen molecule (O_2) contains two atoms of oxygen, we need four electrons to reduce it. We cannot break up the oxygen molecule as suggested earlier by the use of 1/2 O_2; it is written this way for convenience. We can summarize the cytochrome oxidase reaction as it really occurs as follows:

$$4\ \text{cytochrome c-}Fe^{2+} + O_2 + 4\ H^+ \longrightarrow 4\ \text{cytochrome c-}Fe^{3+} + 2\ H_2O$$

The free energy released during electron transfer in cytochrome oxidase (complex IV) results in proton pumping across the inner membrane.

> **KEY POINT**
>
> The four electron transport complexes I, II, III, and IV are located in the inner mitochondrial membrane in such a way that there is direct interaction among them, so that the electrons are passed off to the next complex in the sequence.

Coupled Phosphorylation

Before describing how electron transfer is linked to the formation of ATP, we should summarize what has been covered so far. Some fuel substrates are oxidized by transferring two electrons in the form of a hydride ion ($:H^-$) to NAD^+, making it NADH. Then the electrons on NADH are passed through the respiratory chain to oxygen, using complexes I, III, and IV. During electron transfer, proton pumping occurs. Also, there is a large standard free energy change, shown to be –220 kilojoules/mole.

Other fuel substrates initially transfer electrons to FAD when they get oxidized. These electrons are passed in the form of two hydrogen atoms, each of which contains one electron. The resulting $FADH_2$ then passes the electrons to Q, and then to oxygen. Three complexes are involved: II, III, and IV. However, the standard free energy change is smaller (–200 kJ/mol), and proton pumping occurs only with complexes III and IV.

We intuitively know that water flows downhill but must be pumped uphill. Therefore, the term proton pumping should be clear. The pump is driven by the energy released when the electrons flow from one complex to another that is more easily reduced. The cytosolic side of the membrane has a lower pH (a higher H^+ concentration) and would have a higher positive electrical charge. Thus, protons must be pumped across the membrane against the electrical and chemical gradient. However, like water above a dam, the electrochemical gradient can be exploited by allowing the energy released when protons flow back into the matrix to drive the phosphorylation of ADP, making ATP.

The concept that electron transport is linked to ATP synthesis by way of proton pumping was conceived by Peter Mitchell. His concept is known as the **chemiosmotic hypothesis**, for which Mitchell received a Nobel prize. Although the stoichiometry of protons pumped across the membrane is not firmly established, best estimates are that 10 protons (H^+) are pumped per pair of electrons transferred from NADH to oxygen. Similarly, for each pair of electrons transferred from $FADH_2$ to Q to oxygen, it is widely believed that 6 protons are pumped.

The **ATP synthase**, or complex V, couples proton flow down the gradient into the matrix to phosphorylation of ADP with Pi to make ATP (see figure 5.12). ATP synthase consists of two parts. The F_0 subunit, mainly within the inner membrane, acts as a pore to allow protons to pass into the matrix. Tightly associated with F_0 is the F_1 subunit, which bulges into the matrix. It is composed of a number of peptide subunits (not shown in figure 5.12) and is responsible for combining ADP and Pi into ATP and releasing it into the matrix, although the precise mechanism by which ATP is synthesized is not clear. The substrate for the ATP synthase is $MgADP^-$ and the product is $MgATP^{2-}$, although the role of the magnesium ions is not shown in figure 5.12.

Electron transfer-driven proton translocation and ADP phosphorylation are tightly coupled via complex V—the ATP synthase. The flow of only 3 H^+ down the gradient is enough to drive the phosphorylation of ADP with Pi to make one ATP. Protons can also leak across the inner membrane from the cytosolic side to the matrix side without accompanying ATP formation. Such leakage appears to be most prominent in the liver and more important in small mammals where it can account for as much as 35 to 45% of mitochondrial oxygen consumption. It allows the proton flow to be dissipated as heat, helping small mammals maintain body temperature. This uncoupled electron transport can also be caused by specific chemicals. There is a type of adipose tissue (approximately 1% or less of human body weight) that is rich in mitochondria, giving it a brownish color. This brown adipose tissue (BAT) has an **uncoupling protein** in its mitochondrial inner membranes that allows the electrochemical gradient across the membrane to be dissipated by allowing protons to flow down

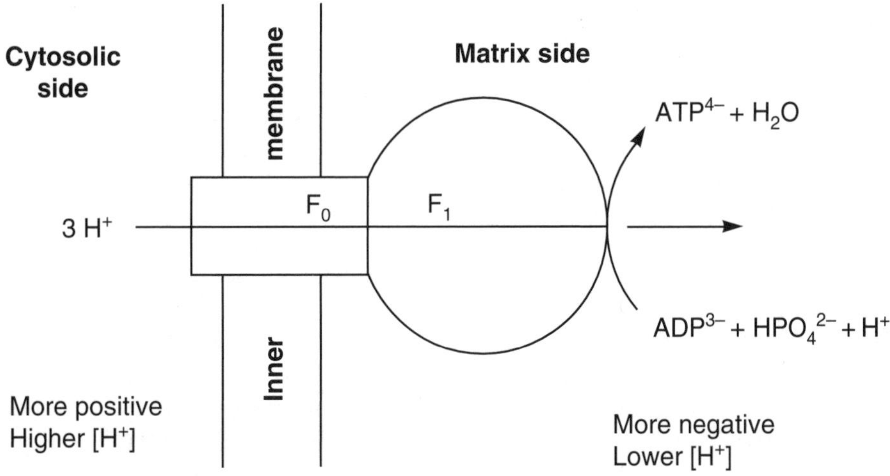

Figure 5.12 ATP synthesis by ATP synthase. The F_0 subunit of ATP synthase acts as a pore, allowing protons to move down their electrochemical gradient from the cytosolic side of the inner membrane into the matrix. Energy released during proton flow is used by the F_1 subunit to drive the phosphorylation of ADP to ATP.

their gradient into the matrix without being coupled to ATP formation using the ATP synthase. In this way, the free energy of the electrochemical gradient is immediately released as heat, warming the animal.

KEY POINT

Some people seem to be able to eat as much as they want and never get fat. Others seem to have a difficult time maintaining a healthy body weight, yet they don't eat a lot. This may be explained by differences in their physical activity. However, another mechanism to account for these individual differences is the extent of uncoupled electron transport.

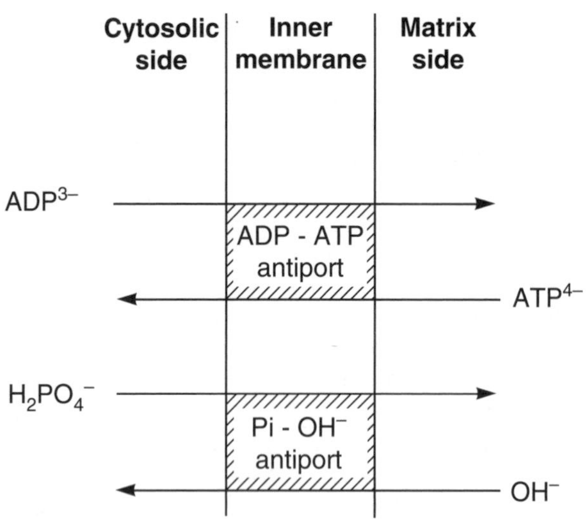

Figure 5.13 Translocation of ATP, ADP, and inorganic phosphate (Pi) across the mitochondrial inner membrane is essential for oxidative phosphorylation. ATP and ADP are transported simultaneously by the ADP-ATP antiport, and the Pi-OH^- antiport allows Pi to enter the matrix. The arrows show the direction of travel.

Mitochondrial Transport of ATP, ADP, and Pi

Most synthesis of ATP occurs in the mitochondria, but most ATP is hydrolyzed in the cytosol. Thus, we must have a way of getting ADP and Pi into the mitochondrion and ATP out of the mitochondrion. ADP and Pi must enter the mitochondrial matrix by crossing the inner membrane, whereas ATP must cross the inner membrane in the opposite direction. Recall that the inner membrane is quite impermeable to most substances and that polar or charged molecules can cross only if they are transported (translocated) by a specialized carrier protein. Figure 5.13 illustrates how ADP and Pi enter the matrix and how ATP crosses to the cytosolic side. Remember we ignore the outer mitochondrial membrane because it is so permeable to small molecules.

The ADP-ATP antiport (also commonly called the ADP-ATP translocase) transports ADP and ATP. The name **antiport** means that two substances are crossing a membrane simultaneously in opposite directions, and they are translocated with a carrier protein. Similarly, the Pi-OH^- antiport transports Pi and OH^- simultaneously across the inner membrane in opposite directions. The terms ADP-ATP translocase and Pi-OH^- translocase also describe these carriers. However, the term antiport is preferable because it gives more information.

The driving force for the ADP-ATP antiport comes from the charge difference between the ADP and ATP. Thus the movement of ATP^{4-} from the more negative matrix region to the more positive region on the cytosolic side of the inner membrane allows the translocation to proceed with a free energy release. The movement of ATP^{4-} out of the matrix, coupled to moving an ADP^{3-} into it, reduces the charge gradient by one but does not alter the proton concentration gradient. The Pi-OH^- antiport is electrically neutral, but acts to dissipate a proton on the cytosolic side of the membrane because OH^- can combine with a proton (H^+) to form water. Thus it is driven by discharge of part of the chemical imbalance of H^+ across the membrane. In summary, moving the constituents to make an ATP molecule into the matrix (e.g., ADP and Pi) costs the equivalent of one of the protons pumped during electron transport from reduced coenzymes to oxygen.

In trying to account for the ATP produced during transfer of a pair of electrons (reducing equivalents), we are assuming that 10 protons are translocated across the inner membrane when NADH is the source of electrons and 6 protons are pumped when the source of electrons is $FADH_2$. We also assume that only three protons are needed to flow down the gradient through the ATP synthase to provide enough energy to phosphorylate one ADP with Pi, making ATP. Further, to transport one ADP and one Pi into the mitochondrial matrix, across the inner membrane, is equivalent to using up one proton. Therefore, we should be able to generate 2.5 ATP from each pair of electrons transported in the respiratory chain from NADH and 1.5 ATP for each pair of electrons transported from $FADH_2$. Nonintegral P/O ratios (e.g., 2.5 and 1.5) may be intuitively disturbing to some. As mentioned, many sources describe P/O ratios of 3 and 2 for NADH and $FADH_2$. Higher P/O ratios could arise if there are more protons pumped per pair of electrons than the 10 and 6 we are assuming. Definitive answers must await more complicated experiments on mitochondria. However, before leaving this issue, it is important to remember that protons do leak across the inner membrane down their electrochemical gradient. As mentioned, this rate of leakage is higher in certain tissues and in smaller animals. Such uncoupled oxidation would serve to lower the P/O ratio because protons would flow back into the matrix without ATP being formed.

Regulation of Oxidative Phosphorylation

Because oxidative phosphorylation generates most of the energy for the cell in the form of ATP, its rate should be precisely connected to the rate of ATP hydrolysis. The TCA cycle is one component of oxidative phosphorylation. As shown in figure 5.8, the TCA cycle and electron transport chain are tightly linked together because the TCA cycle is the major producer of reduced coenzymes needed to funnel electrons into the respiratory chain. Therefore, what regulates the TCA cycle can influence the electron transport chain and vice versa.

If we neglect transfer of electrons from $FADH_2$ to oxygen, a single equation can represent oxidative phosphorylation, in which four electrons reduce a complete oxygen molecule and ADP is phosphorylated with Pi to make ATP:

$$5\ ADP + 5\ Pi + 2\ NADH + 2\ H^+ + O_2 \longrightarrow 2\ NAD^+ + 5\ ATP + 7\ H_2O$$

Two of the seven molecules of water are generated by reduction of a molecule of oxygen and five come about when five ATP are formed.

KEY POINT

FAD is reduced to $FADH_2$ and electrons are transferred to oxygen in only a few reactions. Reduction of NAD^+ to form NADH occurs in many reactions—three are in the TCA cycle. For this reason, we are simplifying our thinking by focusing on NADH and not $FADH_2$.

A simple way to understand the regulation of oxidative phosphorylation is to consider which of the substrates on the left side of the equation (i.e., ADP, Pi, NADH, and O_2) actually limit the overall process. Our discussion focuses on muscle because it has an enormous range of metabolic rate, from complete rest to the vigorous contractions of sprinting. Figure 5.14 aids this discussion. First we must ask, where do the four potentially limiting substrates for oxidative phosphorylation come from?

ADP and Pi are formed mainly in the cytosol when ATP is hydrolyzed to drive endergonic reactions. The increase in Pi is directly proportional to the decline in the concentration of phosphocreatine; that is, PCr decreases and free creatine (Cr) and Pi increase.

NADH comes from the TCA cycle, beta-oxidation of fatty acids in the matrix of the mitochondrion, and the cytosol when pyruvate formed during glycolysis is oxidized in the mitochondrion. NADH is a potential limiting factor because adequate NADH for oxidative phosphorylation requires available substrates for the dehydrogenase reactions that generate it, as well as sufficient activity of the dehydrogenase enzymes to drive the NADH-forming reactions at a sufficient rate.

Oxygen is taken into the lungs when you breathe, diffuses to hemoglobin molecules in red blood cells (i.e., erythrocytes) in the capillaries in the lung, is pumped throughout the body from the heart, and is unloaded from hemoglobin molecules in the capillaries that reach all parts of the body. Oxygen is delivered by way of diffusion from the small capillaries, across the cell membrane, the cytosol, and then to mitochondria.

Figure 5.14 shows that ATP is hydrolyzed to ADP and Pi by the functional ATPases we discussed in chapter 4, whereas ATP is primarily regenerated via oxidative phosphorylation in the mitochondrion. For tightly coupled ATP hydrolysis and ATP

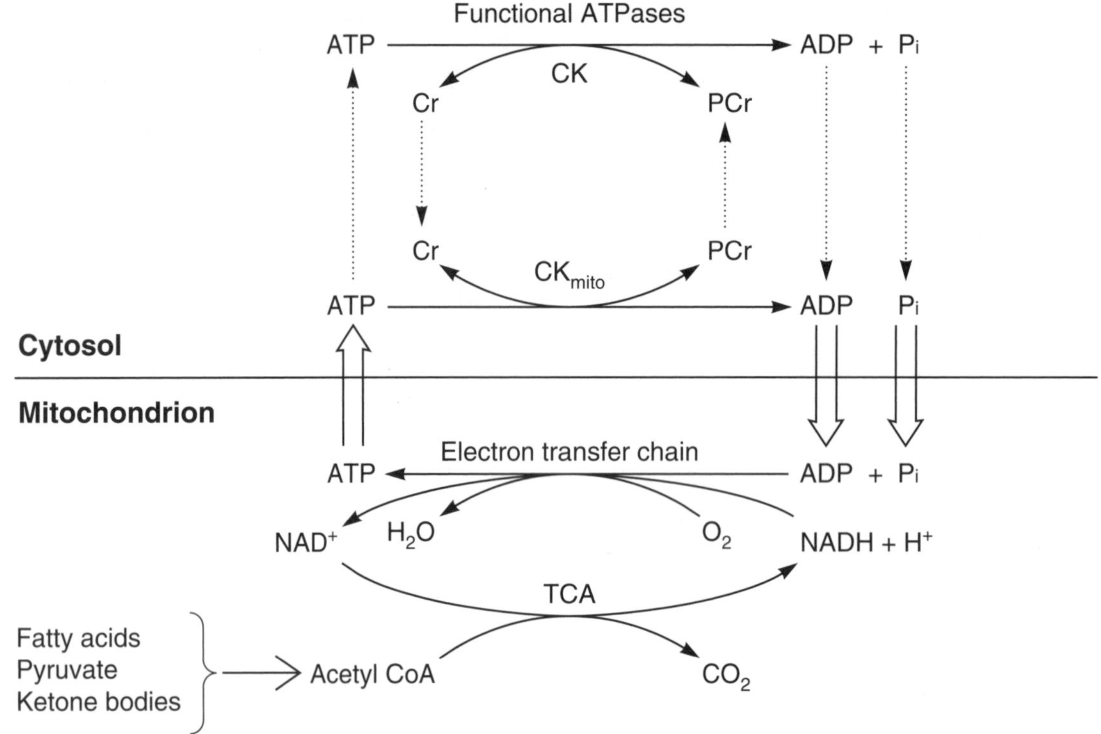

Figure 5.14 Oxidative phosphorylation is coupled to the rate of ATP hydrolysis by ATPases in muscle cytosol. The products of ATP hydrolysis, ADP and inorganic phosphate, Pi, are substrates for oxidative phosphorylation in the mitochondrion, along with electrons from acetyl CoA and the final electron acceptor, oxygen. ADP may diffuse to the mitochondrion and ATP diffuse back to the site of ATP hydrolysis, as indicated by the dotted lines. Also, creatine (Cr) can stimulate ADP formation at the inner membrane by the mitochondrial creatine kinase (CK_{mito}) reaction. Phosphocreatine (PCr) is a source of ATP at the site of ATPases using creatine kinase located there. The movement of Cr and PCr in the energy cycle is called the phosphocreatine shuttle. Double arrows refer to membrane transport. No distinction is shown between the matrix and inner membrane of the mitochondrion.

regeneration to happen, ADP and Pi must cross from the cytosol into the matrix, and ATP must cross from the matrix to the cytosol to be used again. Transport of ADP, Pi, and ATP across the inner mitochondrial membrane is aided by specific carriers that we have already described. It has been generally assumed that ADP diffuses from sites of ATPase activity to mitochondria and ATP diffuses back to the ATPase sites. However, it is widely believed that in skeletal and cardiac muscle much of the cytoplasmic transport of ADP and ATP occurs via Cr and PCr, respectively.

ATP is hydrolyzed in the cytosol of muscle at three major sites: (a) where myosin interacts with actin (the actin-activated ATPase discussed in chapter 4), (b) when calcium ions are pumped back into the sarcoplasmic reticulum (the sarcoplasmic reticulum-Ca^{2+} ATPase), and (c) when sodium ions are pumped out of the cell and potassium ions are pumped back in (the sodium-potassium ATPase). At the site of these ATPases are the creatine kinase enzyme (CK) and PCr. CK catalyzes the phosphorylation of ADP to make ATP, utilizing PCr and producing Cr. At this level, the net direction of the CK reaction is toward ATP formation. Next, Cr can diffuse to the outer side of the inner membrane where it is phosphorylated by an ATP, producing PCr and ADP. A mitochondrial creatine kinase (CK_{mito}) catalyzes this reaction. The net direction of the CK_{mito} reaction is toward PCr formation. The resulting PCr can then diffuse back to the site of the ATPases to rephosphorylate ADP. This process is called the phosphocreatine shuttle (or creatine phosphate shuttle) and is illustrated in figure 5.14.

Use of this shuttle does not prevent ADP from diffusing to the mitochondrion and ATP from diffusing back as shown in figure 5.14. It just means that Cr and PCr can carry out the same process. Since diffusion depends on a concentration gradient, the fact that there are significantly larger changes in PCr/Cr concentrations than ATP/ADP concentrations during muscle work, makes the former molecules more likely candidates in this energy transport process. Moreover, both Cr and PCr are smaller molecules than ADP and ATP, making diffusion easier, plus the normal cellular concentrations of the former are larger than those of ADP and ATP.

The ATP used to phosphorylate Cr at the mitochondrial level, using the mitochondrial form of CK, crosses from the matrix at the same time ADP crosses into the matrix (using the ADP-ATP antiport). The ADP-ATP antiport is the most abundant protein in the inner mitochondrial membrane and is in physical association with mitochondrial creatine kinase, located on the cytosolic face of the inner membrane. ATP is formed via oxidative phosphorylation when electrons are transferred from fuel substrates (pyruvate, fatty acids, ketone bodies, and acetyl CoA) to oxidized coenzymes (such as NAD^+) to make NADH. The electrons on NADH are transferred to oxygen using the electron transport chain.

KEY POINT

For simplicity, we talk about limiting factors as if we have a chain with links that are not equal. It is important to point out that what limits oxidative phosphorylation in one metabolic state may not be the weak link in another condition.

Figure 5.14 shows where all the limiting players (NADH, ADP, Pi, and O_2) are involved in oxidative phosphorylation. Now let us take a more detailed look at what limits the rate of oxidative phosphorylation.

Regulation of the Tricarboxylic Acid Cycle

It has been estimated that in the transition from rest to very intense exercise, the flux through the TCA cycle may increase 100-fold in well-trained humans. Because the TCA cycle is so tightly coupled to the electron transport chain, anything that limits the activity of electron transport to oxygen and ADP phosphorylation will stop the TCA cycle. In other words, the TCA cycle is the primary source of electrons for the electron transport chain. If the flow of electrons from NADH (thus forming NAD^+) is blocked, the TCA cycle will cease to function, because NAD^+ is a substrate for three of the dehydrogenase reactions in the TCA cycle. There are other considerations, since three of the TCA cycle enzymes catalyze reactions that are essentially irreversible. If uncontrolled, these reactions could convert all available substrate into product, thus compromising mitochondrial metabolism.

Citrate synthase (CS) should be controlled because if it were not it could consume the available acetyl CoA and oxaloacetate. CS is inhibited by the allosteric effector NADH, which binds to an allosteric site on CS, increasing the K_m of CS for its substrate acetyl CoA (see the discussion on allosteric enzymes in chapter 2). In addition, citrate is a competitive inhibitor for oxaloacetate at the active site of CS. This means that a rise in citrate can competitively inhibit the binding of oxaloacetate (see chapter 2).

Isocitrate dehydrogenase (ICDH) is likewise inhibited by NADH at a negative allosteric site. Thus, at rest, when NADH concentration is high, ICDH is inhibited. In addition, ICDH is activated by Ca^{2+} ions. Flow of calcium into the matrix through its uniporter increases the matrix [Ca^{2+}]. Calcium ions lower the K_m for the substrate isocitrate, thus increasing enzyme activity for the same isocitrate concentration. The more active a muscle fiber is, the higher and more sustained is the rise in [Ca^{2+}] in both the cytosol and matrix, and the more ICDH is activated.

Alpha-ketoglutarate dehydrogenase is inhibited allosterically by NADH (just like CS and ICDH). Moreover, like ICDH, α-ketoglutarate dehydrogenase is activated by a rise in the concentration of Ca^{2+}, which lowers the K_m for its substrate α-ketoglutarate. In addition, succinyl CoA is a competitive inhibitor for CoA. This means that a rise in the product, succinyl CoA, will competitively inhibit the binding of the normal substrate CoA. Thus, α-ketoglutarate dehydrogenase cannot tie up the available CoA in the matrix.

KEY POINT

The TCA cycle is regulated by substrate concentrations and by allosteric mechanisms. The electron transport process is regulated primarily by the availability of its substrates, reduced coenzymes, ADP, and oxygen.

Control of these three TCA cycle enzymes and pyruvate dehydrogenase (see below) can play a powerful role in controlling the rate of oxidative phosphorylation, again reminding us of the extremely tight relationship between the primary mitochondrial reactions generating NADH and the electron transport system that consumes electrons.

Regulation of Pyruvate Oxidation

We cover carbohydrate metabolism in the next chapter. However, the control of pyruvate oxidation is closely linked with the regulation of oxidative phosphorylation. As shown in figure 5.1, pyruvate is formed during the glycolytic reactions. It may be reduced to lactate in the cytosol or it may enter mitochondria to be oxidized to acetyl CoA. The latter then enters the TCA cycle. Pyruvate is an important source of acetyl CoA, and as we will see shortly, pyruvate availability and its oxidation to acetyl CoA can significantly influence peak rates of oxidative phosphorylation. The reaction for pyruvate oxidation, catalyzed by pyruvate dehydrogenase (PDH) in the mitochondrial matrix is as follows:

$$\text{pyruvate} + NAD^+ + \text{CoA} \xrightarrow{\text{PDH}} \text{acetyl CoA} + \text{NADH} + H^+ + CO_2$$

The PDH complex is composed of five types of enzyme subunits and three tightly bound coenzymes. The coenzymes are lipoic acid (synthesized in the body), thiamine pyrophosphate (TPP), and FAD. Two other coenzymes can also be seen in this reaction, CoA and NAD^+. As mentioned earlier in this chapter, the PDH reaction is similar to that of α-ketoglutarate dehydrogenase in that both use the same five coenzymes, although the subunits of the enzyme complexes are different.

The PDH reaction must be carefully regulated because the irreversible conversion of pyruvate to acetyl CoA means that a potential precursor to make glucose is lost. That is, pyruvate can be converted to glucose in the liver, but acetyl CoA cannot. Because the brain needs glucose and is treated biochemically as the most important tissue in the body, the PDH reaction must be regulated to spare pyruvate from being irreversibly lost. Regulation of PDH occurs via phosphorylation and dephosphorylation and allosteric mechanisms, discussed in chapter 2. To prevent the unnecessary oxidation of pyruvate to acetyl CoA when other fuels such as fat can provide the acetyl CoA, PDH is phosphorylated by PDH kinase into the inactive form, PDH-P (see figure 5.15). The enzyme PDH kinase transfers a phosphate group on to the hydroxyl (OH) group of the amino acid serine. PDH kinase is one component of the PDH complex. It is activated when an excess amount of acetyl CoA and NADH are in the matrix. Under these conditions high intramitochondrial ratios of acetyl CoA/CoA and $NADH/NAD^+$ exist, reflecting an abundance of substrate for the tricarboxylic acid cycle and electron transport chain, respectively.

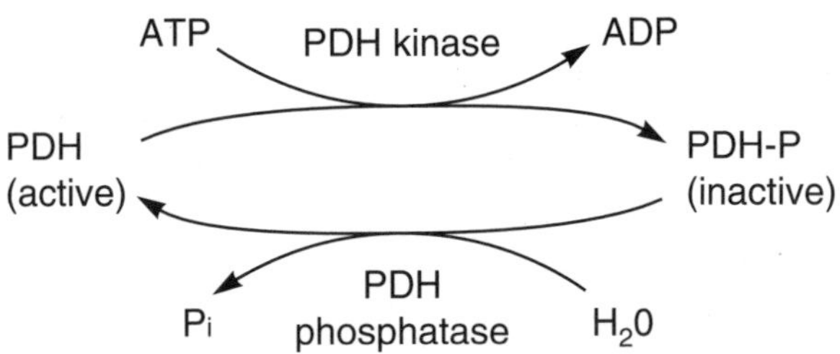

Figure 5.15 The pyruvate dehydrogenase (PDH) reaction is regulated by reversible phosphorylation and dephosphorylation of one subunit. In the dephosphorylated state, PDH is active. When a phosphate group is transferred from ATP to PDH, it becomes inactive. Phosphorylation is catalyzed by PDH kinase, whereas the phosphate group is removed by PDH phosphatase.

Activation of PDH, that is, conversion of pyruvate to acetyl CoA, occurs when PDH is dephosphorylated by PDH phosphatase. The latter is activated by an increase in intramitochondrial Ca^{2+} concentration, which results from a sustained increase in cytosolic calcium ions, for example, by continuous contractile activity in a muscle cell. Aiding the overall activation of PDH is the fact that PDH kinase is inactivated by high intramitochondrial Ca^{2+} concentration. Thus, PDH kinase is inactivated and PDH phosphatase is activated by calcium.

In addition to the phosphorylation control, NADH and acetyl CoA are competitive substrates for NAD^+ and CoA, respectively, in the PDH reaction. In summary, the activity of PDH is based on the relative proportion of unphosphorylated to phosphorylated enzyme. This is affected both by the metabolic state of the muscle fiber and the contractile state of the fiber. In muscle at rest, PDH is primarily inactive, in the phosphorylated state. The onset of muscle contraction leads to dephosphorylation and thus activation. This process does not occur instantaneously since it depends on changes in the concentration ratios of $NADH/NAD^+$ and acetyl CoA/CoA as well as an increase in matrix calcium concentration. In addition, PDH kinase is inhibited by an accumulation of pyruvate, the substrate for PDH. If the phosphatase is activated at the onset of exercise by an increase in [Ca^{2+}] and if the kinase is inhibited by a rise in pyruvate, flux through the PDH reaction will quickly increase. PDH can be activated by a drug, dichloroacetic acid (DCA). Recent experiments using DCA illustrate the significance of PDH activation at the onset of exercise in terms of increasing the contribution of oxidative phosphorylation to ATP provision.

KEY POINT

Rapid activation of pyruvate dehydrogenase at the onset of exercise plays a role in determining the extent of lactate buildup. This is because the glycolytic pathway can produce pyruvate at a much greater rate than it can be oxidized in the mitochondrion even if PDH is maximally activated.

Regulation of Oxidative Phosphorylation in Rested Muscle

In a muscle at rest the rate of energy expenditure is quite low, based mainly on maintaining protein synthesis and normal cell function. Now consider the substrates in the simple equation to describe oxidative phosphorylation (shown earlier in this chapter). Which of these is likely to limit the rate of oxidative phosphorylation?

$$5\ ADP + 5\ Pi + 2\ NADH + 2\ H^{+} + O_2 \longrightarrow 2\ NAD^{+} + 5\ ATP + 7\ H_2O$$

Oxygen is readily available in rested muscle; therefore it cannot be limiting. The concentration of Pi is also high enough to sustain a modestly high rate of oxidative phosphorylation and is therefore not limiting. There is typically sufficient reducing power in the form of NADH in rested muscle. This leaves the availability of ADP to the respiratory chain as the weak link in the chain, limiting oxidative phosphorylation including the TCA cycle. Of course, the availability of ADP depends on the rate of ATP hydrolysis in the cytosol plus the entry of ADP into the mitochondria via the ADP-ATP antiport. Based on studies using isolated mitochondria, some authorities use the term state four to describe the situation of mitochondria in a rested cell where the rate of oxidative phosphorylation is limited by a low availability of ADP.

Many studies support the fact that availability of ADP limits oxidative phosphorylation for a variety of situations in muscle as well as other tissues. For example, if mitochondria are isolated and placed in a well-oxygenated medium, the rate of oxygen consumption (used as an index of the rate of oxidative phosphorylation) is low. If either ADP or creatine is added, the rate of oxygen utilization greatly increases. We have already discussed how addition of ADP stimulates oxidative phosphorylation. However, addition of creatine also effectively increases oxidative phosphorylation by stimulating the mitochondrial creatine kinase enzyme, generating ADP and PCr. The ADP then enters the matrix using the ADP-ATP antiport to stimulate oxidative phosphorylation. When the oxygen utilization of isolated mitochondria is sharply increased by addition of ADP or creatine (the latter generates ADP), we say it is state three respiration.

Overall, the rate of oxidative metabolism for a muscle at rest, and for many other tissues, fits neatly into the concept that a limitation is due primarily to ADP within the mitochondria. This view of the regulation of oxidative phosphorylation is called the kinetic model or **acceptor control model**.

Regulation of Oxidative Phosphorylation in Exercise

When muscle undergoes a transition from rest to moderate exercise or when the intensity of moderate exercise is modestly increased, the rate of ATP hydrolysis undergoes an abrupt step increase to match the new exercise intensity. There will be an increase in the rate of oxidative phosphorylation, measured as $\dot{V}O_2$, but this will follow an exponential time course with a half-time of approximately 20 seconds (see figure 5.16). Two questions can be posed: What is responsible for the increase in the rate of oxidative phosphorylation during step changes in rate of muscle activity? Why is there a lag in oxygen utilization, a measure of the rate of oxidative phosphorylation, before it reaches a new level corresponding to the increased exercise intensity?

The simple kinetic or acceptor control concept explains quite well what keeps oxidative phosphorylation at a low rate in rested muscle. It does not adequately account for the responses that accompany changes in exercise intensity because it is

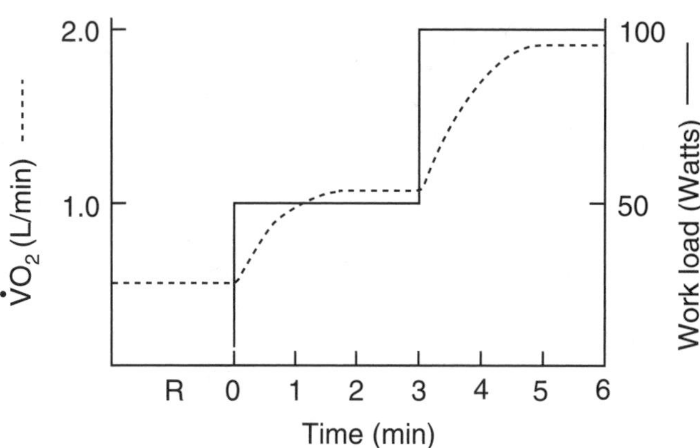

Figure 5.16 The relationship between the rate of oxidative phosphorylation, measured by the $\dot{V}O_2$, and the rate of ATP hydrolysis, which tracks exercise workload for a hypothetical subject. The $\dot{V}O_2$ is determined while the subject sits resting on a cycle ergometer (R). The clock starts and the subject immediately begins pedaling at a workload of 50 watts. After three minutes, the workload is increased to 100 watts. There is a lag of approximately two minutes before the $\dot{V}O_2$ reaches a new steady state for both workloads.

difficult to demonstrate a change in the concentration of any one of the substrates that corresponds to the increase in oxidative phosphorylation. Therefore, exercise biochemists have been more accepting that regulation is based on a combination of factors: changes in the demands of exercise as reflected by the **cytoplasmic energy state** (that is, [ATP]/[ADP] × [Pi], also called the phosphorylation potential), as well as the mitochondrial redox state (the ratio of [NADH]/[NAD^+], also called the mitochondrial redox potential) and the cellular oxygen content (expressed as the partial pressure of oxygen, or PO_2, measured in torr). Let us see how these changes take place.

The phosphorylation potential ([ATP]/[ADP][Pi]) will decrease when there is any increase in exercise intensity. While ATP concentration remains remarkably constant, it will decrease a little. Couple this to a roughly parallel increase in ADP and a much larger increase in Pi due to a decline in PCr and you can see how the phosphorylation potential can vary more widely than any of its constituents. Note also that the phosphorylation potential reflects PCr concentration because the increase in Pi reflects the decline in PCr. The mitochondrial redox potential ([NADH]/[NAD^+]) changes when the rate of electron transfer from NADH to oxygen is not matched by the rate of formation of NADH through dehydrogenase enzymes. Recall that the activities of three of these enzymes (pyruvate, isocitrate, and α-ketoglutarate dehydrogenases) respond to a complex pattern of control factors and mechanisms, including an increase in matrix [Ca^{2+}]. Finally, the oxygen availability to the respiratory chains in each mitochondria ($P_{mito}O_2$) depends on a complicated combination of gas exchange, blood flow, and diffusion in getting from the outside air, through the lungs, into the arterial blood flow, and to the cytosol of each muscle fiber. This long transportation route accounts to a major extent for the observation that $\dot{V}O_2$ measured at the mouth increases in an exponential fashion, whereas the rate of ATP hydrolysis abruptly increases with a step increase in exercise intensity.

The step increase in exercise intensity will cause phosphorylation potential to decline. The extent of the decrease will reflect the increase in exercise intensity. This will stimulate the entry of ADP and Pi into the mitochondrion and electron transport from NADH to oxygen. As NADH is oxidized to NAD^+, the inhibitory effect of NADH on the three irreversible TCA cycle enzymes is reduced and the TCA cycle will speed up. In addition, the gradual rise in matrix calcium as a result of an increased cytosolic concentration in the more active fibers will further stimulate isocitrate and α-ketoglutarate dehydrogenases and activate pyruvate dehydrogenase. Because of the initial mismatch between oxygen utilization by the cytochrome c oxidase complex and oxygen delivery from the air to the fiber, the oxygen tension within the fiber will decline. However, the rate of oxidative phosphorylation can be increased by a combination of a decrease in phosphorylation potential, an increase in mitochondrial redox potential, and a gradual increase in oxygen transport to the muscle mitochondria. The key point is that adjustments in phosphorylation and mitochondrial redox potentials can help to maintain oxidative metabolism in the face of declining oxygen availability to the respiratory chain (i.e., $P_{mito}O_2$). Any mismatch between ATP demand and ATP supplied by oxidative phosphorylation must be provided by PCr and glycolysis. For small step increases in exercise intensity, the former will predominate. This has been demonstrated in a variety of models, showing that any slowing in the delivery of oxygen-rich blood to exercising muscle is compensated by a steeper decline in PCr. Of course this means that phosphorylation potential would be lower compared to the exercise condition with intact blood flow.

Much of the work and exercise we do is performed at a fairly constant rate of energy expenditure. For low to moderate sustained exercise, a steady rate of oxidative phosphorylation can supply virtually all the ATP needs. We call this steady state exercise because the rate of oxidative phosphorylation is precisely matched to the ATP demands by a combination of phosphorylation potential and redox potential adjusted to the steady state content of oxygen in the muscle fibers. The phosphorylation potential would be inversely related to the intensity of exercise while the redox potential should parallel exercise intensity.

So far, we have only looked at control of oxidative phosphorylation when the step increases in muscle activity are relatively small. The situation is in part similar, but more complicated, if we consider transitions to maximum exercise. For a huge and sudden increase in muscle activity, phosphorylation potential will decline even more. Although PCr can partially buffer significant declines in ATP and increases in ADP, both of these will change to a greater extent than in our previous discussion. Further, PCr concentration will decline faster and more precipitously, and Pi will rise in parallel. Oxidative phosphorylation will be stimulated even more than before, with larger changes in mitochondrial redox and matrix calcium concentration. However, we are talking about exercise intensities far beyond what could possibly be supported by even maximal rates of oxidative phosphorylation. Moreover, these supramaximal exercise intensities can be sustained for time periods much less than the time it takes to reach peak rates of oxidative phosphorylation. For these situations, the huge changes in cytoplasmic phosphorylation potential (and other factors we discuss in the next chapter) rapidly turn on glycolysis. With its moderately high power for ATP formation, glycolysis plays a prominent role in providing the ATP needs.

The oxygen available to muscle mitochondria can limit oxidative phosphorylation under certain circumstances. For example, at higher altitude, where the oxygen content of the air is low, the time course in the response of $\dot{V}O_2$ (reflecting active muscle oxygen consumption) to a modest step increase in muscle activity is delayed. This would suggest that the oxygen content available to mitochondria would be lower and slower adjusting with time. Therefore, compared to the same exercise intensity at sea level, we would expect a lower phosphorylation potential and higher redox potential to offset the lower oxygen content of altitude. Oxygen available to the electron transport chain could also be limiting during isometric contractions when intramuscular pressure builds up sufficiently to reduce or completely cut off the blood flow. In situations of reduced mitochondrial oxygen availability due to compression on blood vessel walls, glycolysis becomes extremely important as a source of ATP.

KEY POINT

A keenly debated topic is what limits the maximum rate of oxidative phosphorylation by an individual. Is it limited by oxygen transport from the air to cytochrome c oxidase (i.e., transport limited)? Alternatively, is the limitation based on the production of reducing equivalents for the electron transport chain (i.e., a metabolic limitation)? A program of endurance training generally increases the activities of TCA cycle enzymes and electron transport chain protein components about three to five times more than the ability to transport oxygen as measured by $\dot{V}O_2max$. This is taken to mean that in fit individuals oxygen transport places the upper limit on the maximum rate of oxidative phosphorylation. For the unfit, it is more likely that they cannot generate electrons at a rate sufficient to match their ability to deliver oxygen to the mitochondria of exercising muscle. Athletes engaged in prolonged exercise such as marathons or triathlons can experience a situation where performance falls off near the end. This has been called "hitting the wall," but its cause is easy to understand.

(continued)

As we will see, oxidation of carbohydrates and fats provides the ATP to support submaximal exercise. The higher the exercise intensity, the more carbohydrate is oxidized and the less the reliance on fat. Near the end of a marathon, for example, carbohydrate stores can become severely reduced. This means that the source of acetyl groups for the TCA cycle must come increasingly from the beta-oxidation of fatty acids (discussed in chapter 7). However, numerous studies have revealed that provision of acetyl CoA to the TCA cycle from beta-oxidation of fatty acids alone cannot match that when pyruvate or a mixture of pyruvate and fatty acids is used. This means that the primary supply of reducing equivalents (electrons on NADH) to the electron transport chain is compromised. Accordingly, the athlete must reduce running, cycling, or skiing pace to a level where the rate of ATP demand by the exercise can be met by the new, lower level of oxidative phosphorylation using fatty acids as the primary fuel.

Summary

Oxidative phosphorylation is the synthesis of ATP from ADP and Pi in association with the transfer of electrons from fuel molecules to coenzymes to oxygen. Oxidative phosphorylation, responsible for generating the preponderance of ATP in our bodies, takes place in mitochondria, utilizing the tricarboxylic acid (TCA) cycle and the electron transport (respiratory) chain. The TCA cycle is the primary source of reduced coenzymes. The electron transport system, represented by four protein-lipid complexes in the inner mitochondrial membrane, oxidizes the reduced coenzymes and transports the electrons to oxygen. During the process of electron transport through these complexes free energy is released, which is utilized to transport protons across the mitochondrial inner membrane against a concentration and electrical gradient. Protons are allowed to flow down their electrical and chemical gradient through a specialized inner mitochondrial protein complex known as ATP synthase. Free energy released during proton flow is harnessed to ADP phosphorylation, making ATP. The exact stoichiometry of ATP produced per atom of oxygen consumed (P/O ratio) during electron transport is unclear, but it is likely between 2.5 to 3.0. Although electron transport is normally tightly coupled to ADP phosphorylation, leakage of protons across the inner membrane (i.e., uncoupled oxidation) does occur and is responsible for generating heat, especially for small mammals.

The TCA cycle (Krebs cycle or citric acid cycle) is the pathway that removes the last carbon atoms in the form of CO_2 from all the body's fuels, while electrons associated with the hydrogen atoms of these fuels are used to reduce the coenzymes NAD^+ and FAD. The TCA cycle is a circular pathway catalyzed by eight enzymes, all of which are located in the mitochondrial matrix except succinate dehydrogenase. Acetyl groups containing two carbon atoms enter the pathway attached to CoA and attach to oxaloacetate to form citrate. Because one turn of the cycle yields two CO_2, the tricarboxylic acid cycle neither produces nor consumes oxaloacetate or any other intermediate of the cycle.

Since most of the ATP is consumed in the cytosol of cells, whereas ATP is produced from ADP and Pi in mitochondria, special transporters move ATP from the matrix of the mitochondria while simultaneously bringing ADP in. The rate of ATP hydrolysis in the cytoplasm is closely matched to the TCA cycle, the rate of electron transport, and ADP phosphorylation in mitochondria. Oxidative phosphorylation in skeletal muscle, generating ATP, is tightly regulated to rates of ATP hydrolysis in the cytosol. Rather than focusing on changes in the concentrations of the individual substrates for oxidative phosphorylation, exercise biochemists have adopted two ratios that better account for the coupling of ATP demand to ATP provision. Changes in muscle tissue oxidation rates better match alterations in the energy demand based on the phosphorylation potential ([ATP]/[ADP] [Pi]) and mitochondrial redox potential ([NADH]/[NAD^+]) generated by the regulated dehydrogenases in mitochondria. The oxygen tension at the level of the mitochondria ($P_{mito}O_2$) can limit oxidative phosphorylation. This can occur during exercise at altitude or when blood flow to exercising muscle is partially or completely occluded. The rate of NADH formation in the tricarboxylic acid cycle may limit the overall rate of oxidative phosphorylation in skeletal muscle if carbohydrate available to exercising muscle is depleted, forcing the muscle to use fat as the exclusive fuel. For an athlete, this means reducing the pace of their activity.

Key Terms

acceptor control model 75
antiport 70
ATP synthase 69
beta-oxidation 57
chemiosmotic hypothesis 69
coenzyme A 61
coenzyme Q 64
cristae 57
cytochromes 68
cytoplasmic energy state 76
electrochemical gradient 58
electron transport chain 56
faraday constant (F) 60
intermembrane space 57
lipoic acid 62
matrix 58
NADH 59
P/O ratio 59
redox potential 60
reducing equivalents 60
respiratory exchange ratio (RER) 58
respiratory quotient (RQ) 58
sodium-calcium antiport 58
standard redox potential 60
thiamine pyrophosphate (TPP) 62
ubiquinol 67
ubiquinone 67
uncoupling protein 69
uniporter 58

Review Questions

1. Determine the number of ATP molecules generated from the complete oxidation of a pyruvate if the P/O ratio is 3 for NADH and 2 for $FADH_2$. How does this differ from the numbers we have assumed in this chapter?
2. How many electrons are needed to reduce four molecules of oxygen?
3. A subject exercising at 110 watts, breathing room air (21% O_2) has a $\dot{V}O_2$ of 2.0 liters/minute. What would the $\dot{V}O_2$ be if the subject was doing the exact workload but breathing a gas mixture with only 14% O_2?
4. Using the example in the previous question, explain the adjustments that would likely be made in cytosolic phosphorylation and mitochondrial redox potentials with a lower concentration of inspired oxygen.
5. In figure 5.16, the $\dot{V}O_2$ is approximately 0.4 liters/minute when the subject is merely sitting on the cycle ergometer doing no exercise. Why?
6. Approximately how many kilocalories (or kilojoules) of energy are expended when a subject exercises for two hours at an average $\dot{V}O_2$ of 3.0 liters/minute?
7. You determine the $\dot{V}O_2$max of a group of sedentary students and cross country runners while they are running at constant speed on a treadmill. The treadmill grade is increased every two minutes until the subjects are no longer able to continue. For the untrained subjects, the $\dot{V}O_2$ increases progressively up to the point the subjects can no longer continue. For the cross country runners, the $\dot{V}O_2$ levels off about two minutes before they are forced to stop running. Explain why there are these differences.

CHAPTER

6

Carbohydrate Metabolism

Most active people are aware of the importance of carbohydrates to physical performance. Many others are likely aware that a diet with a greater proportion of carbohydrates to fats is healthier. Few people, however, know much about the chemical reactions in the body involving carbohydrates or how the body, especially skeletal muscle, adjusts its fuel utilization to spare the use of carbohydrates unless stores are high or the need for ATP production is acute. Carbohydrate is brain food, and our carbohydrate chemistry is set to favor scarce carbohydrate stores for brain use.

Our study of carbohydrate metabolism includes a detailed examination of the glycolytic pathway. We look at the metabolism of glycogen and how the synthesis and breakdown of liver and muscle glycogen are tied in to the priorities of carbohydrate storage and utilization. There are important routes to making carbohydrate when dietary sources are inadequate. Gluconeogenesis, or the synthesis of glucose from noncarbohydrate sources, is vitally important. We begin this chapter by looking at the various carbohydrates and how glucose entry into cells is regulated.

Carbohydrates

We classify carbohydrates as monosaccharides, disaccharides, and polysaccharides. Examples of **monosaccharides** are glucose, fructose, and galactose; these are simple sugars called hexoses containing six carbon atoms—*hex* for six carbon atoms and *ose* meaning sugar. We have already mentioned two other monosaccharides that are pentoses. Ribose and deoxyribose (see figure 3.1), each containing five carbon atoms, are found in ribonucleotides and deoxyribonucleotides, respectively. Figure 6.1 shows the structures for the predominant forms of D-glucose, D-galactose, and D-fructose. The D refers to the configuration about carbon atom 5 in each molecule. Only the D forms of the monosaccharides are acceptable to glucose-metabolizing enzymes in animals. In the remainder of this chapter, the D form of each monosaccharide is assumed and thus the letter will be omitted. Recall that only L amino acids can be used by animals to make proteins.

Disaccharides are formed when two monosaccharides join together. The common disaccharide sucrose is composed of the monosaccharides glucose and fructose, while lactose or milk sugar contains glucose and galactose. Maltose, produced during the digestion of dietary starch, contains two glucose molecules joined together. Specific digestive enzymes located on the surface of intestinal cells hydrolyze disaccharides into their monosaccharide constituents. Thus, sucrase, lactase, and maltase digest sucrose, lactose, and maltose, respectively.

Glycogen and starch are polysaccharides, but only starch is significant as a dietary source of

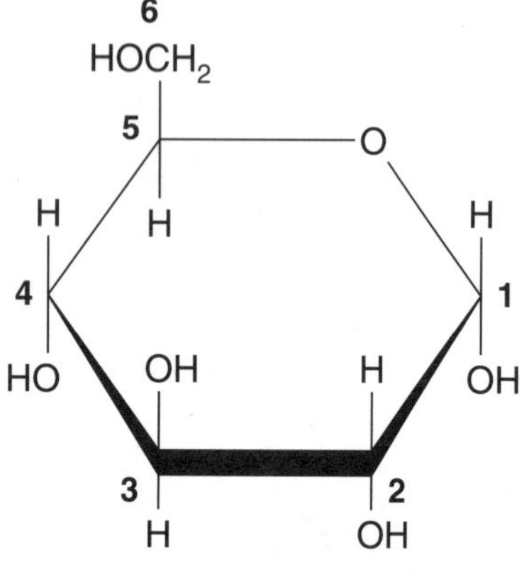

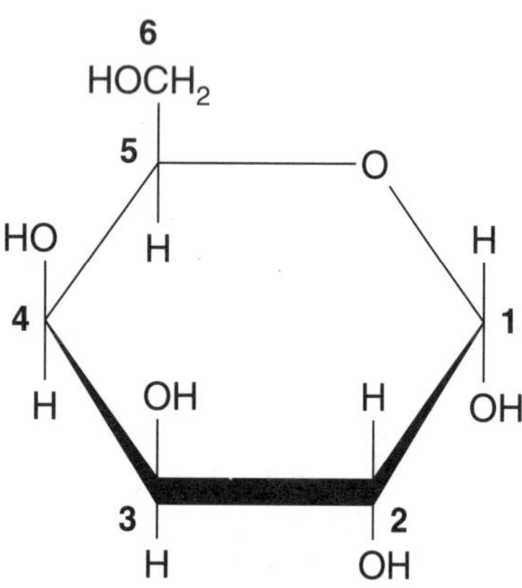

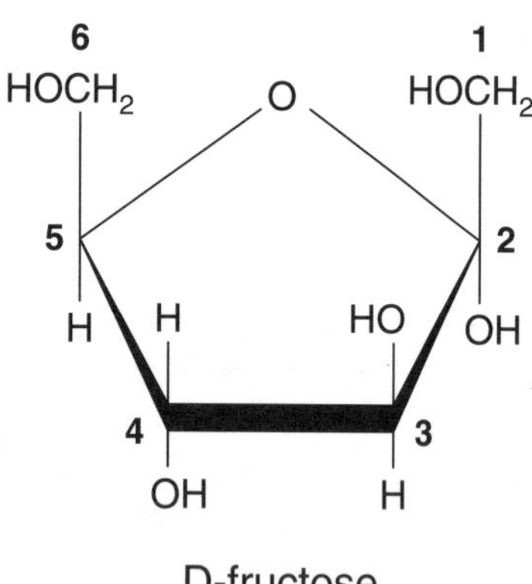

Figure 6.1 The predominant structural formulas for the monosaccharides, D-glucose, D-galactose, and D-fructose. The numbers identify the carbon atoms, and the D refers to the absolute configuration about carbon 5. These molecules are all isomers with the formula $C_6H_{12}O_6$. Note that D-glucose and D-galactose differ only in the configuration about carbon atom 4.

carbohydrate. From a nutritional perspective, starch is a complex carbohydrate. When it is completely digested, its products are glucose molecules. Glycogen is a highly branched polysaccharide, composed only of glucose molecules joined together.

After a meal containing a variety of foods, the main carbohydrate digestion products would be glucose, some fructose, and galactose from milk sugar. These substances are absorbed into the blood and transported to the liver, where galactose is converted to glucose. Fructose can be utilized as a substrate for glycolysis in muscle, liver, and adipose tissue. It enters the pathway at several locations, but in the end, its products are no different from those of glucose.

Cellular Uptake of Glucose

The normal blood glucose concentration, described as **euglycemia**, is approximately 5 millimolar, equivalent to 90 milligrams of glucose per deciliter (100 ml) of blood. Following a meal, blood glucose is elevated above normal and can increase to 9 millimolar (160 mg/dl or more). Uncontrolled insulin-dependent **diabetes mellitus** can result in a glucose concentration over 20 millimolar. We call elevated blood glucose concentrations **hyperglycemia**. Blood glucose concentration well below normal (about 2.5 mM or less, or less than 45 mg/dl) is called **hypoglycemia**. You can become hypoglycemic if you do not eat for a long period of time or if you exercise for hours without ingesting carbohydrate. Individuals who do not pay attention to carbohydrate intake during prolonged competitions or exercise may be forced to stop because of hypoglycemia.

A gradient exists for glucose entry into cells because the glucose concentration in the blood and extracellular fluid is much higher than inside cells. Glucose is a polar molecule with five hydroxy (OH) groups. It is therefore a poor substance for crossing the hydrophobic cell membrane. To get glucose inside cells, a transporter is needed; that is, a protein molecule that will allow glucose entry across the cell membrane. The process of transporting a substance down its concentration gradient across a membrane is known as facilitated diffusion. As mentioned in chapter 2, membrane transport is characterized by simple Michaelis-Menten kinetics with a V_{max} and a K_m. As such, membrane transport will exhibit saturation kinetics similar to the effect of increasing substrate concentration on the activity of an enzyme. A family of membrane transporters exists for glucose. These **glucose transporters** are given designations such as GLUT-1, GLUT-2, GLUT-3, GLUT-4, and GLUT-5. These transporters exhibit tissue-specific locations and unique kinetic parameters.

To be metabolized, glucose must diffuse into a cell through a glucose transporter. Entry into some cells is regulated; entry into others is unregulated. We would expect the unregulated entry of glucose, which depends on the relative concentration gradient of glucose across the membrane, to occur in cells that rely primarily on glucose as an energy source. In fact, this is what happens for red blood cells, brain cells, and kidney cells. Liver cells, which store excess glucose as glycogen, also have unregulated glucose uptake. On the other hand, large tissues, such as skeletal muscle or fat as well as the heart, have regulated glucose uptake. Glucose transport across cell membranes in regulated tissues occurs primarily by GLUT-4 transporter, which, unlike the other transporter isoforms, is regulated by insulin. Skeletal muscle accounts for about 70 to 80% of the insulin-stimulated uptake of glucose.

Insulin, a polypeptide hormone secreted by the beta cells of the pancreas, is the main regulator of glucose transport. When blood glucose concentration is elevated (e.g., following a meal), blood insulin concentration increases to help glucose enter the regulated tissues. Insulin binds to an insulin receptor that spans the cell membrane. Insulin binding results in the phosphorylation of tyrosine residues (see chapter 2) on the cytoplasmic side of the receptor (i.e., auto- or self-phosphorylation), leading to activation of a latent tyrosine kinase activity in the insulin receptor itself. Through a complicated mechanism involving other protein kinases, GLUT-4 transporters are translocated from intracellular storage sites to the cell membrane to aid glucose entry. Thus, insulin increases the V_{max} of glucose transport, but only in those cell types (i.e., muscle and fat) expressing the GLUT-4 transporter gene.

Exercising skeletal muscle also has an increased ability to take up glucose from the blood, independent of the effect of insulin. The exercise effect also involves a stimulation of GLUT-4 transporter translocation to the cell membrane from intracellular storage sites. The mechanism for the exercise effect involves a different signaling pathway, employing the elevated calcium concentration caused by activation of the muscle fiber through its motor neuron. Interestingly, the effects of both exercise and insulin are additive, supporting the fact that the signaling systems are different. This muscle contraction effect persists into the early postexercise period in order to rebuild depleted stores. During prolonged exercise tasks or games, one must ingest glucose to keep blood levels maintained because exercising muscle has an augmented capacity to take up glucose from the blood. Failure to supply enough glucose to the body during prolonged physical activity can lead to problems associated with hypoglycemia.

KEY POINT

People who engage in endurance exercise training programs increase the total content of GLUT-4 transporters in the trained muscle. This means that there is an increased maximal capacity for glucose transport in the trained muscle. Interestingly, when subjects are compared during the same exercise intensity after, compared to before, training, they will have more total muscle GLUT-4 transporters, but fewer in the sarcolemma, to aid in glucose transport. As we will see, one of the adaptations that takes place after training is to use less carbohydrate and more fat to fuel the same exercise level.

Type 1 (I) or insulin dependent diabetes mellitus (IDDM) is a condition in which there is no blood insulin due to an autoimmune destruction of the insulin-secreting beta cells of the pancreas. Individuals with IDDM need exogenous insulin to survive, but because of the exercise effect on glucose uptake, these individuals must carefully balance the type, intensity, and amount of exercise with blood glucose levels, carbohydrate intake, and insulin dose. Noninsulin dependent diabetes mellitus or NIDDM (also known as type 2 diabetes mellitus) is typically characterized by hyperglycemia, hyperinsulinemia (elevated blood insulin, at least in the early stages), and insulin resistance. Unlike type 1 diabetes, insulin is secreted from the pancreatic beta cells in response to an elevation in blood glucose. The problem with NIDDM is that the response of the blood glucose control system is inappropriate. A major source of the problem with NIDDM is resistance to insulin in muscle and adipose tissue, although abnormalities in lipid metabolism also coexist. Type 2 diabetes affects many millions of people and is widely acknowledged to be a major health problem in the wealthy countries. With the understanding of the mechanism for exercise-induced blood glucose transport, health authorities are proposing daily exercise for those predisposed to or suffering from type 2 diabetes. Not only will exercise result in an immediate improvement in glycemic (blood glucose) control by enhancing the effects of insulin to promote glucose uptake in both muscle and adipose tissue, but persistent, daily exercise will increase the total content of GLUT-4 transporters in muscle.

Phosphorylation of Glucose

Once glucose enters a cell, it is covalently modified by transfer of the terminal phosphate from ATP to carbon atom 6 of glucose to make glucose 6-phosphate (glucose 6-P) as shown in the equation that follows. When glucose is phosphorylated, the product, glucose 6-P, is trapped inside the cell.

$$\text{glucose} + \text{ATP} \xrightarrow[\text{}]{\text{hexokinase/ glucokinase}} \text{glucose 6-P} + \text{ADP}$$

This reaction is essentially irreversible because of the large free energy change, as previously shown in chapter 4. Four hexokinase isoenzymes catalyze this reaction, identified as hexokinases I, II, III, and IV (HK I, HK II, HK III, and HK IV). Hexokinase IV is also known by the common name glucokinase and is found in the liver.

The differences between glucokinase (known both by GK and HK IV) and the other hexokinase isozymes (HK I, HK II, and HK III) are as follows:

1. Glucokinase is found only in the liver and pancreas, whereas the other hexokinase isozymes are found in all cells.

2. The amount of hexokinases I, II, and III in cells remains fairly constant; they are thus constitutive enzymes. The amount of glucokinase in liver cells depends on carbohydrate content of the diet. Glucokinase is thus described as an inducible enzyme because a diet high in carbohydrate will induce liver cells to make more. Conversely, diabetes (with its attendant low insulin), starvation, or a low carbohydrate diet will mean less glucokinase. Control of the amount of liver glucokinase is mainly at the transcription level since a 30- to 60-fold increase in gene transcription follows within 30 minutes after injection of insulin into a diabetic rat.

3. Hexokinase isozymes I, II, and III have a low K_m for glucose (0.02–0.13 mM), whereas glucokinase has a high K_m for glucose (~ 5–8 mM). Hexokinases I, II, and III are thus very sensitive to glucose, whereas glucokinase activity only becomes important for phosphorylating glucose when the concentration of blood glucose is slightly elevated. (Figure 2.4 illustrates the responses of the reaction velocities for glucokinase and a hexokinase to changes in glucose concentration.)

4. Hexokinase isozymes I, II, and III, but not glucokinase, can be inhibited by the product of the reaction, glucose 6-P. This is an example of **feedback inhibition**. Thus, if the concentration of glucose 6-P increases inside a cell, it inhibits the activity of hexokinase; glucose will not get phosphorylated, and its concentration will increase in the cell. This increase will reduce the gradient for transport, thus slowing down glucose entry into the cell.

For glucose to be metabolized in a cell, two processes must occur; it must be transported into the cell using a glucose transporter, and it must be phosphorylated to glucose 6-phosphate. Which of these two processes limits subsequent metabolism of glucose can be hotly debated. However, under most normal physiological conditions, the concentration of free glucose inside skeletal muscle cells is very low. This points to glucose transport as being **rate limiting** to glucose metabolism, since glucose is phosphorylated about as fast as it enters the cell.

KEY POINT

The fact that glucose transport is rate limiting to glucose entry into skeletal muscle helps to explain an important point regarding postexercise (competition) feeding. Feeding carbohydrate to athletes after training or competition should take place as soon as possible. One reason for this is the effect of the prior muscle activity on the content of GLUT-4 transporters in the muscle sarcolemma. This exercise effect plus the independent effect of an increase in blood insulin with the carbohydrate feeding will promote glucose transport and hasten the synthesis of muscle glycogen for the next exercise session or competition.

Glycolysis

Glucose entry into a skeletal muscle cell, its phosphorylation to glucose 6-phosphate, and other steps in metabolism of glucose are outlined in figure 6.2. Glucose 6-phosphate has two major fates in a muscle cell. Following a meal, and when the muscle fiber is inactive, glucose 6-phosphate will be directed to the synthesis of glycogen. During exercise, stored glycogen will be broken down to glucose 6-phosphate and metabolized in the glycolytic pathway to pyruvate. As well, any glucose that enters an actively contracting muscle fiber will be broken down to pyruvate using the glycolytic pathway. Therefore, there are two sources of glucose 6-phosphate for glycolysis—glycogen and blood glucose. Glucose 6-phosphate can inhibit hexokinase by feedback as shown in figure 6.2. As we will see, this is important because it prevents unnecessary entry and phosphorylation of blood glucose if stored glycogen can provide glucose 6-phosphate at an adequate rate. One of the interesting questions we will look at later is how important blood glucose is as a source of glucose 6-phosphate to fuel glycolysis during exercise.

As mentioned in chapter 4, the word glycolysis means the splitting of glucose, summarized as follows:

$$\text{glucose} + 2\ \text{ADP} + 2\ \text{Pi} \longrightarrow 2\ \text{lactate}^- + 2\ \text{H}^+ + 2\ \text{ATP}$$

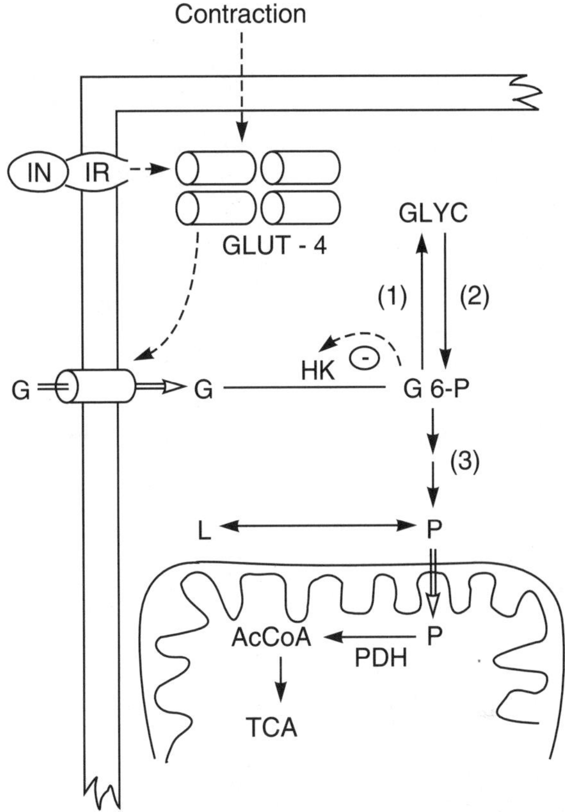

Figure 6.2 Overview of glucose metabolism in a muscle cell. Glucose uptake into muscle through GLUT-4 transporters is greatly facilitated by insulin (IN) binding to the insulin receptor (IR) or by contractile activity. Through two mechanisms, GLUT-4 transporters are then translocated from internal storage sites to the muscle membrane. Once inside the cell, glucose is phosphorylated by hexokinase (HK) to glucose 6-phosphate (G 6-P), which can either be used to make glycogen through pathway (1) or immediately be broken down to pyruvate (P) in the glycolytic pathway (3). Breakdown of glycogen (2) during exercise is accelerated to produce G 6-P for glycolysis. Pyruvate has two major fates, reduction to lactate (L) or entry into the mitochondrion and conversion into acetyl CoA (AcCoA) using pyruvate dehydrogenase (PDH). Acetyl CoA enters the TCA cycle.

Enzymes of Glycolysis

This reaction occurs in the cytoplasm of cells. As the equation above shows, this process is anaerobic. The glucose is split into two lactate ions and protons. From each glucose molecule, two ATP can be formed. This is an example of ATP formation without the use of oxygen and is called substrate-level phosphorylation. The other two forms of substrate-level phosphorylation we have already seen are ATP formation from PCr and the GTP formed in the TCA cycle.

Figure 6.2 shows that the lactate formed in glycolysis is derived from pyruvate. Figure 6.2 also shows that pyruvate can have two major fates: (a) to be reduced to lactate or (b) to enter the mitochondria for oxidation to carbon dioxide and water. Because glycolysis is considered to be an anaerobic process, strictly speaking, glycolysis must end with lactate, since pyruvate oxidation absolutely needs oxygen. To avoid this semantic problem, the term aerobic glycolysis is used for the process in which the pyruvate is oxidized in the mitochondrion. We have already looked at the conversion of pyruvate to acetyl CoA in the mitochondrion, catalyzed by pyruvate dehydrogenase.

In many early textbooks, authors assumed that the enzymes of glycolysis are freely dissolved in the cytosol of the cell. This view is no longer held. Many, if not all, glycolytic enzymes may be bound to other structures in the cell, such as to structural filaments providing shape to the cell, to the endoplasmic reticulum (sarcoplasmic reticulum in muscle), to the outer membrane of the mitochondrion (as is the case for hexokinase), or to contractile proteins such as actin. Some of the enzymes catalyzing the reactions of glycolysis may also be physically linked to each other such that the product of one enzyme is immediately passed to the next enzyme as its substrate. In this case, the actual concentrations of the intermediates in glycolysis, except glucose 6-P, fructose 6-P, pyruvate, and lactate, would not increase much, even if glycolysis were proceeding rapidly.

KEY POINT

We have already used the term flux to describe how fast substrates enter a biochemical pathway and become products at the other end. In terms of glycolysis, its flux may be rapid, but there may not be a large increase in the concentrations of intermediates within the pathway because, as fast as they are formed, they are acted on by the next enzyme in the pathway. For glycolysis, we would expect that the rate of formation of pyruvate and lactate would provide a useful index of the flux of glycolysis since these are the major products.

Glycolysis has two major functions. One is to generate energy in the form of ATP. In fact, in red blood cells, in which glycolysis is the only energy-generating source, making ATP is its only role. The second function is to generate pyruvate for final oxidation in the mitochondrion. Figure 6.3 outlines the major reactions of the glycolytic pathway. If we start from glucose, we can see that there are two reactions early on that require ATP, yet we know that glycolysis is used to produce ATP. This seemingly absurd way of using ATP to make ATP can be understood by using a roller coaster as an analogy. The first part of the roller-coaster ride is a slow ascent to the highest point on the course using a motor that catches the underside of the cars and moves them up. Thereafter, the potential energy of this high elevation is used to create the kinetic energy that makes a roller-coaster ride so thrilling and appealing. Overall, the $\Delta G^{\circ\prime}$ for glycolysis is –120 kilojoules/mole (–29 kcal/mol).

Reactions of Glycolysis

Starting from glucose, glycolysis can be considered to involve three stages. In the first or priming stage, two phosphorylation reactions produce a hexose with two phosphate groups attached. This is called a hexose bisphosphate. Stage two is the splitting stage, conversion of the hexose bisphosphate into two triose phosphates—three-carbon sugar molecules with an attached phosphate group. In stage three, there are oxidation-reduction reactions and the formation of ATP.

Discussion of the individual reactions of glycolysis will start with glucose 6-P, since glucose phosphorylation has already been described. The glucose phosphate isomerase reaction interconverts two hexose phosphates. This is called an isomerase because glucose 6-P and fructose 6-P are isomers.

The next step is the phosphorylation of fructose 6-P, using a phosphate group from ATP and catalyzed by the enzyme phosphofructokinase (abbreviated PFK). This is a priming reaction, increasing the chemical potential energy of the product. The reaction is strongly exergonic with a $\Delta G^{\circ\prime}$ of –19 kilojoules/mole. The phosphate group is attached to carbon 1 of fructose, so the product is known as fructose 1,6-bisphosphate. The term *bis* means two, and two phosphates are attached to separate locations on the same molecule. If there were three phosphates attached to separate places on the same molecule, we would use the term *tris*. PFK catalyzes the committed step of glycolysis, committing the cell to glucose degradation. PFK is under tight regulation, and its activity controls the flux of glycolysis.

Stage two begins with the splitting of a hexose bisphosphate. Aldolase (also called fructose bisphosphate aldolase) splits fructose 1,6-bisphosphate into two triose phosphates, that is, glyceraldehyde 3-phosphate and dihydroxyacetone phosphate. However, only glyceraldehyde 3-phosphate has a further role in glycolysis. Therefore, the enzyme triose phosphate isomerase catalyzes the reversible interconversion of dihydroxyacetone phosphate into glyceraldehyde 3-phosphate. This

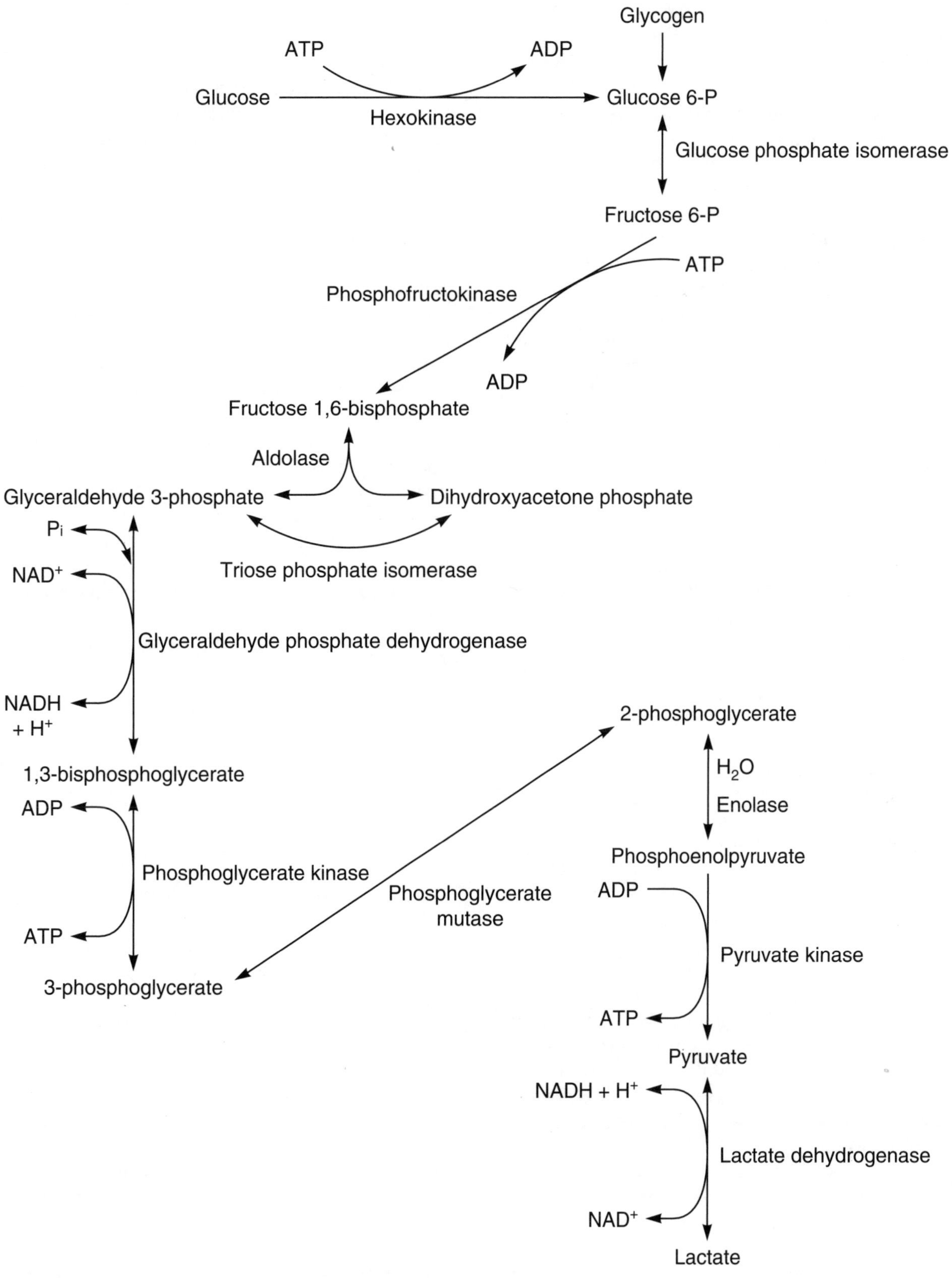

Figure 6.3 The reactions of glycolysis in which glucose is split into two lactate ions. Also shown is the entry of glucose units from glycogen into the glycolytic pathway.

enzyme ensures that all of the carbon atoms in fructose 1,6-bisphosphate are funneled through the glycolytic pathway via glyceraldehyde 3-phosphate.

At this point in the pathway, all of the original carbon atoms in each glucose molecule are in the form of two molecules of glyceraldehyde 3-phosphate.

The third stage begins with a complicated reaction, producing 1,3-bisphosphoglycerate from glyceraldehyde 3-phosphate. Glyceraldehyde phosphate dehydrogenase carries out the following: (a) It oxidizes the aldehyde group of glyceraldehyde to a carboxylic acid group; (b) it reduces NAD^+ to NADH + H^+ when the aldehyde is oxidized; and (c) it reacts the acid group with a phosphate (Pi) to make a mixed anhydride bond between a carboxylic acid and phosphoric acid. This bond is energy rich. A high cytosolic [NAD^+]/[NADH] ratio helps drive this reaction.

The next reaction involves capturing the energy of the mixed anhydride bond in 1, 3-bisphosphoglycerate. Phosphoglycerate kinase catalyzes a substrate-level phosphorylation reaction in which an ATP is generated from ADP by removing the phosphate group from the mixed anhydride, producing 3-phosphoglycerate. After this reaction, the ATP balance scheme starting from glucose is zero (two ATP used and two ATP generated). Remember, two 1,3-bisphosphoglycerates are obtained from each glucose.

Phosphoglycerate mutase catalyzes the movement of a phosphate group from carbon 3 to carbon 2 of the glycerate molecule. Mutases catalyze intramolecular phosphate transfer reactions. There is an intermediate in this phosphate transfer reaction known as 2,3-bisphosphate glycerate (i.e., 2,3-BPG). In red blood cells, 2,3-BPG is important because it can reduce the affinity of hemoglobin for oxygen, thus allowing more oxygen to leave its binding to hemoglobin at the tissue level. When a person moves from lower to higher altitude, one of the adaptations is an increase in the content of 2,3-BPG in red blood cells. This allows the tissues to obtain more oxygen than they would normally.

Enolase catalyzes the dehydration of 2-phosphoglycerate to form the energy-rich molecule phosphoenolpyruvate. Next, pyruvate kinase carries out the second substrate-level phosphorylation reaction, generating ATP from ADP and leaving the product pyruvate. As shown in figure 6.2, pyruvate has two major fates: reduction to lactate or entry into the mitochondrion for complete oxidation. If the former occurs, the NADH + H^+ generated in the glyceraldehyde phosphate dehydrogenase reaction is oxidized to NAD^+ and the pyruvate is reduced to lactate using the enzyme lactate dehydrogenase. If the pyruvate enters the mitochondrion, then the NADH + H^+ generated in the glyceraldehyde phosphate dehydrogenase reaction must be converted back to NAD^+ by one of two shuttle mechanisms. This must occur, or glycolysis will come to a halt due to a lack of NAD^+. We discuss the shuttles later in this chapter.

If the final product of glycolysis is lactate (as opposed to pyruvate, which enters the mitochondrion), the lactate is not a waste product. It can have a number of functions. For example, lactate may leave the cell in which it is formed and enter the blood, where it may be taken up by another type of tissue and oxidized to pyruvate and then converted into acetyl CoA for terminal oxidation in the TCA cycle. Similarly, lactate from one muscle fiber may be transported into an adjacent fiber and oxidized. Blood lactate may be taken up by liver cells and be a precursor for making glucose (i.e., gluconeogenesis), or it may remain in the cell where it is formed and either be used as a source of energy or be changed into glycogen by reversal of glycolysis.

Regulation of Glycolysis

Glycolysis is a pathway that can generate ATP. It also produces cytosolic reducing equivalents in the form of NADH and plays a major role in providing substrate for the TCA cycle as pyruvate. If pyruvate is the main fate of (aerobic) glycolysis, it produces a pair of electrons in the form of NADH for the electron transfer chain. Glycolysis is a major energy-yielding pathway during three general states of muscle. (1) It is important during rapid transitions in muscle activity from rest to exercise or from one exercise intensity to a higher level of intensity. (2) It is important during exercise when oxygen is limited as a substrate for oxidative phosphorylation. This can occur during isometric exercise when intramuscular pressure occludes blood flow. (3) Aerobic glycolysis, producing pyruvate, is important during steady state exercise when the $\dot{V}O_2$ is about 60% or more of $\dot{V}O_2$max.

Glycolysis does not occur haphazardly; its rate is governed by the energy needs of the cell. In part, regulation depends on the concentration of substrates for the various reactions in the pathway:

glucose, glycogen, glucose 6-P, and ADP. The need for glycolysis depends on the rate at which ATP is hydrolyzed to drive endergonic reactions. With ATP hydrolysis, ADP increases in the cytosol and is available for the two substrate-level phosphorylation reactions, the phosphoglycerate kinase and pyruvate kinase reactions. However, the major form of control is in regulating the rate of key reactions.

Normally, regulated enzymes are those that catalyze key irreversible reactions near the beginning of a pathway. There are three irreversible reactions in glycolysis: the hexokinase, the phosphofructokinase, and pyruvate kinase reactions. Of these, only the hexokinase and phosphofructokinase reactions are major points of control, at least for skeletal muscle. In liver, the pyruvate kinase reaction is subject to control, as will be discussed later.

Hexokinase can be regulated by the product of its reaction, glucose 6-P. If the concentration of glucose 6-P increases too much, either because the pathway is being slowed down at another step or because the breakdown of glycogen is producing it at a sufficiently rapid rate, then glucose 6-P binds to a site on the hexokinase enzyme and slows down the reaction rate. We call this process feedback inhibition or product inhibition. As mentioned previously, inhibition of hexokinase results in a sharp decrease in glucose uptake into a cell.

Phosphofructokinase (PFK) is the major regulator of the flux from glucose 6-P to pyruvate and is under complex control. In muscle, PFK activity can be regulated in a variety of ways:

1. PFK exists in dimeric or tetrameric forms (i.e., two or four bound subunits); the tetramer is more active than the dimer. In liver, L4 is the principal active tetramer, while M4 is the principal active tetramer in muscle.

2. PFK may bind to certain proteins in the cell, which can alter its sensitivity to inhibitors.

3. PFK is an allosteric enzyme and is therefore subject to complex modulation by a variety of positive and negative allosteric effectors (allosteric enzymes are discussed in chapter 2). Much of our understanding of the regulation of PFK activity comes from in vitro or test tube simulations. These cannot mimic the in vivo environment of PFK, but they do tell us something about the allosteric effectors and how they might behave.

ATP, citrate, and hydrogen ions have a negative effect on PFK activity when studied in vitro. ATP, a substrate that binds to the active site, can also bind to a negative allosteric site and inhibit the activity of the enzyme. Citrate, from the mitochondrion, changes the more active tetrameric form of PFK to the less active dimeric form. When the concentration of citrate increases in the mitochondrion, some of it passes into the cytosol, indicating that mitochondrial oxidative phosphorylation is working sufficiently fast to supply the energy needs of the cell. In this case, it slows down glycolysis by inhibiting PFK. For example, in a rested muscle, fat oxidation can easily provide the low level of ATP regeneration needed. Citrate will slow down glycolysis because this pathway is not needed; glycolysis uses valuable carbohydrate. Hydrogen ions (an increase in the concentration of H^+ ions, i.e., a decrease in pH) act to promote ATP binding to its negative allosteric site to decrease PFK activity.

In vitro analysis of PFK activity reveals that there are a number of positive effectors. Fructose 1,6-bisphosphate (abbreviated F 1,6-P_2), a product of the PFK reaction, can bind to a positive allosteric site to increase the activity of the enzyme. Fructose 2,6-bisphosphate (F 2,6-P_2) is formed from fructose 6-P by an enzyme known as PFK-2. Epinephrine increases the activity of PFK-2, which increases the F 2,6-P_2 concentration. Like F 1,6-P_2, F 2,6-P_2 also binds to an allosteric site to increase the activity of the PFK. F 2,6-P_2 is not a significant activator of PFK in muscle, but it is in liver. ADP, and to a greater extent AMP, can bind to the same allosteric site to increase the activity of PFK. The concentration of these (particularly AMP) increases in hardworking muscle, thus stimulating glycolysis to provide ATP. Inorganic phosphate (Pi) and ammonium ion (NH_4^+) can bind to a positive allosteric site to increase PFK activity. The concentration of Pi increases in muscle in proportion to the decline in PCr concentration. NH_4^+ concentration will increase in hardworking muscle due to the action of the enzyme AMP deaminase, as discussed in chapter 4.

The ATP concentration in muscle does not change much unless the muscle is working at a very high rate. Buffering of ATP by phosphocreatine is one of the main reasons for this. However, even a small decrease in ATP results in a larger increase in ADP and, especially, AMP because ATP concentration is so much greater due to the high activity of the enzyme AMP kinase (also known as adenylate kinase or myokinase).

KEY POINT

Skeletal muscle represents about 40 to 50% of the total mass of a lean individual. As we have discussed, skeletal muscle can increase its rate of metabolism multifold during exercise. Therefore, controlling glycolysis is an important strategy for an organism since the fuel for glycolysis is carbohydrate—brain food. This becomes particularly important during exercise when body carbohydrate stores are low.

Glycogen Metabolism

Glycogen is a polysaccharide, composed of hundreds of glucose molecules (monosaccharides) joined end to end, with prevalent branches. Glycogen is a convenient way to store glucose inside cells without having an impact on cell **osmotic pressure**. Osmotic pressure depends on the number, not the size, of dissolved substances. For example, one molecule of glycogen may contain 5,000 glucose units, yet produce a minuscule influence on osmotic pressure compared to 5,000 individual glucose molecules. In this section, we look at the synthesis and breakdown of glycogen, integrating the latter with the glycolytic pathway in muscle. Glycogen contains a number of OH groups and therefore interacts with water in the cell. It is estimated that each gram of glycogen has between 2.5 to 3.0 grams of associated water. This means that, on a weight basis, glycogen is a heavy fuel.

Glycogen Storage

The major stores of glycogen are in liver and skeletal muscle. Table 6.1 provides approximate values for the amount of glycogen in liver and skeletal muscle for men and women under three different dietary conditions. As shown, the amount of carbohydrate in the diet influences the amount of stored glycogen. The normal, mixed diet commonly eaten by North American people contains about 45 to 50% carbohydrate. Glycogen is stored in both liver and muscle following a meal. However, after exercise, when muscle glycogen levels are reduced, glycogen is stored preferentially in the exercised muscles.

In glycogen storage or **glycogenesis**, glucose units are added one at a time to existing glycogen molecules, creating long unbranched chains. A branching enzyme then creates the highly branched final structure. Glucose enters the cell and is phosphorylated to glucose 6-P by hexokinase (muscle) or glucokinase (liver). Figure 6.4 illustrates the three-step process of glycogen formation, starting from glucose 6-phosphate. Starting from glucose 1-phosphate, UTP is added, making UDP-glucose, the precursor form of glucose to be added to the glycogen primer. Finally, glycogen synthase adds glucosyl (glucose) units from UDP-glucose to the primer, releasing UDP and producing a glycogen molecule enlarged by an added glucose. Overall, glycogen synthesis is irreversible, due in part to subsequent hydrolysis of inorganic pyrophosphate (PPi) by inorganic pyrophosphatase. The glycogen synthase reaction is also irreversible. Glycogen synthase produces long chains of glucose molecules that the branching enzyme transforms into the treelike structure

Table 6.1 Approximate Glycogen Stores in Liver and Muscle for Adults Following Normal, High Carbohydrate (CHO), and Low CHO Diets

Storage site	Tissue weight (kg)	Total glycogen content (g)		
		Normal diet	High CHO diet	Low CHO diet
Man, 70 kg				
Liver	1.2	40-50	70-90	0-20
Muscle	32.0	350	600	250
Woman, 55 kg				
Liver	1.0	35-45	60-70	0-15
Muscle	22.0	242	410	170

of glycogen found in vivo. The UTP needed to make glycogen comes from the nucleoside diphosphate kinase reaction.

$$\text{UDP} + \text{ATP} \xleftrightarrow[]{\text{nucleoside diphosphate kinase}} \text{UTP} + \text{ADP}$$

The actual primer for glycogen synthesis is a large polypeptide known as **glycogenin**. This is a self-glycosylating protein that transfers glucosyl units to itself from UDP-glucose until there are 7 to 11 glucosyl units. Glycogenin is the primer, involved in initiating glycogen synthesis. Glycogen synthase now adds glucosyl units from UDP-glucose to the glycogenin primer, and branching enzyme creates the branched structure. Two physiologically distinct forms of glycogen have been identified recently. Proglycogen is much smaller than macroglycogen; the difference is due to many more glucosyl units in macroglycogen.

Storage of glycogen following training or competition is an important consideration for athletes because glycogen is such an important fuel for exercising muscle. Moreover, there are limited stores of glycogen in muscle. Research has provided us with some important information. For example, the synthesis of glycogen following exercise is most rapid in the first hour or so after exercise. Since glucose transport into muscle may be the rate-limiting process in glycogen synthesis, it seems that the exercise effect on glucose transport persists past the end of exercise. Glycogen synthesis is stimulated by the provision of glucose. Therefore, athletes are encouraged to eat a rich source of glucose following exercise.

KEY POINT

The dietary sources of glucose to make glycogen can be divided into sugars and starches. The sugars are the monosaccharides glucose and fructose, which are found free in some fruits, and disaccharides, principally sucrose. Monosaccharides are rapidly absorbed because no digestion is needed. Sucrose provides the bulk of our sugar intake. It is naturally found in fruits, but for most people, the richest sucrose source is that which is added to foods during industrial processing. For example, a single 12-ounce (355 ml) serving of a soft drink may contain almost 40 grams of sucrose. Although this must be digested to glucose and fructose, absorption is still rapid. The most important source of glucose for glycogen synthesis is starch, which is found in a variety of grains and vegetables. Besides being a rich source of glucose, although more slowly absorbed, foods containing starch provide a variety of other important nutrients.

The dietary glucose will stimulate an increase in blood insulin, which will further stimulate glucose uptake into the exercised muscle fibers. In general, glycogen synthesis is not a rapid process,

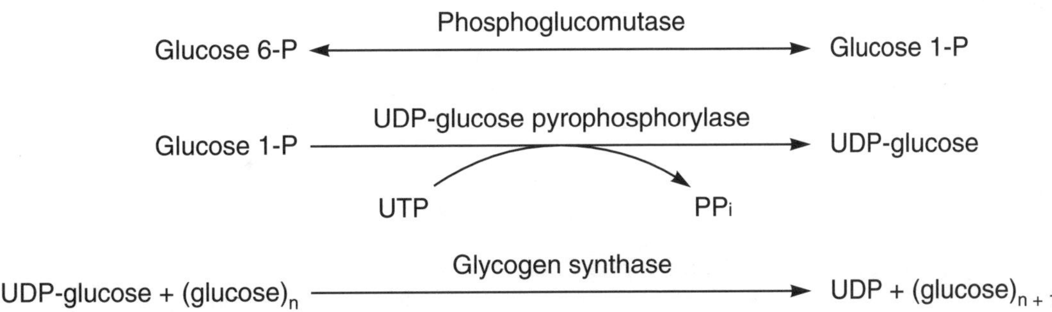

Figure 6.4 Glycogen synthesis involves the addition of glucose units in the form of uridine diphosphate- (UDP-) glucose to a preexisting glycogen primer molecule, shown as (glucose)$_n$. Since one ATP is consumed to make glucose 6-P, the energy cost to store one glucose unit as glycogen would be two energy-rich phosphates: one ATP and one uridine triphosphate (UTP).

and it may take 24 hours or so to replenish depleted stores. Research has also shown that glycogen storage may continue for two to three days after depletion, resulting in what is called glycogen supercompensation. We will look at mechanisms to explain these factors when we discuss the regulation of glycogen metabolism.

KEY POINT

We will see that the supercompensation effect on glycogen synthesis occurs only in the muscles that have been active and thus reduced in their glycogen content. Of course a high carbohydrate diet is necessary to provide the glucose to make the extra glycogen. Athletes engaged in prolonged training or competition use this fact to load muscle glycogen into their muscles for special events.

Mechanism of Glycogen Breakdown

Glycogenolysis, or the breakdown of glycogen, is the phosphorolytic cleavage of glucose units, one at a time, from glycogen molecules through the introduction of Pi. Figure 6.5 shows the process of glycogenolysis. Glycogen phosphorylase, often simply called phosphorylase, acts on chains of glucose units producing glucose 1-P. This is converted into glucose 6-P by phosphoglucomutase. A debranching enzyme is necessary to remove branches because phosphorylase can only cleave off glucosyl units that are joined in a linear fashion. The fact that there are numerous branches in glycogen enhances its exposure to phosphorylase, increasing the rate at which it can be broken down.

In liver, glycogenolysis mainly provides glucose for the blood. Therefore, glucose 6-phosphate from the phosphoglucomutase reaction is most likely dephosphorylated to free glucose, although it may also be used in the glycolytic sequence to make pyruvate. In muscle, glycogen breakdown generates glucose 6-phosphate units for glycolysis.

Regulation of Glycogen Metabolism

In liver and muscle cytoplasm are glycogen particles, glycogenin, and the enzymes both to make and break down glycogen. In muscle, most glycogen particles are found in close association with the sarcoplasmic reticulum, with smaller amounts of glycogen particles found close to the muscle cell membrane (sarcolemma) and in the muscle cytoplasm.

If the processes of synthesis and breakdown are simultaneously active, a futile cycle results that does nothing except use energy-rich phosphates such as UTP. Thus, glycogen phosphorylase should be active when glycogen synthase is inactive or vice versa. In the liver, phosphorylase should be inactive following a meal, but glycogen synthase should be active to store the glucose obtained from food. Between meals, liver phosphorylase should be active to provide glucose for the blood, whereas glycogen synthase should be inactive. In rested muscle, synthase should be active and phosphorylase inactive following a meal, but if the muscle starts to work, the phosphorylase should be active and the synthase inactive. In fact, we might expect the activity of phosphorylase to be graded during exercise; its activity should be most active during very hard exercise when carbohydrate is the most needed fuel and much less active when exercise intensity is low enough for oxidation of fat to maintain ATP levels in the exercising muscle. Control of glycogen phosphorylase can thus be tied to the whole strategy of using carbohydrate for oxidative phosphoryla-

$$(\text{Glucose})_n + \text{Pi} \xleftrightarrow{\text{Glycogen phosphorylase}} \text{Glucose 1-P} + (\text{Glucose})_{n-1}$$

$$\text{Glucose 1-P} \xleftrightarrow{\text{Phosphoglucomutase}} \text{Glucose 6-P}$$

Figure 6.5 Glycogenolysis involves the phosphorolytic cleavage of glucose units from glycogen molecules. The resulting glucose 1-phosphate is converted into glucose 6-P by a specific mutase.

tion when there is plenty available or when the muscle demand for ATP can only be adequately satisfied by carbohydrate degradation.

One enzyme can be active and the other simultaneously inhibited if the two respond in opposite directions to the same stimulus. The main regulation of these enzymes is covalent attachment of a phosphate group, or phosphorylation, which requires a protein kinase. To remove the phosphate groups requires a phosphoprotein phosphatase. The regulation of glycogen metabolism in liver and muscle represents another form of signal transduction, which we discussed in chapter 2. For the control of glycogen metabolism, external signaling molecules (e.g., the hormones epinephrine and glucagon) bind to specific cell membrane receptors and activate protein kinases that phosphorylate specific proteins—phosphorylase and glycogen synthase. Phosphorylation activates phosphorylase and dephosphorylation inactivates it. In contrast, phosphorylation inactivates glycogen synthase but dephosphorylation activates it.

Regulation of Glycogenolysis in Muscle

The unphosphorylated form of phosphorylase is known as phosphorylase *b*, which is normally inactive. The phosphorylated form is phosphorylase *a;* it is active in breaking down glycogen. Figure 6.6 shows how this process is controlled in muscle. Phosphorylation converts phosphorylase *b* (phos *b*) to phosphorylase *a* (phos *a*). The phosphorylation is catalyzed by phosphorylase kinase, which exists in two forms, *a* and *b.* The dephosphorylated or *b* form is inactive unless it binds four calcium ions (Ca^{2+}). As shown in chapter 4, the cytosolic [Ca^{2+}] increases when a nerve activates a muscle fiber, causing the release of calcium ions from the sarcoplasmic reticulum. Phosphorylase kinase *a* catalyzes the phosphorylation of phosphorylase *b.* Phosphorylase kinase *b* becomes active through phosphorylation by a cAMP-dependent protein kinase catalytic subunit, shown as A kinase and known as protein kinase A. The **cyclic AMP** (cAMP) forms when adenylate cyclase (also known as adenylyl cyclase) is activated, converting ATP into cAMP and PPi. In muscle this comes about when **epinephrine** (adrenaline) binds to its receptor on the cell membrane. The blood epinephrine concentration increases during exercise or even the anticipation of exercise or competition. Figure 6.6 shows that protein phosphatase-1 also catalyzes the dephosphorylation of both phosphorylase kinase *a* and phosphorylase *a.*

If we consider a real-life scenario for glycogenolysis, consider a runner just before the start of a 1500-meter race. Anticipation results in increased release of epinephrine, which helps to activate phosphorylase before the race actually begins. The epinephrine activates adenylate cyclase in the muscle membrane. This enzyme changes ATP to cyclic AMP and PPi. Cyclic AMP binds to protein kinase A. This enzyme contains two kinds of subunits, regulatory and catalytic. Binding of cAMP to the regulatory subunit causes it to dissociate from the catalytic subunit, and the latter is now active as a protein kinase. The catalytic subunit phosphorylates phosphorylase kinase *b,* making it the active phosphorylase kinase *a.* Phosphorylase kinase *a* phosphorylates glycogen phosphorylase *b,* making it the active form, phosphorylase *a.* In addition, calcium ion concentration increases in the muscle as soon as the runner begins. The elevated calcium concentration means that inactive phosphorylase kinase *b* can bind four ions to a special regulatory subunit. This makes the phosphorylase kinase *b* active in phosphorylating glycogen phosphorylase *b.* Thus, catalytically active phosphorylase *a* can be created by two different mechanisms.

KEY POINT

Control of the conversion of phosphorylase from the *b* to *a* form occurs through two mechanisms. Epinephrine allows the conversion to take place without significant activity taking place. Excitement due to the anticipation of competition or danger helps to prepare the enzyme to rapidly break down glycogen. Second, the actual muscle contraction activates phosphorylase through an increase in calcium ion concentration. In both cases, the activation can be graded by the extent of elevation in epinephrine in the blood and calcium in the muscle cytoplasm.

Glycogen phosphorylase *b* is influenced by positive (AMP and IMP) and negative (ATP and glucose 6-phosphate) allosteric effectors. It is unclear how important the two positive effectors are, but

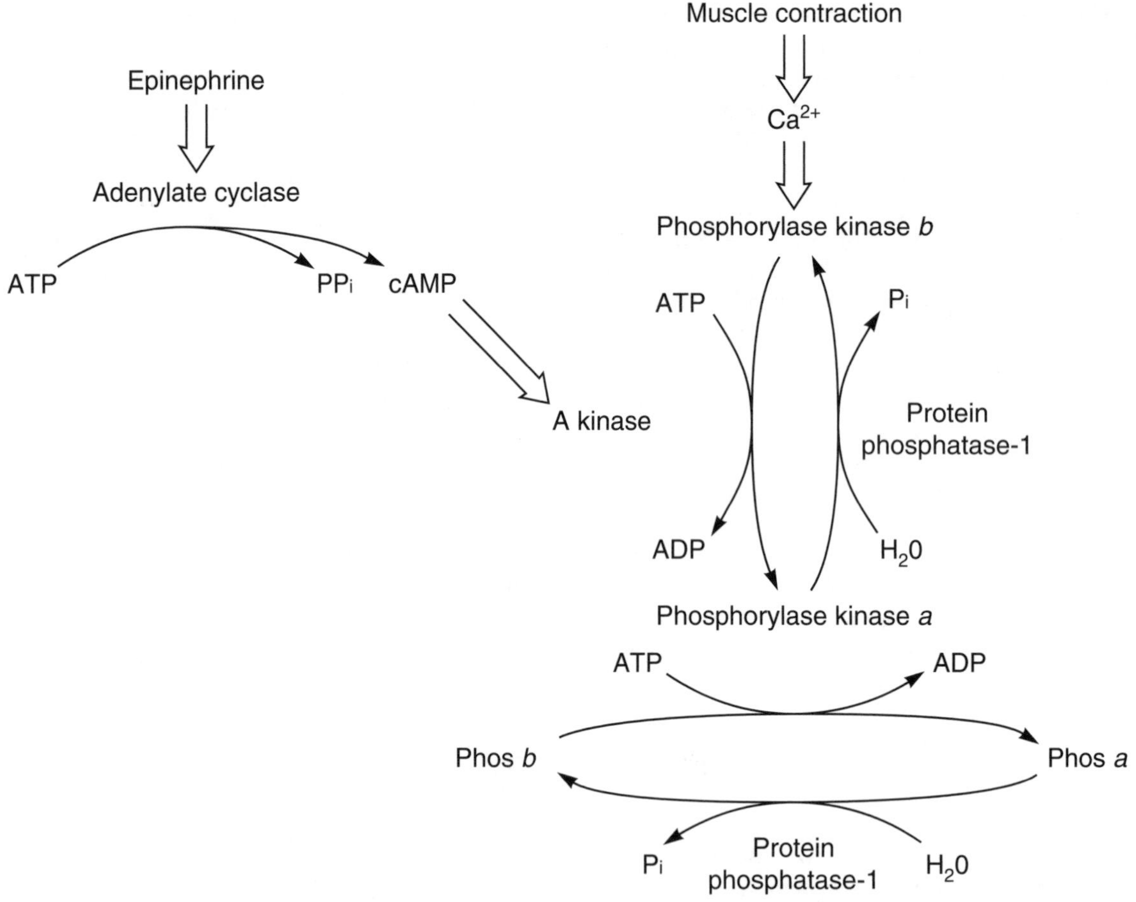

Figure 6.6 Control of glycogen breakdown through epinephrine-directed phosphorylation of glycogen phosphorylase. The dephosphorylated form, phos *b*, is inactive; the phosphorylated form, phos *a*, is active. Phosphorylation of phos *b* is catalyzed by phosphorylase kinase, itself activated by binding four calcium ions (Ca^{2+}) or through phosphorylation catalyzed by cAMP-dependent protein kinase (A kinase). Dephosphorylation of the phosphorylated enzymes is catalyzed by protein phosphatase-1. Thick arrows represent diffusion.

increases in both are necessary to overcome the inhibition of ATP and glucose 6-phosphate. With severe exercise, AMP concentration may triple, sufficient to allosterically activate glycogen phosphorylase *b*. The influence of IMP is unlikely to be great at the beginning of exercise, but with time, the IMP concentration does rise, enough to have a small but potentially minor effect on promoting glycogenolysis. We saw in chapter 4 how IMP is formed from deamination of AMP by AMP deaminase. The amount of IMP formed is directly related to exercise intensity. In addition to the phosphorylation-dephosphorylation and allosteric mechanisms, phosphorylase *a* and *b* both have high K_m (27 mM) values for one of their substrates, Pi. When muscle contracts, phosphocreatine (creatine phosphate) decreases, and the Pi concentration rises from a rest value near 1 millimolar to up to 20 millimolar during severe exercise. Pi must thus increase in the cytosol before significant glycogen is broken down in muscle, even if the active phosphorylase *a* is present. An increase in [AMP] lowers the K_m for Pi of both phosphorylase *a* and phosphorylase *b*. The rate of glycogenolysis in muscle also depends on the amount of glycogen in a muscle fiber. For the same contraction conditions, the higher the glycogen the more rapid the rate of glycogen breakdown. As glycogen decreases during exercise, the protein-glycogen particle releases phosphorylase. As a result, the phosphorylase is less able to be activated by phosphorylase kinase, becoming more available to inactivation through dephosphorylation by protein phosphatase-1.

> **KEY POINT**
>
> Since muscle glycogen is the principle source of substrate for glycolysis, regulation of glycogen phosphorylase is critical to appropriately meet the demands for ATP provision during exercise. This is especially important during exercise at altitude where the oxygen content of the inspired air is reduced, and glycolytic flux is therefore increased.

Regulation of Glycogenolysis in Liver

Regulation of glycogenolysis in liver is similar to that in muscle in that there is an active (phosphorylase *a*) and inactive (phosphorylase *b*) form of liver phosphorylase. There are notable differences. **Glucagon** is the primary hormone that stimulates the rise in liver cAMP concentration and conversion of liver phosphorylase *b* to liver phosphorylase *a*. Glucagon is produced in and released from the alpha cells of the pancreas in response to a decrease in blood glucose. Since liver glycogen is the storage form of glucose that will be released into the blood stream, a rise in glucagon in the blood signals a need to break down liver glycogen to glucose and release this to the blood. Following a meal, when blood glucose and insulin concentrations are elevated, insulin activates protein phosphatase-1, increasing the conversion of liver phosphorylase *a* to liver phosphorylase *b*, while glucose binds to and inactivates any existing liver phosphorylase *a*.

Regulation of Glycogenesis

Like glycogen phosphorylase, glycogen synthase exists in two forms. Glycogen synthase I (GS I) is the unphosphorylated form that is normally active. The *I* stands for independent, that is, GS I is independent of the glucose 6-P concentration. Glycogen synthase D (GS D) is the phosphorylated form that is normally inactive but can become active if the glucose 6-P concentration increases. The *D* means that GS D is dependent on the glucose 6-P concentration, that is, it is allosterically activated by glucose 6-P. Figure 6.7 illustrates this control of glycogen synthase.

Phosphorylation inactivates the active form of glycogen synthase (GS I). A variety of kinases are capable of phosphorylating GS I, including phosphorylase kinase, cAMP-dependent protein kinase (protein kinase A), and a calcium-dependent protein kinase. There are other GS I phosphorylating kinases, but their role in the control of glycogen synthesis in muscle is not clear. GS I can be phosphorylated at 10 sites at least, leading to gradations in glycogen synthesis activity and sensitivity to glucose 6-phosphate. Conversion of inactive GS D to active GS I occurs by dephosphorylation, catalyzed by protein phosphatase-1.

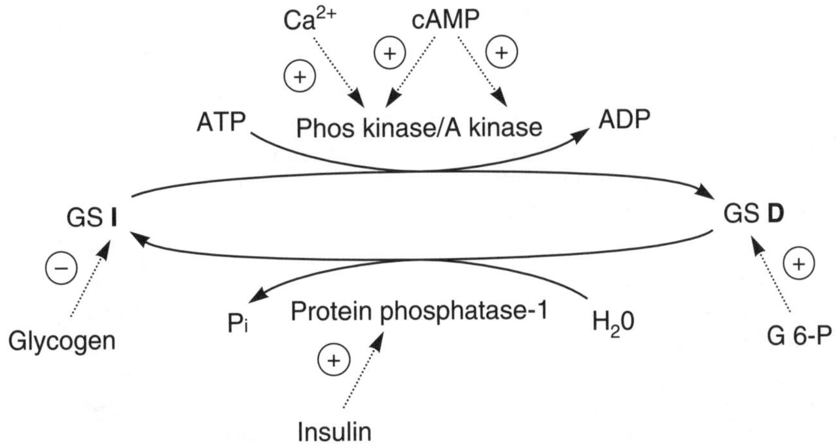

Figure 6.7 Control of glycogen synthase (GS) activity by covalent phosphorylation and allosteric effectors. Phosphorylation of the active form, GS I, generates the less active GS D form. Allosteric effectors and the enzymes that they influence are shown with dotted arrows. Positive and negative signs indicate increase or decrease in activity, respectively.

Insulin enhances the effect of protein phosphatase-1, stimulating the formation of GS I to help make glycogen. Epinephrine increases the concentration of cyclic AMP, leading to phosphorylation and hence inactivation of glycogen synthase in muscle, whereas glucagon has the same effect in liver. A rise in muscle [Ca^{2+}] during activity leads to glycogen synthase phosphorylation. An inverse relationship has been noted between the percentage of glycogen synthase activity as GS I and glycogen concentration for both liver and muscle. This suggests that glycogen concentration can inhibit glycogen synthase activity so that the liver and muscles do not store too much glycogen.

KEY POINT

There is a limit to the amount of glycogen that can be stored in muscle, despite the continued presence of elevated blood glucose and insulin. When some threshold amount of glycogen is reached, glycogen synthase activity is inhibited. Although elevated glucose transport into muscle still occurs, the glucose 6-phosphate is channeled into the glycolytic pathway such that lactate is produced at an accelerated rate even in a rested muscle and pyruvate oxidation is increased at the expense of fat oxidation.

Summary

With the onset of exercise, muscle glycogen breakdown is accelerated, and glycogen synthesis is inhibited. This is accomplished by protein phosphorylation based on a feed forward mechanism due to the increase in epinephrine release prior to exercise, as well as feedback signals within the exercising muscle fibers through increases in calcium concentration. Thus, glycogenolysis is accelerated by phosphorylation of phosphorylase *b*, whereas synthesis is retarded by phosphorylation of GS I. The percentage of phosphorylase in the *a* form increases rapidly from a rest value of less than 10% to more than 50% early in exercise; the increase parallels the relative exercise intensity. However, within minutes the percentage of phos *a* declines to values approaching rest levels even though exercise continues at the same rate. These results tell us that control of glycogen breakdown during exercise is well matched to the fiber's energy needs, less by phosphorylation mechanisms and more by changes in the concentrations of allosteric effectors, substrates, and products. The fact that the concentrations of substrate Pi and allosteric effectors AMP and IMP are related to exercise intensity may play a more important role in grading glycogen breakdown rates to rates of ATP hydrolysis than conversion of glycogen phosphorylase *b* to phosphorylase *a*. Since glycolysis is the only fate for glucose 6-phosphate, created by the breakdown of glycogen, the rate of glycogenolysis must be matched to that of glycolysis. Finally, during submaximal exercise, pyruvate, a product of glycolysis, will be mainly oxidized in the mitochondrion. There is no need to allow pyruvate to be formed in excess of its entry into the mitochondrion and subsequent oxidation except under conditions where the rate of ATP provision by oxidative phosphorylation cannot match the rate of ATP hydrolysis.

Following exercise in which the concentration of glycogen is sharply reduced, muscle glycogen synthesis increases rapidly at first, gradually decreasing over the next 24-hour period. In the first hour after exercise, rapid resynthesis of glycogen can occur in the absence of insulin. However, the presence of glucose and insulin will greatly facilitate the resynthesis rate. What are the factors contributing to the nonlinear pattern of glycogen synthesis after exercise?

In the postexercise condition, the concentrations of Pi and AMP and IMP decrease due to a dramatic drop in the rate of ATP hydrolysis. Together, these changes will sharply decrease the rate of glycogenolysis. Subsequent feeding of carbohydrate-containing foods will raise the blood glucose and insulin concentrations. The combined effects of previous contractile activity, the substrate glucose, and increased insulin mean that the GLUT-4-mediated transport rate of glucose into the fiber is greatly elevated. The glucose will be phosphorylated to glucose 6-phosphate by hexokinase. The glucose 6-phosphate will increase the synthesis of glycogen both by a mass action (push effect as a substrate) and allosteric effect on the GS D form of glycogen synthase. Furthermore, the hormonal milieu of decreased epinephrine and increased insulin will stimulate the dephosphorylation of both glycogen phosphorylase *a* and GS I, depressing the former and activating the latter. Finally, through an incompletely understood mechanism glycogen synthase activity is further enhanced when muscle glycogen concentration is depressed. The net effect is that glycogen synthesis activity is maximal in the first hour or so following exercise.

A number of studies have demonstrated that trained athletes can increase glycogen concentrations to a greater extent and at a faster rate compared to untrained persons. At one time it was believed that this was due to an increase in the activity of glycogen synthase as a result of training. More recent studies reveal that total glycogen synthase activity and the percentage of glycogen synthase in the I form are not different between well-trained and untrained individuals when tested after glycogen-depleting exercise. However, muscle GLUT-4 content is higher in trained muscle, suggesting that an enhanced ability to transport glucose is the primary mechanism that accounts for the faster increase in glycogen replenishment in trained muscle.

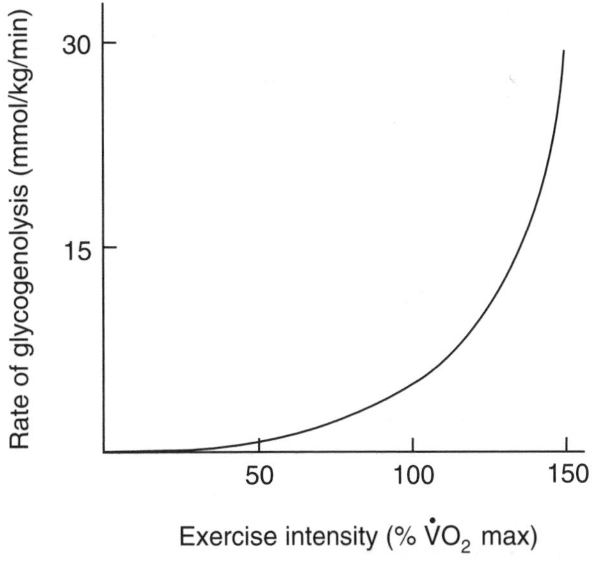

Figure 6.8 Muscle glycogen breakdown rate expressed as millimoles of glucose units per kilogram wet weight of muscle per minute as a function of the relative intensity of dynamic exercise expressed as a percentage of the workload needed to generate $\dot{V}O_2$max.

Carbohydrate Metabolism During Exercise

When we exercise, we place a demand on one or more muscles in terms of ATP hydrolysis. Without being aware how this occurs, the muscles respond in an appropriate manner utilizing the three energy systems we discussed in chapter 4. In the previous chapter we learned how changes in **phosphorylation potential** (i.e., [ATP]/[ADP] [Pi]), reflecting exercise intensity, can help to regulate oxidative phosphorylation. In this section we discuss how exercise demands and diet can influence rates of carbohydrate metabolism.

Glycogen Utilization During Exercise

Figure 6.8 illustrates how glycogen breakdown during exercise increases exponentially with a linear increase in relative exercise intensity, expressed on the basis of $\dot{V}O_2$max. Use of relative as opposed to absolute exercise intensity is helpful in making comparisons with different people because the metabolic response is graded to a large extent by the relative effect of the exercise on a person. For example, an elite cross-country skier weighing 90 kilograms (198 lb) with a $\dot{V}O_2$max of 7.2 liters/minute (80 ml/min/kg) may be able to comfortably cycle for an hour at a workload of 300 watts. An untrained adult male weighing 60 kilograms with a $\dot{V}O_2$max of 2.4 liters/minute (40 ml/kg/min) may be able to cycle for the same length of time at a workload of only 100 watts. Despite the threefold difference in work intensity, the metabolic response should be fairly similar between these two men.

As figure 6.8 reveals, glycogen becomes increasingly important as a fuel as the intensity of exercise increases. There are two major metabolic explanations for this. First, we have seen in chapter 4 that more ATP can be generated per unit of oxygen consumed when carbohydrate is the fuel, compared to fat. In other words, the P/O ratio is higher for carbohydrate. As we will discover in the next chapter, important control mechanisms regulate fuel use during exercise, sparing carbohydrate during low to moderate intensity exercise when oxidative phosphorylation using fat can meet ATP demands. As the relative intensity of exercise increases, placing extra demands on oxygen transport to mitochondria, carbohydrate use is favored. The net effect is that at workloads approaching $\dot{V}O_2$max, carbohydrate is virtually the exclusive fuel.

Second, at workloads beyond $\dot{V}O_2$max, when oxidative phosphorylation cannot increase further, additional ATP demands must be met by anaerobic glycolysis to lactate. Starting from muscle glycogen, each glucose unit degraded to two lactate through glycogenolysis and glycolysis yields 3 ATP. Complete oxidation of the same glucose unit, including glycogenolysis, glycolysis to pyruvate, pyruvate oxidation to acetyl CoA, and oxidation of acetyl CoA in the TCA cycle would yield at least 33 ATP. As mentioned in the previous chapter, the exact number of ATP per glucose depends on the P/O ratios for NADH and $FADH_2$. Therefore, approximately 11 times as many glucose units in glycogen are broken down to lactate to produce the same number of ATP produced if that same glucose unit were completely oxidized. Moreover, with higher intensities of exercise, there is a greater reliance on fast twitch (type II) muscle fibers, which have a lower capacity for oxidative phosphorylation and a greater capacity for glycolysis. Finally, as the intensity of exercise increases, the concentration of epinephrine in the blood increases. This hormone has a potent effect, as we have seen, on promoting the conversion of phosphorylase *b* to the more active phosphorylase *a* form.

When all other factors are the same, the concentration of glycogen will play a direct role in how much is used. In other words, glycogen promotes its own use. In the hypothetical exercise condition shown in figure 6.9, 80 millimoles of glucose units in glycogen were consumed per kilogram of muscle to fuel the exercise when glycogen concentration was higher. This is twice the amount of glycogen that was used for the identical exercise task when glycogen stores were only half as great. Reinforcing this point is the fact that the peak rate of glycogen utilization shown in figure 6.9 occurred at the beginning of exercise when glycogen concentration was higher. This is a classic example of what is called a mass action effect. This means that for the same enzyme activity, a higher substrate concentration will mean more substrate is consumed. (This point can be understood by reviewing figure 2.3.)

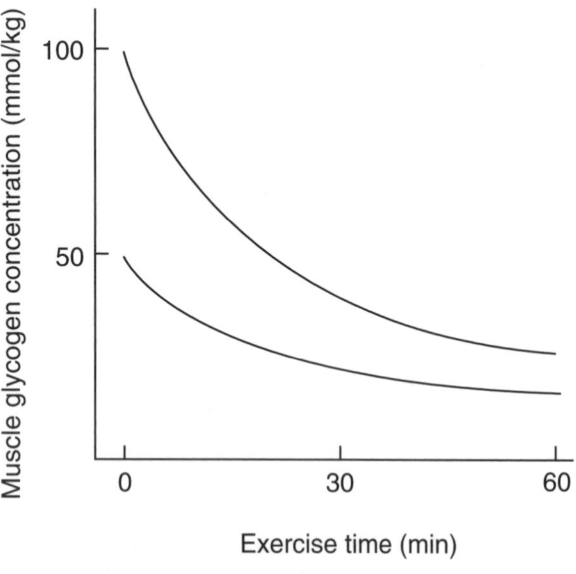

Figure 6.9 Glycogen utilization in the vastus lateralis muscle of the same individual on two occasions, performing 60 minutes of exercise at 70% of $\dot{V}O_2$max on a cycle ergometer. The top curve shows the situation when the initial glycogen concentration is twice as high as the lower curve.

KEY POINT

With a higher glycogen concentration, there is more glycogen used compared to the identical condition with a lower glycogen concentration. This points to an accelerated flux through glycolysis in the glycogen-loaded condition, and with this, more lactate will be formed. Therefore, both muscle and blood lactate concentrations will be elevated more in a glycogen-loaded exercise condition.

One of the explanations for increased glycogen use when glycogen concentration is higher is related to the size of the glycogen molecule. When glycogen concentration is high, each glycogen molecule has a near maximal number of glucose units arranged in a highly branched fashion. In a larger glycogen molecule there are more branches, making the glucose units more accessible to phosphorylase, thus increasing the rate of breakdown to glucose 1-phosphate.

Besides directing its own utilization, the amount of glycogen in skeletal muscle can modulate the rate of uptake of glucose from the blood. As we have seen, skeletal muscle glucose uptake is controlled by the content of GLUT-4 transporters in skeletal muscle cell membranes. During exercise when glycogen content is high, the number of membrane-bound GLUT-4 transporters is reduced compared to the same exercise when glycogen content is lower. This suggests that the actual concentration of glycogen plays a role in modulating GLUT-4 translocation from intracellular sites to the muscle cell membrane, thereby regulating glucose uptake into the contracting muscle fiber.

Fates of Pyruvate in Muscle

In addition to being oxidized in the mitochondrion or reduced to lactate in the cytosol, pyruvate produced during glycolysis can accept an amino group from the amino acid glutamic acid in a **transamination** reaction. The pyruvate is then changed to the amino acid alanine. Exercising muscle releases alanine to the blood, a topic we consider in chapter 8.

If we focus on the two major fates of pyruvate in muscle, reduction to lactate in the cytosol or pyruvate oxidation in the mitochondrion, we can assess the factors that potentially determine which process predominates.

1. The rate of pyruvate formation or the glycolytic flux rate. This may depend both on the intensity of the activity as well as the availability of glycogen, as we have discussed.

2. The cytosolic redox state, which is the concentration ratio of [NADH]/[NAD^+]. NADH and NAD^+ participate in oxidation-reduction reactions as substrates; thus major relative changes in their concentration influence the direction of reversible redox reactions.

3. The number and size of mitochondria, which reflect the potential for oxidative phosphorylation.

4. The availability of oxygen, which is the final acceptor of electrons in oxidative phosphorylation.

5. The total activity of lactate dehydrogenase (LDH) and the LDH isozyme type, which is important because the greater the activity of LDH and the more the isozyme type M_4 or M_3H, the greater the chance that pyruvate will be reduced to lactate, all other factors being similar.

If we compare slow twitch (ST) and two types of fast twitch (FTA and FTX) skeletal muscle fibers with heart (cardiac) muscle fibers, we learn some important metabolic lessons. First, their typical activities differ greatly. Heart muscle is continuously active because the heart is always working. Slow twitch (ST or type I) muscle fibers are involved in low intensity activity and are also active during strong contractions. For the same cross-sectional area, these fibers generate the same tension as fast twitch muscle fibers, but they cannot shorten as rapidly. Fast twitch (FT or type II) fibers are usually active during intense activity but are not normally active during low intensity activity. They can shorten faster than slow twitch muscle fibers primarily because they have fast myosin isozymes. We generally identify two major fast twitch muscle fibers in humans based on their myosin heavy chain composition, yielding two different myosin isozymes and thus different fiber types (i.e., FTA and FTX or IIA and IIX—see Quaternary Structure of Proteins in chapter 1).

Table 6.2 summarizes important differences between the heart muscle, ST, FTA, and FTX skeletal muscle fibers. Both FT fibers can hydrolyze ATP at a maximum rate that exceeds ST fibers and heart muscle. Thus, the regeneration of ATP must also be faster in FT fibers. However, ATP regeneration in FT fibers will involve more glycolysis and less fuel oxidation compared to the ST and heart muscle fibers because the FT fibers have a greater capacity for glycolysis and a poorer blood supply—thus less oxygen and a lower capacity for oxidative phosphorylation. As table 6.2 summarizes, there are also subtle differences in the metabolic capacities of the FT fibers, with FTA fibers having a higher oxidative capacity but a lower glycolytic capacity. In summary, either FT fiber would produce more lactate and consume less oxygen when regenerating ATP compared to an ST fiber or heart muscle.

If a person begins an endurance training program, such as cycling, running, or swimming,

Table 6.2 Relative Comparisons for Heart Muscle (HM) and Slow Twitch (ST) and Two Fast Twitch Skeletal Muscle Fibers (FTA and FTX) for a Variety of Metabolic Factors

Factor	Fiber type comparison
Maximum rate of ATP hydrolysis	FTX > FTA > ST > HM
Maximum glycolytic flux rate	FTX > FTA > ST > HM
Blood supply or availability of oxygen	HM > ST > FTA > FTX
Fiber size	FTX > FTA > ST > HM
Maximum oxidative capacity	HM > ST > FTA > FTX
Percentage of LDH as the isozyme M_4	FTX > FTA > ST > HM

significant changes take place in the muscles that participate in the training. The capacity for glycolysis generally does not increase and may even decrease. However, the number and size of the mitochondria increase. This means that the activities of the enzymes of the TCA cycle, electron transport chain, and beta-oxidation of fatty acids all increase. Because of increases in the numbers of capillaries around fibers (capillarization), the ability to deliver oxygen to mitochondria increases. We should therefore expect that for the same rate of ATP hydrolysis after training compared to before, less lactate would be produced and more fuel oxidized to generate ATP. In addition to the changes in mitochondria, endurance training increases the total content of muscle GLUT-4 transporters and the activity of hexokinase.

KEY POINT

The adaptations that take place with training allow the trained person to carry out the same exercise task as before but with less carbohydrate being used, compensated by a greater utilization of fat.

Lactate Transport

During exercise with the blood flow occluded, as in modestly strong isometric contractions, glycolysis is the major source of ATP. As we have seen, glycolysis is also significant during exercise at intensities beyond $\dot{V}O_2max$. For these two exercise conditions lactate concentration will increase in the muscle fiber, and lactate will appear in the blood, well beyond the normal one millimolar level. Lactate can diffuse across cell membranes as the undissociated acid, lactic acid. However, with a low pK_a (3.9) the fraction of undissociated lactic acid in a muscle fiber with a cytosolic pH of 7.0 would be very low. Therefore, lactate must cross the cell membrane down its concentration gradient as the lactate ion. Being charged, this means that there should be a lactate transporter. A lactate transporter has been discovered as a **monocarboxylate transporter** (MCT), since this can transfer other singly charged anions such as pyruvate. Recently, isoforms of the MCT have been observed and identified with numbers, much like the numbering system for the glucose transporters (e.g., MCT-1 and MCT-4 are in skeletal muscle). The MCT acts as a symport, transferring lactate down its gradient, accompanied by a proton (H^+).

Lactate transport across the muscle sarcolemma can occur in both directions; the favored direction depends only on the lactate and proton gradient. Thus we might expect that during exercise at an intensity near $\dot{V}O_2max$, an active FT fiber may be producing significant lactate and H^+, which are transported together through MCT-4 and MCT-1 into the extracellular fluid surrounding the fiber. A neighboring ST fiber with a low lactate concentration may take up the lactate and a proton using an MCT isoform because the lactate concentration outside this fiber is higher than inside. Recent studies have also demonstrated that endurance training can increase the content of MCT isoforms in skeletal muscle. Such an effect, as noted for the training-induced increase in GLUT-4 content, would increase the V_{max} for lactate transport in the trained muscle. Recent research has demonstrated the presence of MCT-1 in mitochondrial inner membranes from animal and human muscle. In particular, MCT-1 is found in both sarcolemmal and mitochondrial membranes, whereas MCT-4 is localized to the sarcolemma. Moreover, it is known that the heart-type isozyme of LDH (H_4) is present in the mitochondrial matrix. This means that cytosolic lactate can flow down its gradient into the mitochondrial matrix through MCT-1, be oxidized to pyruvate producing NADH, and the pyruvate can be oxidized within the mitochondrion using pyruvate dehydrogenase. Since the mitochondrial density and therefore the content of the MCT-1 transporter and matrix LDH are higher in type I muscle fibers, these fibers can consume lactate produced and released by neighboring type II fibers. As well, lactate produced within the cytosol of any fiber can theoretically be oxidized by the mitochondria of that same fiber. With endurance training, the content of total MCT is increased, but more recent studies demonstrate that it is the MCT-1 isoform that increases its content in both the sarcolemma and mitochondrial membranes.

Lactate Formation—Exercise Intensity

In rested muscle, the activity of the glycolytic pathway is quite low, based on the observation that the RQ of rested muscle is close to 0.75, indicating the preponderance of fat as the fuel for oxidative phosphorylation. During moderate intensity dynamic exercise, the rate of glycolysis may increase by 30- to 40-fold or more above the rest level. At this exercise intensity, some lactate will be formed, but the concentration of lactate in muscle and blood may not increase significantly, as much of the pyruvate will be oxidized in the mitochondrion. As exercise intensity increases linearly beyond 50 to 60% of $\dot{V}O_2max$, the rate of glycolysis, expressed as the rate of pyruvate formation, increases at an even faster rate. Although exercise may still be described as submaximal, that is, at an intensity less than $\dot{V}O_2max$, pyruvate can be formed at such a rate that a significant fraction will be reduced to lactate. This does not mean that there is a lack of oxygen to act as the electron acceptor, because recent studies using sophisticated instruments that can measure cytosolic PO_2 reveal that there is sufficient O_2 to meet the needs of the electron transport chain at exercise intensities approaching $\dot{V}O_2max$.

In a previous section, it was pointed out that the rate of glycogenolysis during submaximal exercise depends on the concentration of glycogen. Since the glucose 6-phosphate produced during glycogen breakdown will feed into the glycolytic pathway, we might expect that the amount of lactate formed during submaximal exercise will also be directly related to muscle glycogen concentration. Indeed, a number of experiments have demonstrated that the same person doing exactly the same submaximal exercise task can have a muscle lactate concentration at least three times higher if starting muscle glycogen stores are high, compared to low.

Lactate is produced constantly in the body by the red blood cells (erythrocytes) and resting muscle. All lactate will eventually be oxidized, either directly, by a tissue such as the heart, or after conversion to glucose in the liver and subsequent oxidation in another tissue. We follow the conversion of lactate to glucose in a later section of this chapter.

KEY POINT

In an earlier section in this chapter, we pointed out that when glycogen stores are maximal, carbohydrate feeding will promote glycolysis and lactate formation even in a muscle at rest. Glucose taken into the rested muscle, promoted by elevated blood insulin, is phosphorylated to glucose 6-phosphate. With filled glycogen stores, the glucose 6-phosphate is directed to glycolysis. When glycolysis takes place, some lactate will be produced. This is a further example of the point that carbohydrate promotes its own oxidation.

In summary, the formation of lactate by a muscle means that pyruvate is being produced in the cytosol by the glycolytic pathway. It may or may not mean a lack of oxygen because pyruvate formed in glycolysis can either be reduced to lactate or enter a nearby mitochondrion and be oxidized to acetyl CoA. In a sense, the path pyruvate takes involves a chance encounter: It meets a mitochondrial membrane or a lactate dehydrogenase enzyme. Since most muscle has a higher capacity for pyruvate reduction, some lactate should be formed whenever glycolysis is even slightly active.

Oxidation of Cytoplasmic NADH

We learned that pyruvate, a product of glycolysis, has two major fates. First, it is reduced to lactate, using the enzyme lactate dehydrogenase and NADH, which is generated in the glyceraldehyde phosphate dehydrogenase reaction and is converted back to NAD^+ when pyruvate is reduced. Second, it enters the mitochondria for terminal oxidation, using the pyruvate dehydrogenase reaction and oxidation of the resulting acetyl CoA in the TCA cycle. If the latter occurs, the NADH formed during the glyceraldehyde phosphate dehydrogenase reaction is not oxidized by the lactate dehydrogenase reaction. If NADH is not oxidized to NAD^+, glycolysis will thus stop due to a lack of NAD^+ as a substrate for glyceraldehyde phosphate dehydrogenase.

The simplest solution to this metabolic problem would be for cytoplasmic NADH to enter the mitochondrial matrix and be oxidized in the electron transport chain. However, the inner mitochondrial membrane is impermeable to NADH. To get around this roadblock, there are two shuttle systems that transfer electrons on cytoplasmic NADH into the mitochondrion without actual inner membrane crossing by NADH.

Glycerol Phosphate Shuttle

The glycerol phosphate shuttle (see figure 6.10) transfers electrons on cytosolic NADH to FAD, then to ubiquinone (coenzyme Q) in the mitochondrial inner membrane. The cytosolic NADH forms during the glyceraldehyde phosphate dehydrogenase reaction of glycolysis (reaction 1 in figure 6.10). This and subsequent reactions in glycolysis are not affected by this shuttle system. In reaction 2, the cytosolic NADH, not reoxidized back to NAD^+ by the lactate dehydrogenase reaction, transfers a hydride ion and proton to dihydroxyacetone phosphate, changing it to glycerol 3-phosphate (glycerol 3-P) and oxidizing NADH to NAD^+. The enzyme catalyzing this reaction (reaction 2 in figure 6.10) is cytosolic glycerol phosphate dehydrogenase. Glycerol 3-P diffuses to the outer side of the inner mitochondrial membrane and is oxidized by mitochondrial glycerol phosphate dehydrogenase (reaction 3). This enzyme is located in the inner membrane but faces the intermembrane space so that glycerol 3-P need not penetrate the inner membrane. Instead of using the NAD^+ coenzyme, the mitochondrial form of glycerol phosphate dehydrogenase uses FAD. FAD binds tightly to the mitochondrial form of the glycerol phosphate dehydrogenase and is another flavoprotein dehydrogenase described as complex II in oxidative phosphorylation. In reaction 3, the products are $FADH_2$ and dihydroxyacetone phosphate. The electrons on $FADH_2$ transfer to ubiquinone (coenzyme Q) in the electron transport chain and then to oxygen. The dihydroxyacetone phosphate then diffuses back to again accept electrons from cytoplasmic NADH.

The glycerol phosphate shuttle exhibits no net consumption or production of dihydroxyacetone phosphate or glycerol 3-P. It simply transfers electrons from cytosolic NADH to the electron transport chain. This shuttle will generate approximately 1.5 ATP per pair of electrons transferred from cytosolic NADH to oxygen and is irreversible because reaction 3 goes in only one direction. Although not the most prevalent of the two major shuttle systems, it is easier to explain. The glycerol phosphate shuttle operates to a minor extent in a variety of tissues such as brain but is very important in fast twitch (type II) skeletal muscle fibers.

Malate-Aspartate Shuttle

The malate-aspartate shuttle is more complicated and is reversible (see figure 6.11). It is the dominant shuttle for the liver, the heart, and slow twitch (type I) muscle fibers. This shuttle system transfers electrons on NADH in the cytosol to NAD^+ in the mitochondria (as opposed to electron transfer to FAD in the glycerol phosphate shuttle). As a result, for each two cytosolic electrons on NADH transferred to oxygen in the mitochondria, we get approximately 2.5 ATP.

Cytosolic NADH is converted to NAD^+ by reducing oxaloacetate to malate. This allows the NAD^+ to be used in the glyceraldehyde phosphate dehydrogenase reaction in glycolysis. The malate is transported into the matrix using an antiport mechanism with α-ketoglutarate. In the matrix, the malate now transfers its electrons to NAD^+, generating NADH and oxaloacetate. In this way, cytosolic reducing equivalents become matrix electrons on NADH. The complication of the malate-aspartate shuttle is that there

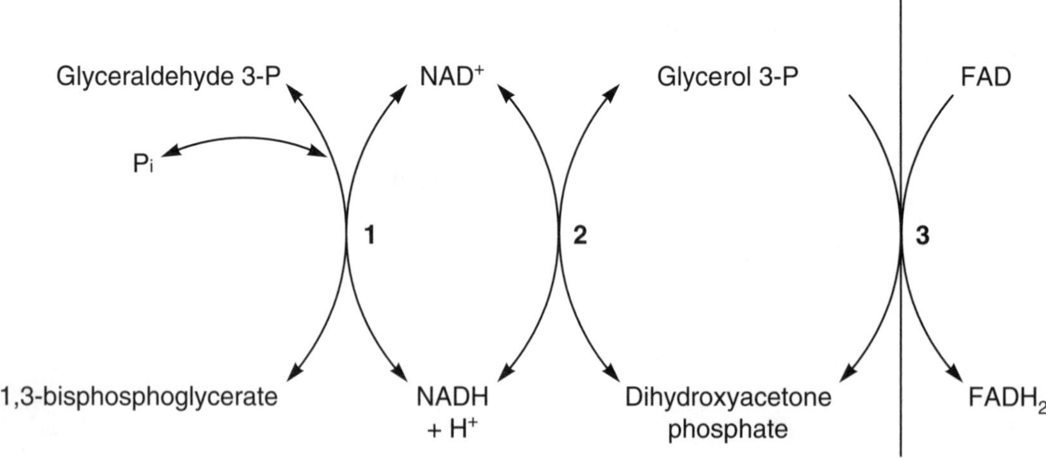

Figure 6.10 The glycerol phosphate shuttle transfers electrons from cytosolic NADH to FAD in the inner mitochondrial membrane, generating $FADH_2$. The NADH generated in reaction 1, catalyzed by glyceraldehyde phosphate dehydrogenase, is used to reduce dihydroxyacetone phosphate in reaction 2, catalyzed by cytoplasmic glycerol phosphate dehydrogenase. Glycerol 3-P diffuses to the inner mitochondrial membrane and transfers a pair of electrons to FAD in reaction 3, catalyzed by a mitochondrial glycerol phosphate dehydrogenase. Electrons on $FADH_2$ are subsequently transferred to ubiquinone. The dihydroxyacetone phosphate created in reaction 3 is now available to accept another pair of electrons from NADH. The shuttle is irreversible because reaction 3 is irreversible.

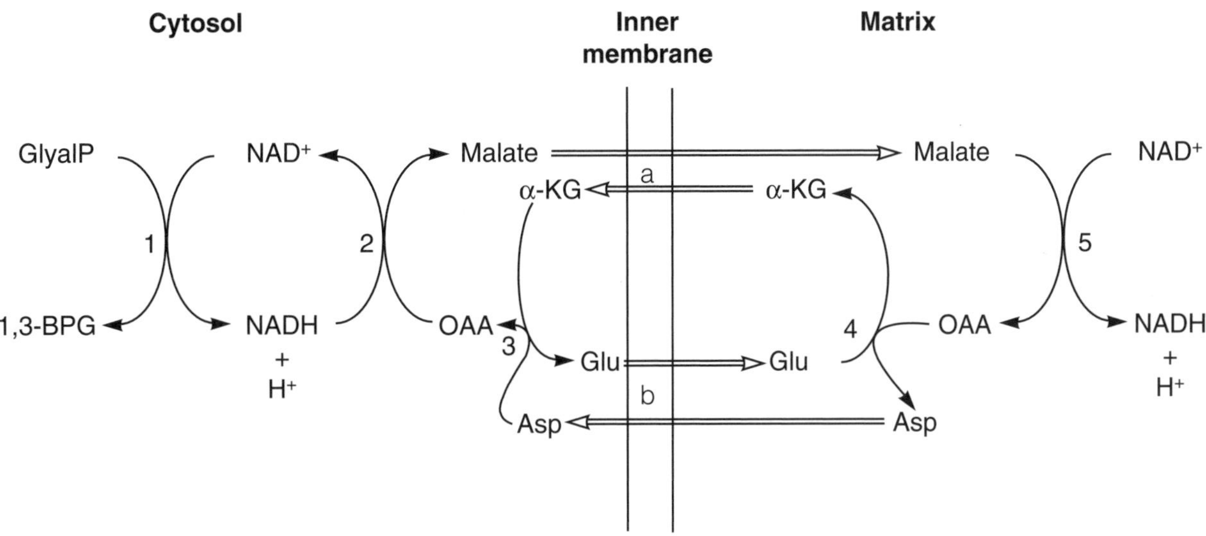

Figure 6.11 The malate-aspartate shuttle transfers electrons from cytoplasmic NADH to mitochondrial NADH. Reduction of NAD^+ in the cytosol by electrons on glyceraldehyde phosphate (GlyalP) is catalyzed by glyceraldehyde phosphate dehydrogenase (1). The resulting NADH is used to reduce oxaloacetate (OAA), producing malate and NAD^+ catalyzed by cytosolic malate dehydrogenase (2). Malate is transported into the matrix using an antiport (a) that simultaneously transports α-ketoglutarate (α-KG) from the matrix to the cytosol. The α-KG is converted into glutamate (Glu) by accepting an amino group from aspartate (Asp); the latter is converted into oxaloacetate using cytosolic aspartate aminotransferase (3). Glutamate is transferred from the cytosol to the matrix using the antiport (b) that simultaneously transfers Asp to the cytosol. In the matrix, NADH is produced by the oxidation of malate to OAA using mitochondrial malate dehydrogenase (5). The oxaloacetate is converted to aspartate using mitochondrial aspartate aminotransferase (4).

is no mechanism to directly transport oxaloacetate from the matrix to the cytosol across the inner membrane. Instead, oxaloacetate is converted into the amino acid aspartic acid (aspartate) by moving the amino group on glutamic acid (glutamate) to oxaloacetate. Amino group transfers are catalyzed by a class of enzymes known as aminotransferases (formerly known as transaminases). The resulting aspartate is transported across the inner membrane by an antiport that simultaneously transports glutamate from the cytosol to the matrix. As in the glycerol phosphate shuttle, there is no net production or consumption of any of the intermediates in this shuttle. Direction is provided to this potentially reversible shuttle by the formation of NADH in the cytosol and its oxidation in the electron transport chain of the mitochondria.

Gluconeogenesis

As mentioned previously, glucose is the principal fuel for the brain and also erythrocytes (red blood cells) and some other tissues. In the brain, the glucose is almost exclusively oxidized. In red blood cells, anaerobic glycolysis is the fate of the glucose since these cells lack mitochondria. The amount of glucose in the blood is very small, and it is used constantly by a variety of tissues. To maintain blood glucose levels, liver glycogen is broken down through glycogenolysis, and the glucose is released to the blood. At rest, in the postabsorptive state, glucose release from the liver is about 10 to 15 micromoles/kilogram/minute (about 150 mg/min for a 70 kg person). This R_a (rate of appearance) of glucose can increase up to eightfold during hard exercise. However, there are limited supplies of glycogen in liver to maintain blood glucose. For example, an overnight fast can markedly reduce liver glycogen, while a 24-hour fast can deplete liver glycogen completely. Consider what can happen with exercise and no food intake. Without exogenous glucose to help maintain blood glucose, and in the face of an increased uptake by exercising muscle, blood concentration could fall to hypoglycemic levels during prolonged exercise, seriously impairing performance. Obviously, there must be another mechanism to make glucose for the body when glycogenolysis of liver glycogen becomes limited.

As the name suggests, gluconeogenesis is the formation of new glucose from noncarbohydrate precursors such as lactate, pyruvate, glycerol, propionic acid, and particularly the carbon skeletons of amino acids. The carbon skeletons of amino acids are what is left of the amino acids when the amino group is removed. Humans cannot oxidize amino groups on amino acids. These are mainly converted to the molecule **urea**, and the urea is excreted in the urine. Gluconeogenesis occurs mainly in the liver and to a much smaller extent in the kidney.

Gluconeogenesis becomes important when liver glycogenolysis cannot produce glucose for the blood at a necessary rate. The central nervous system (principally the brain) needs about 125 grams of glucose each day, and other tissues that rely exclusively on glucose need an additional 30 to 40 grams of glucose a day. This means we need about 160 grams or so of glucose just for the glucose-dependent tissues. If we exercise or do work, we will use blood glucose in the contracting muscles. Fat cells need glucose to act as a source of glycerol (as glycerol 3-phosphate) to make triglyceride molecules.

Whenever there is an inadequate supply of dietary carbohydrate (glucose, fructose, and galactose) to supply the body with the glucose it needs, gluconeogenesis makes up the difference. This becomes particularly significant during fasting or starvation, when eating a low carbohydrate diet, or during prolonged exercise when the working muscles use a lot of blood glucose. Gluconeogenesis will also occur whenever the blood lactate concentration rises, such as during moderate to severe exercise. Although lactate will be used as a fuel by other skeletal muscles and the heart, a considerable amount will be extracted from the blood by the liver and used to make glucose. The **Cori cycle** describes the process by which carbon cycles between liver and muscle. Lactate released from muscle circulates to the liver where it is converted to glucose. The glucose is released to the blood and taken up by active muscle. Through glycolysis, glucose is converted to lactate, released from the muscle, and recycled again by the liver.

KEY POINT

During prolonged exercise, gluconeogenesis gradually increases in activity to compensate for the diminishing ability of liver glycogenolysis to produce glucose as the liver glycogen particles decrease in size.

Reactions of Gluconeogenesis

For the most part, new glucose is made from simple precursors by the reversal of the glycolytic pathway, as figure 6.12 reveals. Most of the reactions of glycolysis are reversible and so from this perspective can go in either direction, driven by mass action. However, there are three irreversible (nonequilibrium) glycolytic reactions: the pyruvate kinase reaction, the phosphofructokinase reaction, and the hexokinase or glucokinase reaction. These irreversible reactions act as one-way valves, making the glycolytic pathway flow only to pyruvate and lactate. In principle, glycolysis can go backward if there are alternate ways of getting around the three irreversible reactions.

As figure 6.12 shows, if we start from pyruvate and proceed toward glucose there is an immediate block because pyruvate kinase is a nonequilibrium reaction. Therefore, to convert pyruvate to phosphoenolpyruvate an alternate route is needed. This requires two new reactions with two new enzymes, pyruvate carboxylase and phosphoenolpyruvate carboxykinase. Pyruvate is first converted to oxaloacetate using pyruvate carboxylase, then the oxaloacetate is converted to phosphoenolpyruvate, catalyzed by phosphoenolpyruvate (PEP) carboxykinase. The irreversible pyruvate kinase reaction is bypassed, but it costs the equivalent of two ATP to do so.

Pyruvate carboxylase catalyzes a carboxylation reaction, using the tightly bound coenzyme **biotin**. This is a B vitamin that acts to bind and transfer CO_2 in the form of bicarbonate ion (HCO_3^-), which is needed in this reaction to change a molecule with three carbon atoms into one with four carbon atoms.

$$\text{ATP} + \text{pyruvate} + HCO_3^- \xrightarrow{\text{pyruvate carboxylase}} \text{oxaloacetate} + \text{ADP} + \text{Pi}$$

This reaction takes place in the mitochondrial matrix.

The oxaloacetate is converted into phosphoenolpyruvate by phosphoenolpyruvate carboxykinase, often designated PEP carboxykinase or just PEPCK.

$$\text{oxaloacetate} + \text{GTP} \xleftrightarrow{\text{PEPCK}} \text{phosphoenolpyruvate} + \text{GDP} + CO_2$$

PEPCK is located in both the cytosol and mitochondrial matrix. If the oxaloacetate is converted

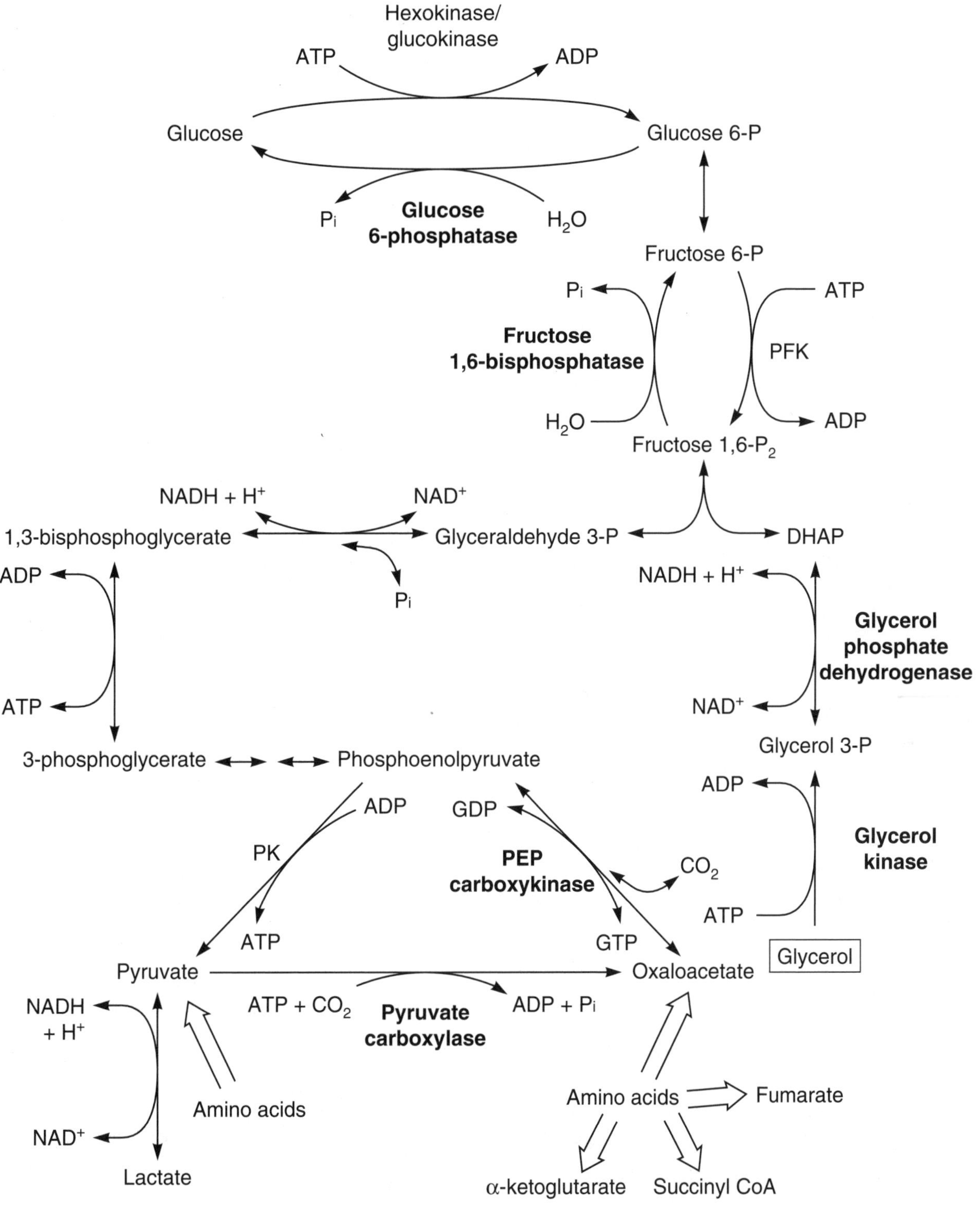

Figure 6.12 Except for the irreversible reactions catalyzed by glucokinase or hexokinase, phosphofructokinase (PFK), and pyruvate kinase (PK), glycolysis in the liver could go backward from lactate to glucose. New reactions, catalyzed by the enzymes shown in bold, allow glycerol, released from fat cells, and amino acid carbon skeletons to be used as a source of new glucose. DHAP is dihydroxyacetone phosphate, and PEP is phosphoenolpyruvate.

to phosphoenolpyruvate in the matrix, the latter is then transported across the mitochondrial inner membrane where it will continue toward glucose. If oxaloacetate is not converted to phosphoenolpyruvate by the matrix PEP carboxykinase, it must be a substrate for the cytosolic PEP carboxykinase enzyme. However, as noted earlier, oxaloacetate cannot be transported across the inner mitochondrial membrane. Therefore, oxaloacetate must be converted into something that can cross the inner membrane. It can be transaminated to aspartate, which can use the aspartate-glutamate antiport. Alternatively, it can be reduced to malate, and the malate can be transported by an antiport with α-ketoglutarate.

Once the pyruvate kinase reaction is circumvented, glycolysis can run backward up to the fructose 1,6-bisphosphate step. At this point, an enzyme known as fructose 1,6-bisphosphatase removes one phosphate group, creating fructose 6-P. After this latter compound is converted to glucose 6-phosphate, glucose 6-phosphatase removes the phosphate group and the product free glucose is ready to be released by the liver (and, to a very small extent, the kidney) to the blood stream.

Figure 6.12 also illustrates where glycerol comes in. Glycerol is released from fat cells when triglyceride molecules are broken down in a process known as lipolysis. Two glycerol molecules are capable of making one glucose molecule. In liver, the glycerol is first phosphorylated by glycerol kinase to make glycerol 3-P, which is oxidized to form dihydroxyacetone phosphate. The enzyme that catalyzes this latter reaction is the cytosolic form of glycerol 3-P dehydrogenase that we saw in the glycerol phosphate shuttle. During the oxidation of glycerol 3-P, NAD^+ is reduced to NADH. During starvation, glycerol produced during lipolysis of stored fat can act as a significant gluconeogenic precursor, providing a third or more of the glucose produced by the liver.

KEY POINT

If you start from pyruvate and make a molecule of glucose, you can see that this is an energy-requiring process. Estimate the number of ATP (or equivalent) to make one glucose molecule starting from pyruvate.

The Role of Amino Acids

Chapter 8 contains a detailed description of amino acid metabolism. For now, we will take a simple overview. Figure 6.12 reveals that amino acids can be used as a source of carbon atoms to make glucose. In fact, of the 20 common amino acids, 18 may have all or part of their carbon atoms directed to the formation of glucose; we call these **glucogenic amino acids**. The exceptions, in mammalian tissues, are leucine and lysine; these are said to be **ketogenic amino acids**.

Most adults are in a state we call **protein balance** or, as it is also known, **nitrogen balance**. The former term means that, over a long period of time, adults will maintain a constant level of protein in their bodies. The latter term means that if the entire nitrogen intake into the body is measured (mainly in the form of amino acid amino groups) and all of the nitrogen lost by the body is measured (this is measured in urine, feces, sweat, etc.), these measures will balance. Whether we use the term protein balance or nitrogen balance, we are essentially talking about the same thing. The average adult takes in about 1 gram or more of amino acids in dietary protein each day per kilogram of body weight. If they are in protein balance (or nitrogen equilibrium), they must get rid of the same amount of amino acids. We do not excrete these amino acids. Rather, the nitrogen in the form of the amino groups is removed and then excreted, mainly in the form of urea, while the remaining parts of the amino acids, which can be called the carbon skeletons, are used for

1. immediate oxidation to generate energy,
2. conversion to glucose and later oxidation, or
3. conversion into fat, which is stored and then oxidized later.

When the amino groups on most of the amino acids are removed, we get TCA cycle intermediates such as oxaloacetate, α-ketoglutarate, fumarate, succinyl CoA, as well as pyruvate. Of course, these four tricarboxylic acid cycle intermediates and pyruvate can be used to make glucose, as figure 6.12 reveals. If you review the TCA cycle, you can see that α-ketoglutarate, fumarate, and succinyl CoA can all produce oxaloacetate, which is the precursor to forming phosphoenolpyruvate. The formation of TCA cycle intermediates through removal of the amino groups from amino acids is known as **anaplerosis**. While our focus here is on the formation of glucose in liver, it is appropriate to mention that these TCA cycle intermediates can be formed in exercising muscle where they might increase the flux capacity for the cycle.

The carbon skeletons of amino acids, including the hydrogen and oxygen, are not wasted but treated as fuel. The amino groups can be removed from most of the amino acids by transferring them to other molecules. An example of this is shown in figure 6.13 for the **branched chain amino acids** (leucine, isoleucine, and valine). The amino groups on these amino acids are removed, primarily in muscle, by transferring them to α-ketoglutarate to make branched chain keto acids (BCKAs) from the branched chain amino acids plus the amino acid glutamic acid (glutamate). Then, the amino group on glutamic acid is transferred to pyruvate, regenerating the α-ketoglutarate and forming alanine.

The alanine is released from muscle to the blood. Liver extracts the alanine and removes the amino group by transferring it to oxaloacetate, making aspartic acid. The carbon skeleton remaining, which is pyruvate, can then be used to make glucose. Figure 6.14 shows the cycling of amino

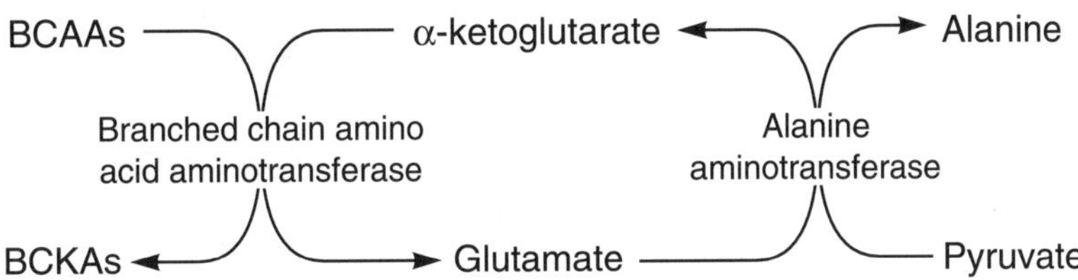

Figure 6.13 In muscle, amino groups on the branched chain amino acids (BCAAs) such as leucine, isoleucine, and valine can be transferred to α-ketoglutarate, making glutamate and branched chain keto acids (BCKAs). The amino group on glutamate can be transferred to pyruvate, making alanine and regenerating α-ketoglutarate. Alanine can exit the muscle cell and travel through the blood to the liver.

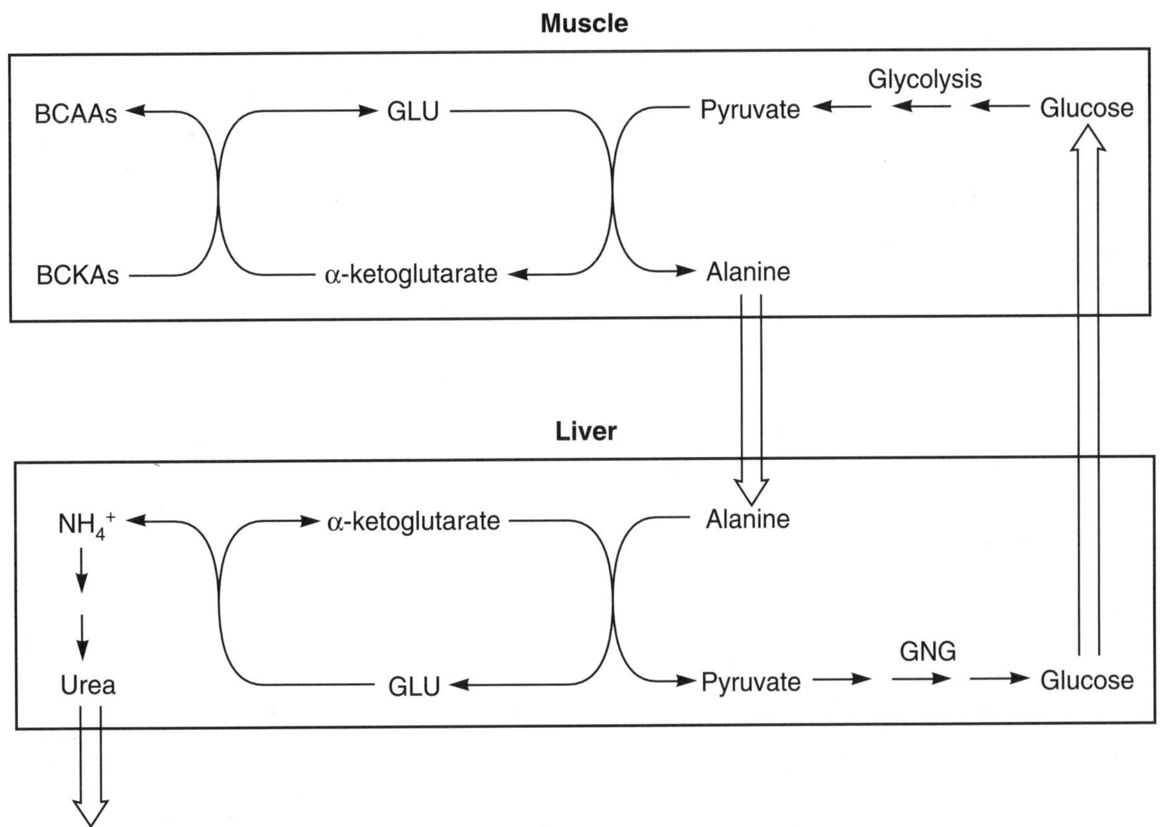

Figure 6.14 The glucose-alanine cycle transfers amino groups (NH_2) and carbon in the form of alanine from muscle to liver. In the liver the amino group is used to make urea, whereas the carbon skeleton is used to make glucose by gluconeogenesis (GNG). The glucose leaves the liver and is a carbon source for pyruvate in muscle glycolysis. BCAAs are the branched chain amino acids; BCKAs are the corresponding branched chain keto acids that result from the loss of amino groups. GLU is glutamic acid.

groups from muscle to liver by way of alanine. The amino group ends up as urea and the carbon skeleton (pyruvate) is converted into glucose in the liver. This has been described as the glucose-alanine cycle, and during exercise the rate of this cycle increases. The glucose-alanine cycle shows the importance of amino acids as a source of glucose. It also reinforces the fact that muscle protein can be broken down more rapidly under certain conditions, generating amino acids that can be a source of glucose. Muscle, the primary protein-containing tissue, is wasted to make glucose if there is not enough food energy in the diet.

KEY POINT

Diets high in protein are favored by a number of athletes. Because there is a limit to how much protein can be synthesized in skeletal muscle and other tissues, excess amino acids are oxidized or used to make glucose. When the need for gluconeogenesis is minimal, the excess amino acids are oxidized. Thus like carbohydrate, which promotes its own oxidation when there is an excess, a diet too high in protein promotes protein oxidation.

Regulation of Gluconeogenesis

The conditions that favor the onset of gluconeogenesis are those where blood glucose levels are threatened because liver glycogenolysis is declining as a result of low liver glycogen stores. As shown in figure 6.12, gluconeogenesis and glycolysis share the same enzymes in the middle portion of the glycolytic pathway, with the flux of gluconeogenesis going from pyruvate toward glucose and glycolysis going in the opposite way. Like a rail route going east to west and west to east and sharing the same section of track, traffic cannot be going in both directions simultaneously. Thus, the control of gluconeogenesis and glycolysis is integrated, insuring that one direction is favored, while the other is inhibited. Regulation is complex and involves expression of genes for key enzymes in liver, control by protein phosphorylation, and allosteric regulation of enzymes. The net effect is to depress glycolysis and stimulate gluconeogenesis simultaneously under circumstances where blood glucose levels are threatened and to stimulate glycolysis and inhibit gluconeogenesis under carbohydrate feeding conditions.

At rest, in the postabsorptive period, the brain accounts for 60% or more of the glucose output from the liver. Of the remainder, most can be accounted for by skeletal and heart muscle. During prolonged exercise, utilization of glucose by skeletal muscle can increase almost 20-fold. This helps to explain the need for careful control of liver glucose output between meals, particularly under exercise conditions.

Role of Hormones

Under conditions where gluconeogenesis is needed, specific hormones play a direct or participatory role in turning on and increasing the rate of gluconeogenesis. Glucagon is a polypeptide hormone secreted by the alpha cells of the pancreas. Its concentration rises as the blood glucose concentration decreases below normal (< 5 mM). It acts to stimulate the breakdown of liver glycogen in the postabsorptive period. When liver glycogen concentration is low, glucagon stimulates gluconeogenesis. Glucagon has a variety of actions in liver to promote gluconeogenesis. It stimulates the formation of gluconeogenic enzymes. It promotes gluconeogenesis while inhibiting glycolysis by protein phosphorylation and allosteric mechanisms.

As mentioned earlier in this chapter, insulin is released by the beta cells of the pancreas. Its concentration rises when blood glucose concentration increases beyond normal levels (> 5 mM). Insulin promotes glucose uptake into insulin-sensitive tissues and thereby helps lower blood glucose concentration. Insulin has a negative effect on gluconeogenesis, opposing not only the secretion of glucagon, but also its biochemical effects on liver.

Under more extreme conditions of physical stress, the glucocorticoid cortisol may play an important role in gluconeogenesis. Cortisol is a steroid hormone secreted by the adrenal cortex under catabolic conditions. We may suspect that its concentration would increase under severe or prolonged exercise conditions, particularly in the fasted state. Cortisol has a catabolic effect in that it promotes net protein breakdown in skeletal muscle. In this way, it increases the availability of amino acids to act as gluconeogenic precursors. In the liver, cortisol stimulates the expression of genes coding for gluconeogenic enzymes. If cortisol levels do not increase, the presence of even a low level of cortisol in the blood acts to assist the process of gluconeogenesis since persons with cortisol deficiency can develop hypoglycemia more easily.

Control by Altering Gene Expression

Glycolysis and gluconeogenesis share the same equilibrium (reversible) enzymes in the central portion of the glycolytic pathway. Control of gluconeogenesis and glycolysis takes place outside of this central portion through three substrate cycles that regulate the overall rates of glycolysis and gluconeogenesis in the liver (see figure 6.12). These substrate cycles involve the intermediates glucose and glucose 6-phosphate, fructose 6-phosphate and fructose 1,6-bisphosphate, and phosphoenolpyruvate and pyruvate. If you focus on these substrate cycles, you should be able to see that if the enzymes involved are simultaneously active, the net effect is no flux in either direction, but there would be a loss of energy-rich phosphates. The enzymes involved in these substrate cycles are different and nonequilibrium (irreversible), and the reactions they catalyze are highly exergonic.

Glycolysis can be favored by increasing the activity of glucokinase, phosphofructokinase, and pyruvate kinase by stimulating the transcription of their genes while simultaneously depressing transcription of genes for the gluconeogenic enzymes. On the other hand, gluconeogenesis can be favored by increasing transcription of glucose 6-phosphatase, fructose 1,6-bisphosphatase, pyruvate carboxylase, and PEPCK while depressing transcription of the three nonequilibrium glycolytic enzymes. The hormones insulin and glucagon play a major role in controlling gene expression, with glucagon promoting gluconeogenic enzymes and insulin promoting glycolytic enzyme formation. We discuss the regulation of transcription in chapter 9. An interesting element to the transcription control of genes coding for enzymes of gluconeogenesis is the role of endurance exercise training. It has been shown in a number of studies that endurance training improves the maintenance of blood glucose concentration levels during exercise by inducing higher levels of hepatic (liver) gluconeogenic enzymes.

Control of Enzyme Activity

Control of gluconeogenesis and glycolysis by regulating the transcription of genes for key enzymes plays a powerful role. However, there is a considerable lag in time before changing the expression of specific genes can lead to significant changes in the amounts of functional enzymes. Therefore, rapid-acting mechanisms must be available to control glycolysis and gluconeogenesis.

Pyruvate kinase, the nonequilibrium glycolytic enzyme, can be inhibited by phosphorylation. The increase in glucagon under conditions favoring hepatic glucose output results in an increase in cyclic AMP (see the section in this chapter on control of liver glycogenolysis). In addition to stimulating glycogenolysis, cAMP activates a protein kinase that results in the phosphorylation of pyruvate kinase. This effectively blocks glycolysis at this step, yet has no influence on gluconeogenesis. A glucagon-induced rise in hepatic cell cAMP concentration results in the phosphorylation of another enzyme known as phosphofructokinase-2 (PFK-2). This enzyme is distinct from phosphofructokinase (PFK), the glycolytic enzyme, in that PKF-2 adds a phosphate group to fructose 6-phosphate at carbon atom 2, creating fructose 2,6-bisphosphate (F 2,6-P_2). Phosphorylation of PFK-2 depresses its activity so that it does not produce F 2,6-P_2, which is an allosteric activator of PFK. The net effect of this enzyme phosphorylation reaction is to slow down glycolysis. In summary, phosphorylation of two key enzymes by a cAMP-dependent protein kinase acts in switchlike fashion to inhibit glycolysis in liver. If glycolysis is inhibited, gluconeogenesis can proceed.

Allosteric regulation of PFK and fructose 1,6-bisphosphatase (FBPase) plays a very rapid-acting role in regulation of the substrate cycle around fructose 6-P and fructose 1,6-bisphosphate. Figure 6.15 highlights the role of the allosteric effectors and their control of PFK and FBPase. In liver, PFK is activated by AMP and F 2,6-P_2 and inhibited by citrate. The opposing enzyme, FBPase, is inhibited by F 2,6-P_2 and AMP and activated by citrate. Therefore, the two opposing enzymes in this substrate cycle respond in opposite ways to the same allosteric effectors.

Under conditions where gluconeogenesis needs to be stimulated, there is a rise in the acetyl CoA level. This arises from an increase in beta-oxidation of fatty acids, a topic we cover in the next chapter. Acetyl CoA from fatty acids is a substrate for the TCA cycle. As well, acetyl CoA can act as an allosteric effector in two ways. First, an increase in cytosolic acetyl CoA inhibits pyruvate kinase at a negative allosteric site. Secondly, acetyl CoA is an obligatory activator for pyruvate carboxylase,

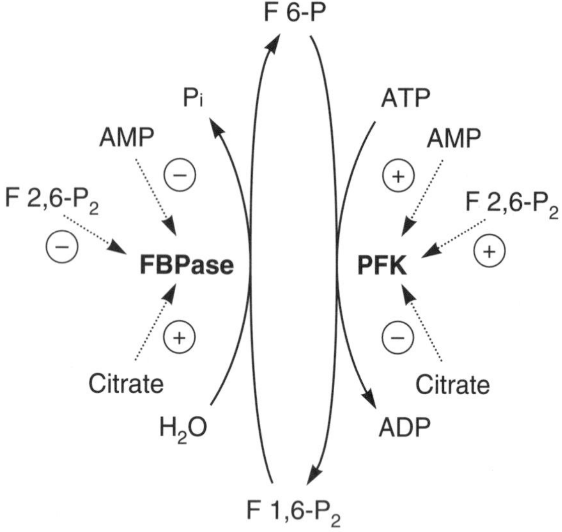

Figure 6.15 To prevent wasteful cycling between fructose 6-P (F 6-P) and fructose 1,6-bisphosphate (F 1,6-P_2), phosphofructokinase (PFK) and fructose 1,6-bisphosphatase (FBPase) are regulated in the opposite direction by the same allosteric effectors. The dotted arrows and the positive and negative signs show the influence of the effectors on each enzyme.

which carboxylates pyruvate to make oxaloacetate. The rise in acetyl CoA signals that there is plenty of fuel for the TCA cycle, so carbon should be shunted to make glucose.

KEY POINT

From a biochemical perspective, the brain could be described as the body's most important tissue. It is hard to argue with this statement given the complex, yet complementary, mechanisms to regulate the level of glucose in the blood to act as a fuel for the brain.

The Pentose Phosphate Pathway

As we will see in the next chapter, most cells require a constant source of reducing equivalents for biosynthesis. These reducing equivalents are found in NADPH, a molecule differing from NADH by the fact that the former contains an additional phosphate group. As we have seen, electrons on NADH are used in the electron transport chain to reduce oxygen and generate ATP. Electrons on NADPH are used to synthesize new molecules such as fatty acids, amino acids, and steroids. The pentose phosphate pathway is the most common name, but it is also known as the hexose monophosphate shunt or the phosphogluconate pathway. Besides NADPH, the pentose phosphate pathway generates another important substance used in biosynthetic reactions. Ribose 5-phosphate is used in nucleotide biosynthesis and thus is needed to make ATP, DNA, RNA, FAD, Coenzyme A, and NAD^+ and $NADP^+$.

A detailed discussion of this pathway is beyond the scope of this book. However, reactions of the pentose phosphate pathway are important in tissues such as the liver, adipose tissue, adrenals, testes, and ovaries, but not skeletal muscle. The reaction begins with glucose 6-phosphate, which is also a beginning point for glycolysis. However, there are huge differences. In the pentose phosphate pathway, the glucose 6-phosphate first undergoes an oxidation, catalyzed by glucose 6-phosphate dehydrogenase, generating NADPH plus H^+. This first reaction is the key reaction in the pentose phosphate pathway, and it is under tight regulation by one of its products, NADPH. When the ratio of the reduced to oxidized form of this coenzyme is elevated, glucose 6-phosphate dehydrogenase is inhibited. Subsequent reactions of the pentose phosphate pathway generate ribose 5-phosphate and even glycolytic intermediates such as fructose 6-phosphate and glyceraldehyde 3-phosphate.

Summary

Most dietary carbohydrate is broken down in the intestinal tract to glucose, fructose, and galactose, but glucose is by far the most important carbohydrate. In the form of blood glucose and liver and muscle glycogen, body carbohydrate is a major fuel, particularly for the brain. The concentration of blood glucose is carefully regulated so that its concentration is approximately 5-6 millimolar (90–110 mg/100 ml of blood). Glucose from the blood enters cells down a concentration gradient through a specific glucose transporter. The GLUT-4 transporter is found in muscle and fat cell membranes, and its content in these membranes is greatly increased by insulin. Exercise also increases the uptake of glucose into muscle. The principle metabolic pathway for breaking down carbohydrate is glycolysis. Pyruvate, a product of glycolysis, has two major fates. It can be reduced to lactate by the enzyme lactate dehydrogenase, or it can enter the mitochondrion, to be oxidized to acetyl CoA to enter the TCA cycle.

Glycolysis is a carefully regulated process with most of the control exerted by the enzyme phosphofructokinase (PFK). This enzyme is sensitive to the binding of a number of positive and negative modulators to allosteric sites. Rest or low level muscle activity is associated with a low activity of glycolysis, whereas vigorous exercise greatly increases its rate in order to produce ATP. When pyruvate is oxidized in the mitochondrion and not reduced to lactate during glycolysis, NADH can potentially build up to the extent that there is too little NAD^+ to permit glycolysis to continue. Oxidation of cytoplasmic NADH, other than by lactate dehydrogenase, occurs through the action of two shuttle systems. The glycerol phosphate shuttle and the malate-aspartate shuttle transfer electrons from the cytosol to the mitochondria for use by the electron transport chain.

Glycogen, the storage form of glucose in liver and muscle tissue, is synthesized from glucose taken up from the blood in a process that is greatly accelerated following a carbohydrate-rich meal. Liver glycogen is a reservoir of glucose for the blood. In muscle, the glycogen is ready to be fed into the glycolytic pathway when the muscle fiber becomes active. Since glycogen synthesis and glycogenolysis are opposing processes taking place in the cytosol of cells, one process must be inactive when the other is active to avoid a futile cycle. Two regulatory enzymes, glycogen phosphorylase and glycogen synthase, are thus affected in opposite ways by the same signal, which leads to their simultaneous phosphorylation. In addition to covalent modification, the activities of both synthase and phosphorylase are influenced by the binding of regulatory molecules. Following a meal, when blood insulin concentration is increased, the synthase is active and the phosphorylase inactive in both liver and muscle. Between meals, liver phosphorylase is activated by a rise in glucagon and a fall in insulin so that the glycogen is slowly broken down to glucose. During exercise, when epinephrine concentration is increased, muscle phosphorylase is active and synthase inhibited. The tight regulation of glycogen metabolism emphasizes the importance of controlling body carbohydrate stores.

When body carbohydrate content is decreased and glucose output from the liver is threatened, glucose is made from a variety of noncarbohydrate sources such as lactate, pyruvate, and the carbon skeletons of amino acids. This process, gluconeogenesis, takes place in liver and kidney and involves the reversible reactions of glycolysis. New reactions that bypass irreversible glycolytic reactions are catalyzed by pyruvate carboxylase, phosphoenolpyruvate carboxykinase, fructose 1,6-bisphosphatase, and glucose 6-phosphatase. These reactions allow the liver to produce glucose during times of need while glycolysis is virtually shut down. The pancreatic hormone glucagon promotes gluconeogenesis. Glycerol, produced when triglycerides are broken down, is an important gluconeogenic precursor in liver. In addition, when the amino group of the 20 common amino acids is removed, the carbon skeletons of 18 of these can be converted to glucose. The fact that skeletal muscle can be severely wasted during diets providing inadequate food energy demonstrates the importance of the brain as a tissue, since it can be a significant user of glucose derived from amino acids obtained from muscle proteins. The pentose phosphate pathway is not important in muscle, but in breaking down glucose 6-phosphate in a variety of tissues, this pathway produces reducing equivalents in the form of NADPH and precursors for making nucleosides.

Key Terms

anaplerosis 106
biotin 104
branched chain amino acids 107
Cori cycle 104
cyclic AMP 93
diabetes mellitus 82
disaccharides 81
epinephrine 93
euglycemia 82
feedback inhibition 84
glucagon 95
glucogenic amino acids 106
glucose transporters 83
glycogenesis 90
glycogenin 91
glycogenolysis 92
hyperglycemia 82
hypoglycemia 82
insulin 83
ketogenic amino acids 106

Review Questions

1. Stored fat is considered an ideal fuel from an efficiency perspective because, being stored in an anhydrous form, it provides about 9 kilocalories (38 kJ) per gram. From the same perspective, we could say that glycogen is not an efficient fuel because it is stored with water and has a lower fuel value. Calculate the approximate fuel value of glycogen in kcal and kJ per gram weight as stored.
2. We have identified a variety of exercise and competition activities where elevations in muscle glycogen concentration can enhance performance. Can you think of activities where glycogen is an essential fuel but where supercompensated levels may be detrimental to performance?
3. The concentrations of glucose and lactate in the blood may be described in units such as mM or in units of mg/dl of blood. If the molecular weight of glucose is 180 and lactate is 90, convert the following concentrations of glucose and lactate to mM units: lactate concentration is 990 mg/dl and blood glucose concentration is 108 mg/dl.
4. From RER and other measures, it is determined that a distance runner is oxidizing carbohydrate at the rate of 4 grams per minute and fat at the rate of 0.5 grams per minute. If he is running at 4 minutes per kilometer, what percentage of his fuel is derived from fat if his total energy expenditure is 3.43 megajoules (MJ)? How far does he run?

CHAPTER 7

Lipid Metabolism

Most people are well aware that fats (or lipids) are very important fuels in the body. They are aware, maybe painfully so, of the storage of lipid throughout their bodies. They may also know that lipids are not soluble in water. Few are aware, however, that lipids are also important structural elements and are involved in cell signaling. Fat is stored as triglyceride (triacylglycerol) in fat cells. This is considered a long-term energy store, in contrast to glycogen, which is considered a short-term energy store. To understand the importance of lipids one must learn about the various kinds of lipids and their metabolism.

Types of Lipids

Lipids are categorized as fatty acids, triglycerides, **phospholipids**, and sterols, whose roles in the body are largely determined by their chemical structure. Lipids are hydrophobic, containing only nonpolar components, or they may be amphipathic, meaning that they have some polar groups, although they are largely nonpolar. We have already mentioned fatty acids in chapter 5, when it was disclosed that the beta-oxidation of fatty acids provides acetyl CoA for the TCA cycle. Fatty acids are therefore important fuel molecules.

Fatty Acids

Fatty acids are **long chain carboxylic acids**, which means they have a long hydrocarbon tail and a carboxyl group for the head. They are usually known by their trivial names. At neutral pH, fatty acids exist in anion form as carboxylates. Normally they have an even number of carbon atoms and can be saturated, with no carbon-to-carbon double bonds, or unsaturated, with one or more carbon-to-carbon double bonds. Figure 7.1 gives examples of saturated and unsaturated fatty acids. Palmitic acid and stearic acid are saturated fatty acids. A shorthand notation for describing their structure appears in the square brackets. The first number reveals the number of carbon atoms and the second indicates the number of double bonds. Oleic acid and linoleic acid are unsaturated fatty acids. The numbers in parentheses indicate the positions of the double bonds, beginning with the carboxyl group as carbon 1. The configuration of the double bonds in naturally occurring fatty acids is cis, as is shown in figure 7.1. Most of the fatty acids in your body come from your diet. We have a limited capacity to synthesize fatty acids. From a human nutrition perspective, the three most common fatty acids are palmitic acid (palmitate at pH 7), oleic acid (oleate), and stearic acid (stearate).

Linoleic acid is an **essential fatty acid** that we must get in the diet. It is also known as a polyunsaturated fatty acid or PUFA. Linolenic acid is [18:3 (9, 12, 15)] and is called an omega-three or n-3 fatty acid, that is, the last double bond begins three carbons from the end carbon, or carbon 18. Note that linoleic acid is an omega-six or n-6 fatty acid. The n-3 (or omega-three) fatty acids are reputed to offer special protection to people by lowering blood lipid concentration. Linolenic acid is also a dietary essential fatty acid. The essential fatty acids are found in cell membranes in phospholipids. They are also precursors for the synthesis of a family of hormonelike compounds known as the eicosanoids. Eicosanoids are synthesized in cells from the fatty acid arachidonic acid [20:4 (5, 8, 11, 14)]. The eicosanoids are known as local hormones because they are released from cells where they may have an effect on the secreting cell itself (**autocrine effect**) or on the cells in the immediate environment of the secreting cell (paracrine effect). Examples of eicosanoids are prostaglandins and leukotrienes.

Triglycerides

Although they should be called **triacylglycerols**, the term **triglyceride** is more commonly used. An older term, neutral fat, is also used. Figure 7.2 shows these molecules as triesters, made from the combination of a three-carbon molecule with three alcohol groups, known as glycerol, and three fatty acids. Remember that the combination of an acid with an alcohol is known as an ester. In the body, stored triglycerides rarely have the same fatty acid attached at all three positions on the glycerol. Stored fats have both saturated and unsaturated fatty acids. Diglycerides or, more correctly, diacylglycerols have only two fatty acids attached to the glycerol. Monoglycerides or monoacylglycerols have only one.

The physical state of a triglyceride depends on the length of the carbon chain in the fatty acids and the number of double bonds. Shorter length fatty acids as well as more double bonds lower the melting point. Thus, vegetable fats with polyunsaturated fatty acids are liquids, whereas the fat on the side of a steak, which contains more

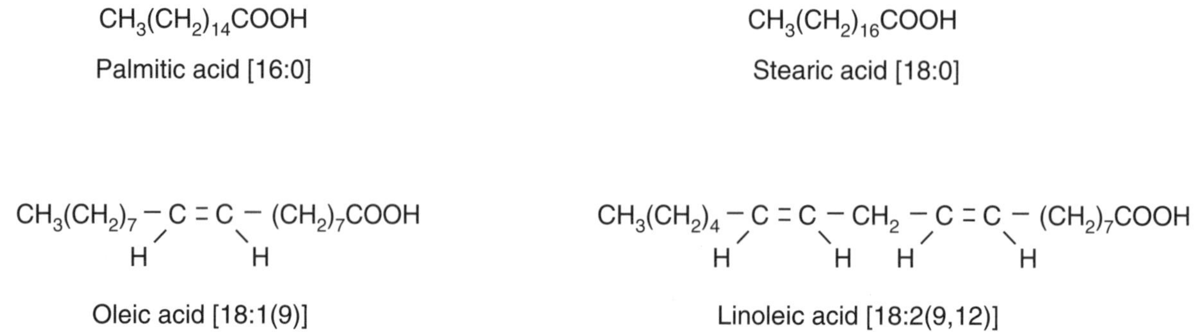

$CH_3(CH_2)_{14}COOH$

Palmitic acid [16:0]

$CH_3(CH_2)_{16}COOH$

Stearic acid [18:0]

$CH_3(CH_2)_7-\underset{H}{C}=\underset{H}{C}-(CH_2)_7COOH$

Oleic acid [18:1(9)]

$CH_3(CH_2)_4-\underset{H}{C}=\underset{H}{C}-CH_2-\underset{H}{C}=\underset{H}{C}-(CH_2)_7COOH$

Linoleic acid [18:2(9,12)]

Figure 7.1 Two examples of saturated (upper) and unsaturated (lower) fatty acids. The square brackets show a shorthand way to describe fatty acid structure.

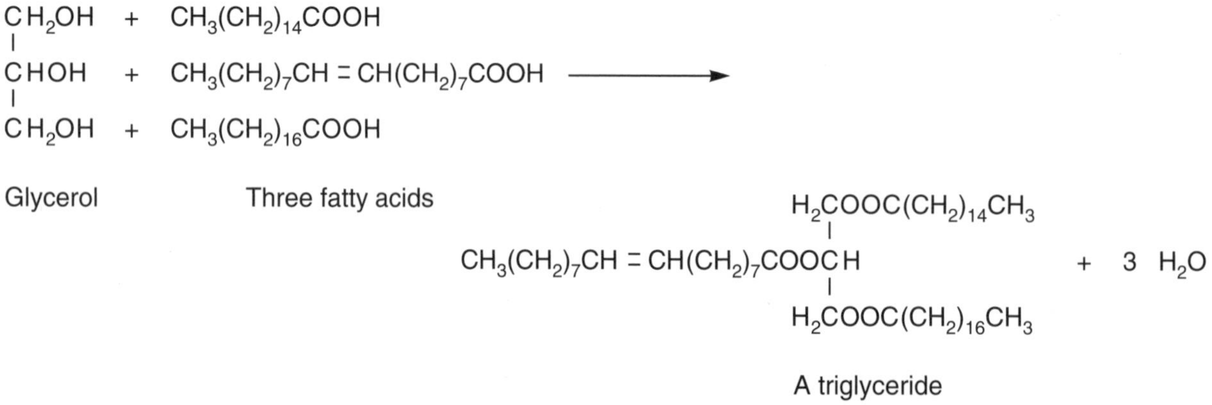

$$\begin{array}{lcl} CH_2OH & + & CH_3(CH_2)_{14}COOH \\ | & & \\ CHOH & + & CH_3(CH_2)_7CH=CH(CH_2)_7COOH \\ | & & \\ CH_2OH & + & CH_3(CH_2)_{16}COOH \end{array} \longrightarrow$$

Glycerol — Three fatty acids

$$\begin{array}{r} H_2COOC(CH_2)_{14}CH_3 \\ | \\ CH_3(CH_2)_7CH=CH(CH_2)_7COOCH \\ | \\ H_2COOC(CH_2)_{16}CH_3 \end{array} \quad + \quad 3\ H_2O$$

A triglyceride

Figure 7.2 Triglycerides are formed by the union of glycerol and three fatty acid molecules.

saturated fatty acids, is a solid. Palm oil is a liquid, despite having mainly saturated fatty acids, because the fatty acids have 10 to 12 carbon atoms as opposed to 16 to 18 in most other triglycerides. Triglycerides and the fatty acids found in triglycerides are insoluble in water due to the large degree of hydrophobic hydrocarbon components. The hydrophobic nature of triglycerides makes them ideal for storing energy because, by weight, they have more chemical potential energy than other fuel molecules such as carbohydrates or protein. In fact, because the fatty acids in triglycerides are so reduced, the oxidation of one gram of triglyceride has a standard free energy change of more than 9 kilocalories/gram (38 kJ/g).

Phospholipids

Many phospholipids are derivatives of **phosphatidic acid**, whose structure is illustrated in figure 7.3. Different groups can be attached to the phosphate in phosphatidic acid. If choline is attached, the molecule is called phosphatidyl choline or, commonly, lecithin (see figure 7.3). Phospholipids have a **hydrophilic** (water liking) part and a hydrophobic part. The hydrophilic part is due to polar chemical bonds and charged groups on the phospholipid. The hydrophobic part is the long hydrocarbon tail of the fatty acids, which can contain more than 16 carbon atoms. Phospholipids are major components of cell membranes.

Phosphatidyl inositols are found in membranes, where they play an important role in cellular regulation. Figure 7.4 shows the structure of inositol. When phosphorylated at carbons 4 and 5 and attached to phosphatidic acid, it is known as phosphatidyl inositol 4,5-bisphosphate, abbreviated Ptd Ins 4,5-P_2. Hydrolysis of Ptd Ins 4,5-P_2 produces inositol 1,4,5-trisphosphate, abbreviated Ins 1,4,5-P_3, and a **diacylglycerol** (DG).

Ptd Ins 4,5-P_2 and Ins 1,4,5-P_3 are important in signal transduction. This is the name given to a process of communication whereby a signal external to the cell results in specific action taking place within the cell. For example, certain intercellular signals such as hormones or neurotransmitters

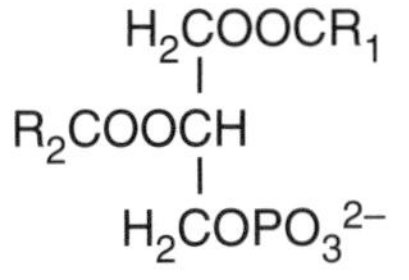

H_2COOCR_1
R_2COOCH
$H_2COPO_3^{2-}$

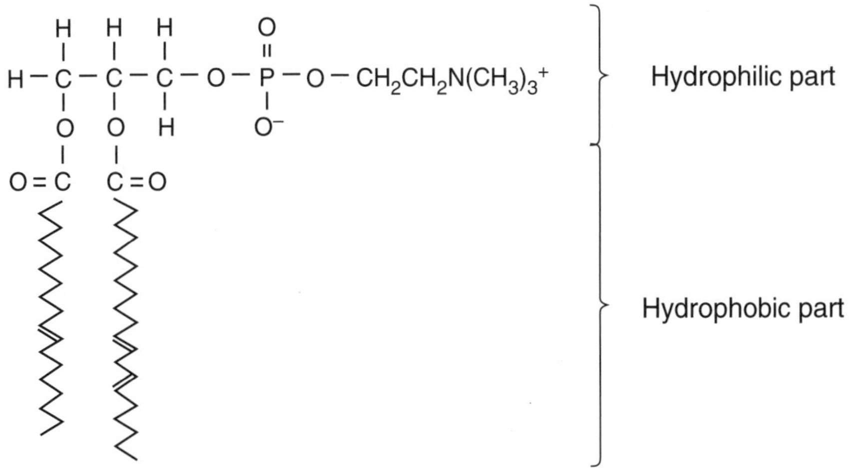

$H-C(-O-C{=}O)-C(-O-C{=}O)-CH_2-O-P({=}O)(O^-)-O-CH_2CH_2N(CH_3)_3^+$

Hydrophilic part

Hydrophobic part

Figure 7.3 The top structure shows the structure of phosphatidic acid. R_1 and R_2 represent long chain fatty acids attached to glycerol. The lower structure illustrates phosphatidyl choline, formed when phosphatidic acid combines with choline. The structure has been rotated to illustrate a hydrophilic part with polar bonds and charged groups and a long hydrophobic part, due to the tails of the fatty acids, shown in chemical shorthand.

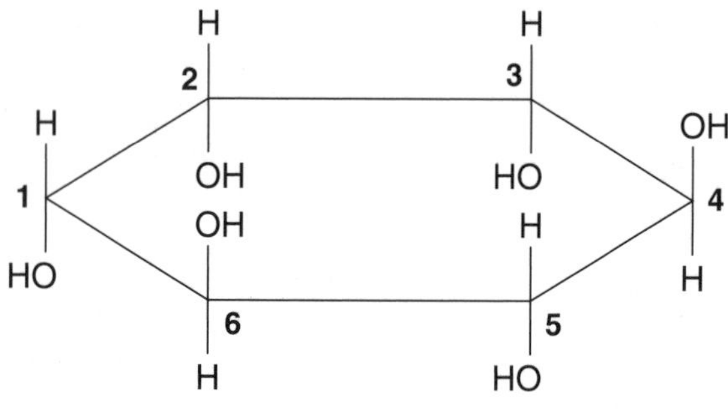

Figure 7.4 The structure of inositol. The numbers identify the six carbon atoms.

(i.e., an agonist) can activate a cell membrane enzyme known as phospholipase C on binding to its membrane receptor. This enzyme can then hydrolyze cell membrane phospholipids (Ptd Ins 4,5-P_2) to produce Ins 1,4,5-P_3 and a diacylglycerol. The Ins 1,4,5-P_3 can act on the endoplasmic reticulum of the cell to release calcium ions, which can then activate other enzymes. The diacylglycerol can also activate an enzyme known as protein kinase C, which in turn can phosphorylate other enzymes. In this way, a neural signal or hormone can cause specific changes in a cell. We call this process signal transduction because the nerve impulse or hormone (called the first messenger) acts at the cell membrane by binding to specific receptors, resulting in the formation of second messengers inside the cell. Second messengers, such as Ins 1,4,5-P_3, diacylglycerol, Ca^{2+}, and, as we saw in the previous chapter, cyclic AMP, can promote the phosphorylation of certain proteins. Phosphorylation of an enzyme protein can lead to dramatic changes in activity, generating specific responses within the cell. Malfunctions in the signal transduction process underlie many harmful diseases in the body. Figure 7.5 provides an overview of the phosphatidyl inositol signaling pathway.

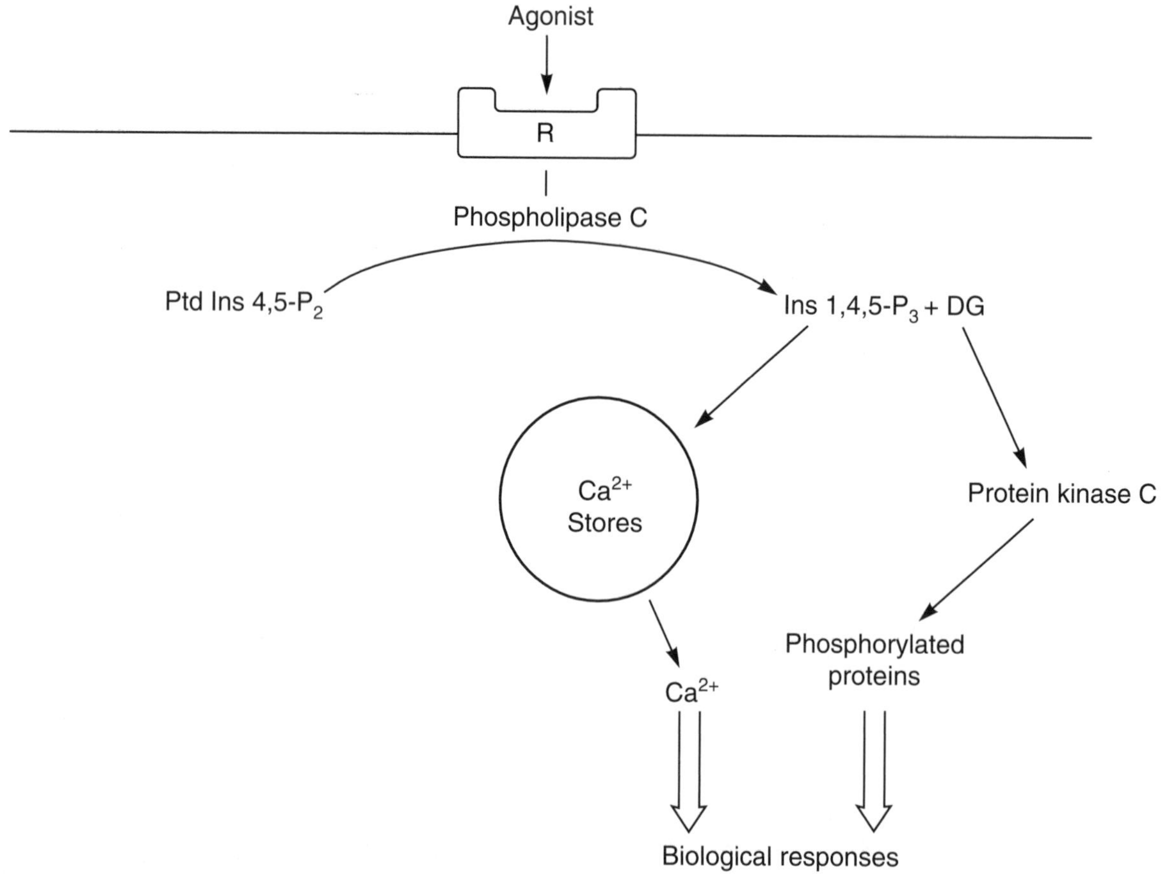

Figure 7.5 The phosphatidyl inositol signaling pathway. An agonist such as a hormone or neurotransmitter binds to a specific receptor (R) activating phospholipase C, which splits phosphatidyl inositol 4,5-bisphosphate (Ptd Ins 4,5-P_2) into two products. The inositol 1,4,5-trisphosphate (Ins 1,4,5-P_3) acts to release calcium ions (Ca^{2+}) from internal stores. This Ca^{2+} release has a number of biological effects. The other product, diacylglycerol (DG), activates protein kinase C, which phosphorylates proteins with biological effects.

KEY POINT

In chapter 2 we introduced the concept of protein phosphorylation and dephosphorylation in signal transduction. In that section, we mentioned MAP (mitogen-activated protein) kinases, which in humans play a major role in determining a variety of cellular responses, including morphological changes, stress responses, and cell death (apoptosis).

Fat Stores

Fat is stored as triacylglycerol (triglyceride) in fat cells. It is considered a long-term energy store in contrast to glycogen, which is considered a short-term energy store. A normal adult can store about 8,000 to 12,000 kJ (2,000 to 3,000 kcal) of glycogen in liver and muscle, but even a lean male or female has more than 300,000 kJ (75,000 kcal) stored as triglyceride in fat cells. The major storage form of triglycerides is a liquid droplet, occupying much of the volume of the adipocyte (fat cell).

Fatty acids are obtained mainly from food fat. After the fatty meal has been digested and the fatty acids absorbed, triglycerides are formed inside intestinal cells, combined with some protein and phospholipid, and released into the lymphatic system as lipoprotein particles known as **chylomicrons**. Another source of triglyceride is from the liver. The liver makes and secretes triglyceride-rich lipoproteins known as **very low density lipoproteins** (VLDLs). The fatty acids in chylomicrons and VLDLs are released from their triglycerides through hydrolysis by an enzyme known as lipoprotein lipase. This enzyme is synthesized in adjacent fat cells, secreted from the cell, and attached to the endothelial lining of a nearby capillary. Fatty acids freed in the capillaries by lipoprotein lipase can flow down a concentration gradient into adjacent adipocytes with the aid of a specific carrier. Lipoprotein lipase is also present in capillaries in skeletal muscle where it can generate fatty acids for muscle fibers. We discuss this in a later section of this chapter.

Formation of Triacylglycerols

Fat synthesis is favored following a meal when chylomicrons and blood glucose levels are elevated. An elevated insulin concentration in the blood facilitates entry of glucose into fat cells using the GLUT-4 transporter. Formation of triacylglycerol is a simple process, illustrated as a brief sequence of reactions in figure 7.6. This process is also known as esterification because it forms the triester, triglyceride. The fatty acid substrates are attached to CoA and each is known as a fatty acyl CoA. The initial precursor is glycerol 3-phosphate, which is produced by the partial breakdown of glucose to dihydroxyacetone phosphate (discussed under the section on glycolysis in the previous chapter). Reduction of dihydroxyacetone phosphate by glycerol phosphate dehydrogenase using NADH generates glycerol 3-P. The enzyme fatty acyl transferase moves the fatty acid from the CoA to glycerol 3-phosphate. The 1-lysophosphatidic acid is simply phosphatidic acid without a fatty acid attached to carbon 2 of glycerol. The formation of fatty acyl CoA is shown in the reaction below.

$$\text{fatty acid} + \text{ATP} + \text{CoA} \xrightarrow[\text{synthetase}]{\text{acyl CoA}} \text{fatty acyl CoA} + \text{AMP} + \text{PPi}$$

Therefore, to make one triglyceride molecule, we need one glycerol 3-P (from glucose) and three fatty acids attached to CoA.

Mobilization of Fat

This process is also known as triglyceride hydrolysis but is better known as **lipolysis**. In this process, hydrolysis of a triglyceride molecule yields glycerol and three fatty acid molecules. This process is favored under conditions of exercise, low calorie dieting, fasting (starvation), or when the body is cold. An enzyme inside fat cells and other fat storing cells (e.g., muscle) known as hormone-sensitive lipase is responsible for this. Lipolysis involves three hydrolysis reactions, each catalyzed by hormone-sensitive lipase (see figure 7.7). The last hydrolysis reaction is also assisted by a monoglyceride lipase. The numbers refer to the carbon atoms of glycerol. For example, a 2-monoglyceride (2-monoacylglycerol) has a single fatty acid attached to the glycerol molecule at the middle carbon atom.

Lipolysis takes place in locations other than in fat cells. For example, hydrolysis (digestion) of dietary triglyceride occurs in the small intestine, catalyzed by **pancreatic lipase**. As mentioned, hydrolysis of triglyceride in blood lipoproteins is catalyzed by lipoprotein lipase. Finally, triglyceride is stored inside

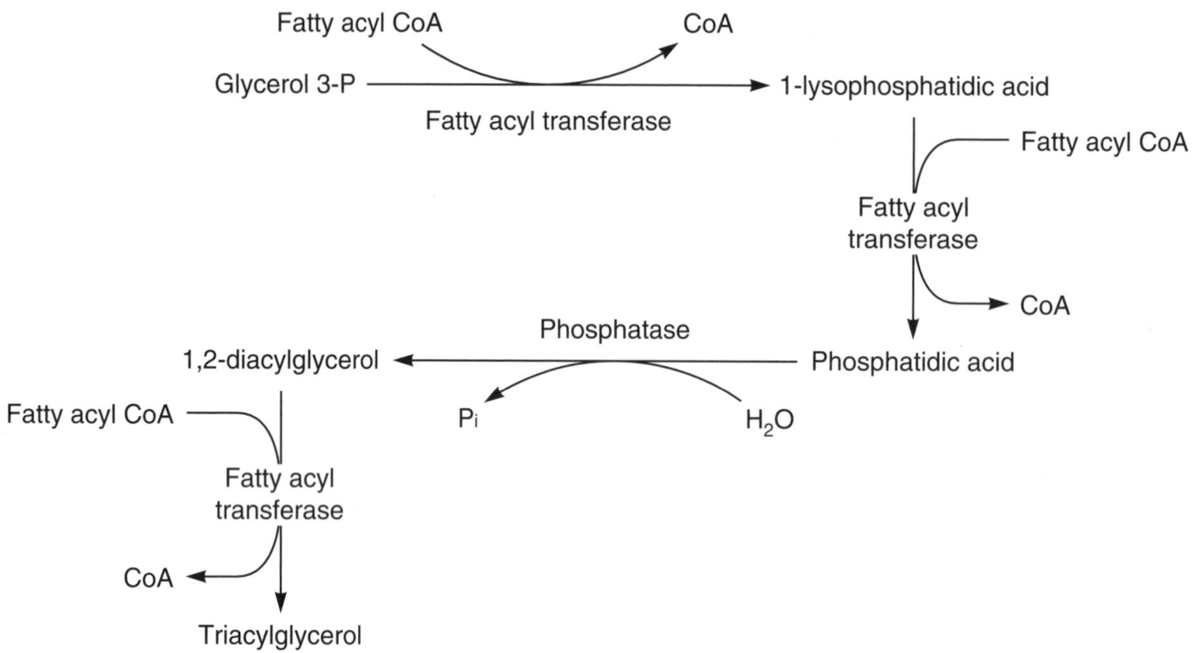

Figure 7.6 The formation of triacylglycerol in a sequence of reactions in which fatty acids attached to CoA are transferred to glycerol, which starts out as glycerol 3-phosphate.

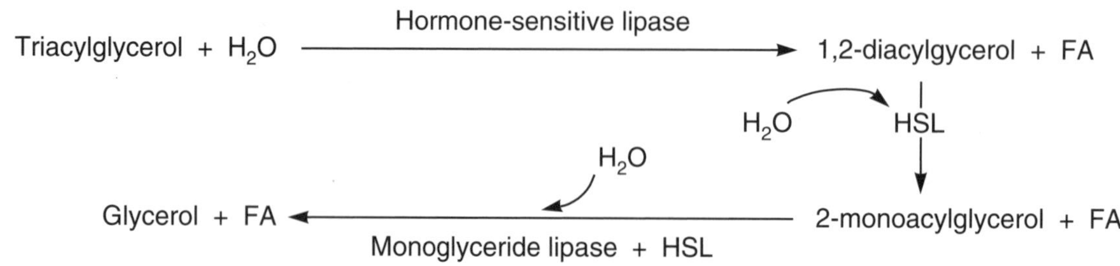

Figure 7.7 The breakdown of stored triacylglycerol in three steps known as lipolysis or fat mobilization. HSL represents hormone-sensitive lipase.

of muscle fibers. Hydrolysis of this intramuscular triglyceride (IMTG) produces fatty acids that are immediately available as a fuel for that fiber.

KEY POINT

The processes of triglyceride synthesis (esterification) and breakdown (lipolysis) take place in the cytosol of cells. This might suggest to you that since these are opposing processes occurring in the same location they should not be simultaneously active. Second, the amount of triglyceride stored in a fat cell is based on the long-term net balance between esterification and lipolysis.

Fate of Fatty Acids and Glycerol

The glycerol released when a triglyceride is hydrolyzed cannot be reused by the fat cell. Instead, the glycerol leaves the fat cell and circulates in the blood. Since glycerol has three OH groups, it is quite soluble in the blood. It is metabolized by tissues that contain a glycerol kinase, an enzyme that phosphorylates the glycerol to form glycerol 3-phosphate. The major tissue where this occurs is the liver. As we saw in the previous chapter, glycerol is an important source of glucose during periods of fasting or starvation. Because glycerol is not reused by the fat cell, whereas fatty acids may be esterified, release of glycerol is taken as the index of lipolysis.

As shown in figure 7.8, fatty acids released when triglyceride molecules are hydrolyzed have two major fates:

1. Reesterification to a triglyceride, following attachment of CoA to the fatty acid
2. Leaving the fat cell, becoming a free fatty acid in the blood stream

Using the energy from ATP, the enzyme acyl CoA synthetase can attach a CoA to the fatty acid forming fatty acyl CoA. This is now a substrate for triglyceride formation. If all the fatty acids released during lipolysis exit the adipocyte, then there should be a FFA to glycerol ratio of 3:1. However, because of the reutilization of some of the fatty acids in the formation of new triglyceride molecules, this three to one ratio is seldom observed. Indeed, during inactivity, nearly two thirds of the fatty acids produced during lipolysis are esterified. On the other hand, during light to moderate intensity exercise, most fatty acids produced by lipolysis leave the adipocyte.

When fatty acids leave a fat cell and enter the blood, they become attached to the blood protein albumin since the fatty acids are not soluble in the aqueous plasma. We call fatty acids attached to albumin in the blood **free fatty acids** or **FFA**. They are also known as nonesterified fatty acids or NEFA. The FFA circulate in the blood and may enter other cells where they can be used as an energy substrate, as we shall soon see. Transfer of FFA from **adipose tissue** to skeletal muscle during exercise requires sufficient perfusion of adipose tissue by blood. Therefore, adipose tissue blood flow (ATBF) could pose a limitation to the delivery of fatty acids from fat tissue to exercising skeletal muscle.

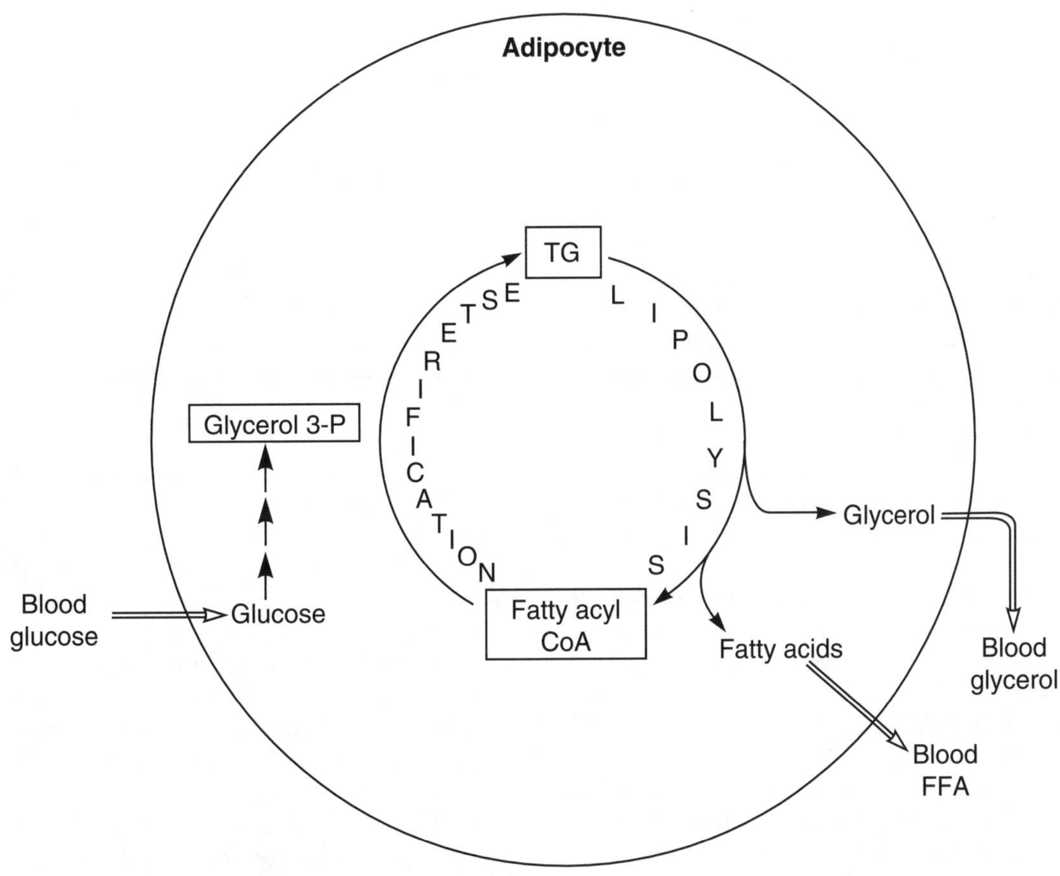

Figure 7.8 The fatty acids released from triglyceride molecules during lipolysis have two fates. They can be converted into fatty acyl CoA and used to make a new triglyceride molecule in the process of esterification, or they can leave the fat cell. Because of their insolubility in water, fatty acids in the blood are attached to the protein albumin and transported in the blood as free fatty acids (FFA). Esterification, or the formation of a triglyceride, requires glycerol 3-phosphate (glycerol 3-P), which is made by the partial breakdown of glucose taken up into the fat cell from the blood.

Regulation of Triglyceride Turnover

Formation of triglyceride molecules and their degradation take place in the cytosol of adipose tissue, although different enzymes are used. Theoretically, both processes could be active at the same time. The net effect of a continuous process of lipolysis and esterification would be wasteful ATP hydrolysis. To prevent this futile cycling, regulatory mechanisms ensure that one process is dominant and the other depressed. For example, during exercise, esterification is inhibited, whereas lipolysis is accelerated. Following a meal, lipolysis is inhibited, whereas esterification is accelerated. Regulation takes place primarily at the level of hormone-sensitive lipase (HSL), controlled by phosphorylation-dephosphorylation, although other mechanisms operate.

When HSL is phosphorylated, it is active; when it is dephosphorylated, it is inactive. Figure 7.9 shows the mechanism for regulating metabolism in the fat cell. The principal mechanism for controlling the activity of hormone-sensitive lipase is mediated through the binding of specific agonists to receptors on the cell membrane. There are two classes of receptors, shown in figure 7.9 as R_β and $R_{\alpha 2}$. These are **adrenergic receptors** that respond to the catecholamine agonists epinephrine and **norepinephrine**. Epinephrine is synthesized in and released from the adrenal medulla. Norepinephrine is the principal neurotransmitter released from sympathetic nerve endings. There are three subgroups of beta adrenergic receptors in adipose tissue, known as β_1, β_2, and β_3. These differ in their sensitivity to the agonists epinephrine and norepinephrine. In human skeletal muscle, it is the β_2 receptor that has the greatest influence on lipolysis. When the agonists epinephrine or norepinephrine bind to the beta adrenergic receptors, they activate the enzyme adenylate cyclase (also known as adenylyl cyclase) via a stimulatory guanine nucleotide binding protein complex (G protein) shown in figure 7.9 as G_s. When these same agonists bind to the α_2 adrenergic receptor, they inhibit the activity of adenylate cyclase via an inhibitory G protein (G_i).

When activated, adenylate cyclase converts an ATP molecule into cyclic AMP (cAMP) and inorganic pyrophosphate (PPi). The cAMP has a single phosphate group that is attached both to the 5' and the 3' carbon atoms of the sugar ribose, thus the term cyclic. Cyclic AMP binds to protein kinase A and activates it as follows. Normally, protein kinase A exists as a tetramer, with two regulatory (R) and two catalytic (C) subunits held together by noncovalent bonds. In this form, protein kinase A is inactive. Binding of cyclic AMP to the R subunits causes them to dissociate from the catalytic (C) subunits. Thus freed, the catalytic subunits attach a phosphate group from ATP to hormone-sensitive lipase (HSL), forming a phosphorylated enzyme (HSL-P). The phosphorylated form is active in breaking down triglyceride molecules, as shown previously.

The cAMP has a finite lifetime within the cell. An enzyme known as cyclic AMP phosphodiesterase (PDE) breaks the bond that attaches the phosphate to the 3' carbon of ribose, producing 5'-AMP, thus stopping the cAMP signal. The PPi formed when ATP is converted into cAMP is hydrolyzed by inorganic pyrophosphatase to form two Pi. Cyclic AMP is also known as a **second messenger**, the first messenger being the hormone epinephrine or the neurotransmitter norepinephrine. The balance between the number of stimulatory (R_β) and inhibitory ($R_{\alpha 2}$) receptors helps determine the final response of fat cells to the agonists epinephrine and norepinephrine.

> ***KEY POINT***
>
> Triglyceride is also stored in muscle fibers, especially the ST (type II) fibers. Regulation of intramuscular triglyceride lipolysis follows a similar pattern as discussed above with epinephrine playing a key role.

There are other factors besides the catecholamines that can influence lipolysis. **Growth hormone**, released from the anterior pituitary gland, strongly facilitates lipolysis. This hormone is released in pulses at the onset of sleep and following meals. The pulsatile bursts in growth hormone are followed one to two hours later by measurable increases in both free fatty acids and glycerol in the blood. Although the precise mechanism of growth hormone action in vivo is not clear, it does increase the activity of adenylate cyclase. Since there is a delay of at least an hour, this suggests that growth hormone is influencing the expression of one or more genes, leading to subsequent changes in the content of protein with lipolytic activity. The effect of growth hormone on lipolysis has been known for years, which is reflected in its use by athletes wishing to increase muscle mass and to decrease fat mass of their bodies.

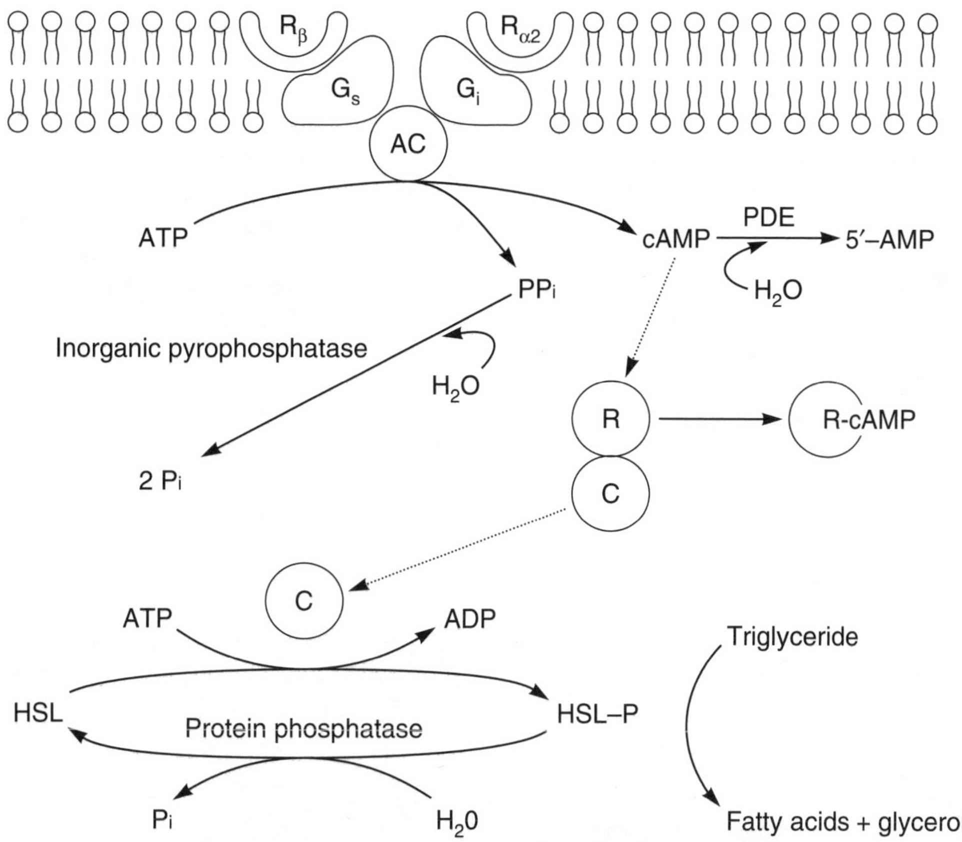

Figure 7.9 A highly schematic view of the regulation of hormone-sensitive lipase (HSL) in a fat cell. An agonist binds to its specific receptor, indicated as R_β or $R_{\alpha2}$. These in turn are connected to the enzyme adenylate cyclase (AC) by either a stimulatory (G_s) or inhibitory (G_i) G protein. Receptors, G protein, and AC are found in the cell membrane, shown as the lipid bilayer. Agonists binding to R_β activate AC. Cyclic AMP (cAMP), formed by AC, activates a protein kinase consisting of regulatory (R) and catalytic (C) subunits. The catalytic subunit activates the inactive form of HSL by attaching a phosphate group donated by ATP. The active or HSL-P form now breaks down triacylglycerol to fatty acids and glycerol. The cAMP signal ceases when cAMP phosphodiesterase (PDE) breaks it down to 5'-AMP.

Thyroid stimulating hormone (TSH), also from the anterior pituitary, has a positive influence on lipolysis. TSH acts on the thyroid gland to stimulate release of thyroid hormones. Hypothyroid humans have a blunted lipolytic response to catecholamines, whereas individuals with hyperthyroidism have a potentiated response. It is believed that the thyroid hormones play their role at the level of the protein kinase A or the hormone-sensitive lipase, rather than on influencing the number of alpha or beta adrenergic receptors.

Male sex hormones, androgens (e.g., testosterone), increase the sensitivity of the catecholamines (adrenaline and noradrenaline), so that androgens may be said to have a permissive effect on lipolysis; the effect of androgens is mild. Castration is associated with a blunted adrenergic-dependent response. This can be reversed by testosterone administration. Androgens influence fat cell metabolism by binding to intracellular receptors. Glucocorticoids such as cortisol, released from the adrenal cortex, can potentiate lipolysis. As mentioned in the previous chapter, cortisol concentration increases with exercise duration and intensity.

There are negative effectors for lipolysis. Adenosine, produced locally from ATP, can bind to its own receptor on the adipocyte membrane. This receptor is coupled to the G_i, the inhibitory G protein that depresses the activity of adenylate cyclase, resulting in inhibition of lipolysis.

Insulin is the most potent inhibitor of lipolysis. We have already seen that it stimulates triglyceride synthesis, promoting esterification of fatty acids released during lipolysis by enhancing the availability of glycerol 3-phosphate. Insulin depresses lipolysis at several sites. One of the effects of insulin on lipolysis is to stimulate the activity of cyclic AMP phosphodiesterase (PDE), which degrades cyclic AMP, by promoting the phosphorylation of a specific serine residue in PDE. Insulin also promotes the dephosphorylation of the active (phosphorylated) form of HSL. In addition to decreasing the rate of lipolysis, insulin also promotes triacylglycerol synthesis by promoting glucose uptake into fat cells.

KEY POINT

Lipolysis is accelerated as exercise intensity increases to moderate submaximal levels. However, there is not a good correspondence for all exercise intensities since the rate of appearance of fatty acids and fat oxidation are depressed during higher intensities of exercise even below $\dot{V}O_2$max. Since epinephrine increases continuously as exercise intensity increases, the factors regulating the rate of lipolysis and release of fatty acids from adipose tissue are complicated.

Estrogens also function by binding to intracellular receptors. The number of these receptors varies in fat depots from different sites, pointing to the role of estrogen in promoting fat deposition at specific locations. For example, the gluteofemoral fat deposits are augmented with the rise in estrogen in pubescent females. Estrogens have a relatively mild effect to reduce the lipolytic effect of catecholamines on fat cells.

KEY POINT

A number of studies have demonstrated that there is a greater proportion of fat oxidized during the luteal phase of the menstrual cycle compared to other phases when women are studied at the same exercise intensity. This period coincides with the end of the peak of progesterone release, about one week following the peak of estrogen release. Obviously, accounting for the mechanisms favoring lipid oxidation in the luteal phase are complicated.

Gender Differences in Lipolysis

Some studies have demonstrated that during exercise at the same relative intensity women have higher glycerol concentrations than men, suggesting an increased rate of lipolysis. Several recent studies have demonstrated that glycerol and FFA levels are significantly higher in women compared to men during exercise at the same relative intensity. Although FFA and glycerol concentrations in women during exercise are up to double those of men, adrenaline concentration was higher in men while plasma noradrenaline and insulin concentrations were similar. These results suggest that lipolysis is more sensitive to adrenergic agents in women.

KEY POINT

Gender comparisons for metabolic responses to exercise can be complicated by a variety of factors, including differences in the training state of subjects, generally larger relative fat deposits in women compared to men, and menstrual cycle influences on fuel utilization. One recent study by Romijn showed that gender differences in fuel use during moderate and high intensity submaximal exercise were not observed when trained women were compared with trained men.

Oxidation of Fatty Acids

To obtain the free energy stored in fatty acid molecules, they must be transported from the blood, across the cell membrane of the utilizing cell, through the cytosol of the cell, across the inner mitochondrial membrane, and into the matrix. There, the fatty acids will be broken down into two-carbon acetyl groups attached to CoA by beta-oxidation. We will look at these steps in detail.

Intracellular Transport of Fatty Acids

Fatty acids enter cells with the aid of transporters that allow the fatty acids to be moved down their concentration gradient into the cytosol of fat utilizing cells. With its ability to increase its metabolic rate to high levels, muscle is of primary importance for oxidizing fatty acids. A variety of transporters have been identified in skeletal muscle, including a plasma membrane fatty acid binding protein ($FABP_{pm}$), a fatty acid translocase (FAT), and a fatty acid transport protein (FATP).

Because of their insolubility, fatty acids in the cell cytoplasm of many tissues are attached to a cytosolic fatty acid binding protein ($FABP_c$). In muscle, fatty acids attached to $FABP_c$ can be those originating in adipose tissue and taken up from the blood or those released when intramuscular triglyceride is hydrolyzed. To be oxidized in the mitochondria, fatty acids must first be activated, forming a fatty acyl CoA, which we have seen earlier in this chapter. This reaction is shown below:

$$\text{fatty acid} + \text{ATP} + \text{CoA} \xrightarrow[\text{synthetase}]{\text{acyl CoA}} \text{fatty acyl CoA} + \text{AMP} + \text{PPi}$$

This reaction is essentially irreversible because the PPi is hydrolyzed by inorganic pyrophosphatase to two Pi (not shown), which drives the reaction to the right. The fatty acyl CoA that is formed is an energy-rich molecule, much like acetyl CoA. Formation of a fatty acyl CoA costs two ATP because two phosphates are removed from ATP.

Transport as Acyl Carnitine

Formation of fatty acyl CoA (often called simply acyl CoA) occurs in the cytosol, whereas oxidation of the fatty acyl CoA occurs in the mitochondrial matrix. However, the mitochondrial inner membrane is impermeable to CoA and its derivatives, which permits separate regulation of CoA compounds in mitochondrial and cytosolic compartments. Transport of fatty acyl CoA into the mitochondrial matrix occurs using three different proteins and the small molecule **carnitine** (see figure 7.10). Only fatty acids attached to carnitine are able to cross the inner mitochondrial membrane.

Carnitine, formed from the amino acids lysine and methionine, can cross the inner membrane, and a fatty acyl form of carnitine can also cross in the opposite direction. The carnitine acyl carnitine translocase is called the carnitine acyl carnitine antiport because this membrane protein transfers two different substances across the inner membrane in opposite directions. The enzyme carnitine acyl transferase exists in two forms, located on

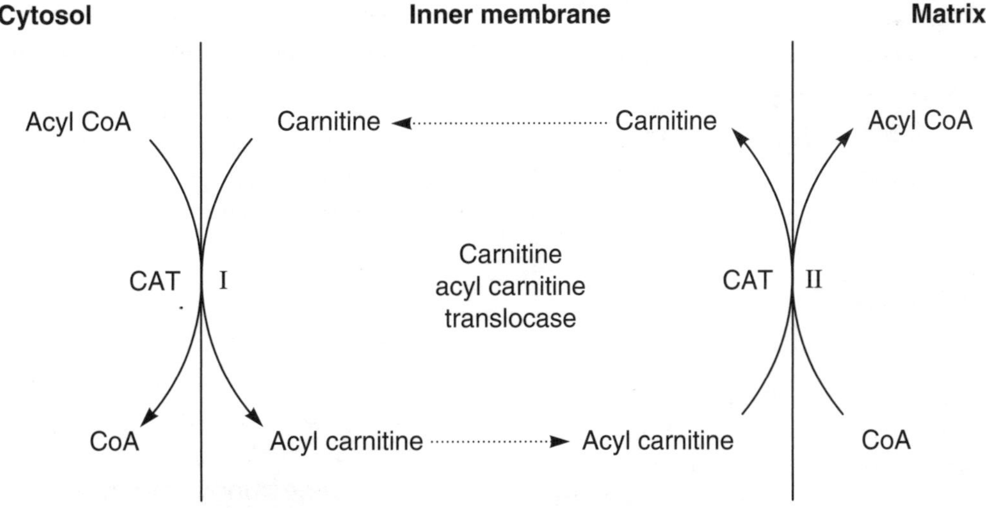

Figure 7.10 Fatty acyl groups are transported across the inner mitochondrial membrane for oxidation when attached to the small carrier molecule carnitine. The translocase moves the carnitine and acyl carnitine in the direction shown by the dotted arrows. Attachment from coenzyme A (CoA) to carnitine and from carnitine back to CoA is catalyzed by the enzyme carnitine acyl transferase, located on the outer (represented by CAT I) and the matrix side of the inner membrane (represented by CAT II).

the cytosolic side (CAT I) and matrix side of the inner membrane (CAT II). CAT I transfers the fatty acyl group from CoA to carnitine on the cytosolic side of the inner membrane, while CAT II transfers the fatty acyl group from carnitine to CoA. In some sources, CAT is identified as carnitine palmitoyl transferase (CPT), as the specificity for the 16-carbon palmitate is high. As we will see later, CAT I can be inhibited by malonyl CoA, which helps to regulate fatty acid oxidation.

Carnitine deficiency, due to the body's inability to convert lysine into carnitine, is not an unusual metabolic disease. Patients have muscle weakness and poor exercise tolerance due to accumulation of triglyceride in muscle and the inability to oxidize fatty acids. Carnitine supplementation by some athletes has been tried to enhance lipid oxidation and spare carbohydrate utilization.

KEY POINT

The carnitine transport system is necessary for the long chain fatty acids (16–20 carbon atoms) that are typically stored as triglyceride in adipose tissue and skeletal muscle. Medium chain triglycerides (MCT) contain fatty acids that have a length of 6 to 10 carbon atoms. MCT are more rapidly digested and their fatty acids absorbed because these are not incorporated into chylomicrons. In addition, the medium chain fatty acids can enter mitochondria for oxidation without the need for carnitine.

Beta-Oxidation of Saturated Fatty Acids

The initial process in the oxidation of fatty acids is known as beta-oxidation (β-oxidation), which occurs in the matrix of the mitochondrion. Beta-oxidation begins as soon as the fatty acyl CoA appears in the matrix, using repeated cycles of four steps, so that with each cycle the fatty acyl CoA is broken down to form a new fatty acyl CoA, shortened by two carbon atoms, plus an acetyl CoA. Figure 7.11 shows the four steps in β-oxidation.

The first reaction is catalyzed by the enzyme acyl CoA dehydrogenase, located on the matrix side of the inner membrane of the mitochondrion. The asterisk (*) in figure 7.11 indicates the β-carbon atom of the fatty acyl CoA. This is also carbon 3, numbering from the acyl carbon, or carbon 1 of the fatty acid. In this reaction, two hydrogen atoms, each with an electron, are removed and transferred to FAD, making it $FADH_2$. These electrons are then passed through an electron transfer from flavoprotein to coenzyme Q. The removal of the two hydrogen atoms results in the formation of a carbon-to-carbon double bond between the α-carbon (or carbon 2) and the β-carbon (or carbon 3) of the fatty acyl group. The double bond is trans because the two hydrogens are on opposite sides.

In the second reaction trans-enoyl CoA is hydrated, accepting a water molecule. The OH part of the water is added to the β-carbon while the other hydrogen atom of water is added to the α-carbon. The product is a 3-hydroxyacyl CoA (or β-hydroxyacyl CoA). The enzyme, enoyl CoA hydratase, is located in the matrix. In the third reaction, catalyzed by the matrix enzyme 3-hydroxyacyl CoA dehydrogenase, the β-carbon (carbon 3) of the 3-hydroxyacyl CoA is oxidized, losing a hydride ion and a proton. Such oxidation reactions always use the nicotinamide coenzymes, NAD^+ in this case. This reaction changes the OH group to a ketone and is therefore a β-oxidation. The enzyme 3-hydroxyacyl CoA dehydrogenase could also be named β-hydroxyacyl CoA dehydrogenase.

In the last step CoA attaches to the β-carbon (i.e., carbon 3), which allows carbons 1 and 2 to come off as an acetyl CoA, leaving a new acyl CoA shortened by two carbon atoms. The enzyme responsible for the fourth reaction is acyl CoA thiolase (called thiolase because the CoA contains a terminal SH [thiol] group, as discussed in the section on the beginning of the TCA cycle).

KEY POINT

The steps in the TCA cycle from succinate to oxaloacetate are remarkably similar in nature to the first three steps in beta-oxidation. For both, there is a dehydrogenation, generating a carbon-to-carbon double bond (fumarate in the TCA cycle), a hydration (forming malate), and then a further dehydrogenation generating the keto group in oxaloacetate.

The new acyl CoA now undergoes the same four enzyme-catalyzed steps, creating another acetyl CoA and a new acyl CoA, again shortened by two

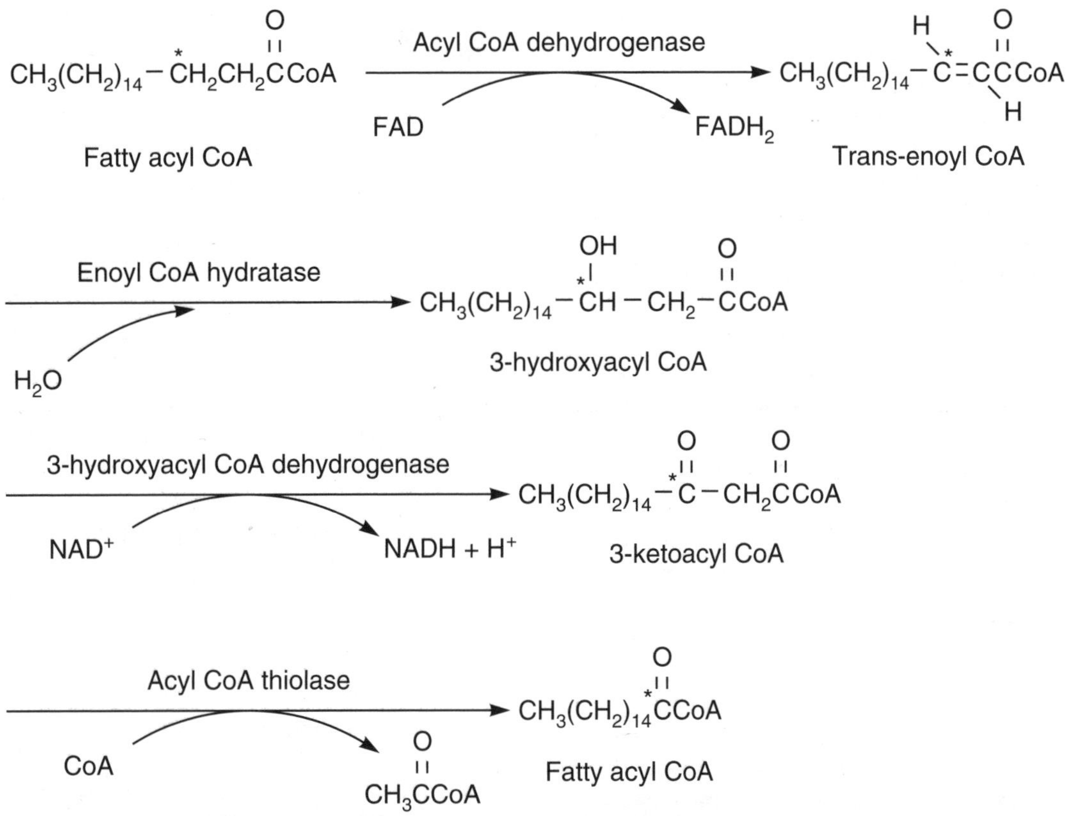

$$\underset{\text{Fatty acyl CoA}}{CH_3(CH_2)_{14}-\overset{*}{C}H_2CH_2\overset{O}{\overset{\|}{C}}CoA} \xrightarrow[FAD \;\rightarrow\; FADH_2]{\text{Acyl CoA dehydrogenase}} \underset{\text{Trans-enoyl CoA}}{CH_3(CH_2)_{14}-\overset{*}{C}H=CH\overset{O}{\overset{\|}{C}}CoA}$$

$$\xrightarrow[H_2O]{\text{Enoyl CoA hydratase}} \underset{\text{3-hydroxyacyl CoA}}{CH_3(CH_2)_{14}-\overset{*}{C}H(OH)-CH_2-\overset{O}{\overset{\|}{C}}CoA}$$

$$\xrightarrow[NAD^+ \;\rightarrow\; NADH + H^+]{\text{3-hydroxyacyl CoA dehydrogenase}} \underset{\text{3-ketoacyl CoA}}{CH_3(CH_2)_{14}-\overset{*}{\overset{O}{\overset{\|}{C}}}-CH_2\overset{O}{\overset{\|}{C}}CoA}$$

$$\xrightarrow[CoA \;\rightarrow\; CH_3\overset{O}{\overset{\|}{C}}CoA]{\text{Acyl CoA thiolase}} \underset{\text{Fatty acyl CoA}}{CH_3(CH_2)_{14}\overset{*}{\overset{O}{\overset{\|}{C}}}CoA}$$

Figure 7.11 The four enzyme-catalyzed reactions that comprise the beta-oxidation process in which fatty acyl coenzyme A (acyl CoA) is reduced to acetyl CoA units. In this example, the starting material is the 18-carbon fatty acyl CoA known as stearoyl CoA. Eight cycles of beta-oxidation reduce stearoyl CoA to nine acetyl CoA units. The * identifies the beta carbon (carbon atom 3) that is oxidized.

carbon atoms. Eventually the original fatty acyl CoA, which in figure 7.11 contained 18 carbon atoms, is reduced to nine acetyl CoA units by eight cycles of β-oxidation. The acetyl CoA units can each feed into the TCA cycle. The reactions in figure 7.11 are shown as if they are irreversible. Actually, the free energy changes overall are modest, and the reactions could rightfully be described as equilibrium (reversible). Direction is given to β-oxidation by the fact that the acetyl groups feed into the TCA cycle. There is a considerable energy yield from β-oxidation of fatty acids. For example, from the 18-carbon stearic acid, the yield is 9 acetyl CoA, 8 $FADH_2$, and 8 NADH.

KEY POINT

When calculating the ATP yield from the oxidation of a particular fatty acid, one must account for the ATP generated with electron transport from the $FADH_2$ and NADH produced during beta-oxidation as well as the acetyl CoA units that will be substrate for the TCA cycle. Finally, remember to account for the fact that formation of fatty acyl CoA costs the equivalent of two ATP.

Oxidation of Unsaturated Fatty Acids

Many fatty acids stored in body fat are unsaturated. These fatty acids are also sources of energy in the form of acetyl CoA units and reduced coenzymes during β-oxidation. For oxidation of oleic acid an additional step is required when the β-oxidation process reaches the double bond because this carbon-to-carbon double bond is the cis configuration, whereas enoyl CoA hydratase can act only on trans double bonds. A new enzyme, enoyl CoA isomerase, converts the cis to a trans carbon-to-carbon double bond and β-oxidation continues.

Other unsaturated fatty acids, such as linoleic acid, have their double bonds in the wrong position as well as the wrong (cis) configuration for β-oxidation. For these, enoyl CoA isomerase converts the cis to a trans double bond. An enzyme known as a reductase converts the carbon-to-carbon double bond in the wrong position into a carbon-to-carbon single bond by the addition of hydrogen in the form of NADPH and H^+. As we have seen, NADPH is closely related to NADH and serves to reduce double bonds during synthetic reactions.

Oxidation of Odd Carbon Fatty Acids

Most of the fatty acids found in land animals contain an even number of carbon atoms. However, plants and marine animals contain fatty acids with an odd number of carbon atoms. Oxidation of these odd carbon fatty acids occurs through the normal route of β-oxidation. In the last cycle of β-oxidation, the starting substrate is a five-carbon acyl CoA. When this is split by the acyl CoA thiolase enzyme, the result is acetyl CoA and propionyl CoA, which contains the three-carbon propionyl group attached to CoA. Propionyl CoA is not treated as a fat. In fact, it is converted into the TCA intermediate succinyl CoA in a three-step process. This means that propionyl CoA generates a gluconeogenic precursor (see chapter 6).

Oxidation of Ketone Bodies

Figure 7.12 shows the three ketone bodies. D-3-hydroxybutyrate and acetoacetate are formed primarily in the matrix of liver mitochondria and to a minor extent in the kidneys. Acetoacetate undergoes slow, spontaneous decarboxylation to yield acetone and carbon dioxide. Both 3-hydroxybutyrate and acetoacetate are water soluble. A small amount of acetone is formed by decarboxylation of acetoacetate, and because it is volatile, its presence, like that of alcohol, can be smelled in the breath.

Formation of Ketone Bodies

Ketone body formation accelerates in normal individuals when the body carbohydrate content is extremely low, as during starvation or self-controlled fasting, when extremely low carbohydrate diets are eaten, and during prolonged exercise with insufficient carbohydrate ingestion. In all of these conditions, carbohydrate content in the body is low, and thus carbohydrate utilization and blood insulin concentration are low. Accelerated ketone body formation also occurs during uncontrolled diabetes mellitus. Although blood glucose concentration is high because insulin is lacking, carbohydrate utilization by insulin-dependent tissues is low.

With starvation or fasting, low carbohydrate diets, prolonged exercise, or uncontrolled diabetes mellitus, the adipose tissue releases large quantities of fatty acids due to an imbalance between triglyceride formation and lipolysis caused by low blood insulin concentration. Recall that insulin promotes triglyceride formation in fat cells and inhibits lipolysis. Thus the net effect of low blood insulin is that lipolysis greatly exceeds triglyceride formation, resulting in a large increase in blood free fatty acid (FFA) concentration. In uncontrolled diabetes mellitus, for example, blood FFA levels are extremely high.

Under normal conditions, the liver is able to extract about 30% of the FFA that pass through it. With high blood FFA concentration, the liver extracts even more. The fate of the extracted fatty acids by liver is formation of fatty acyl CoA,

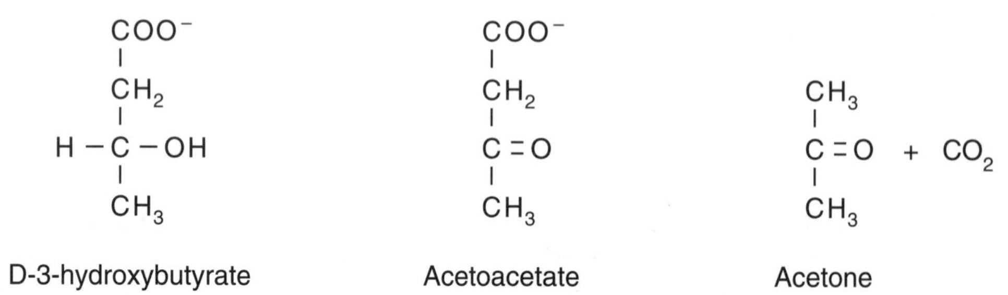

$$\begin{array}{ccc} COO^- & COO^- & \\ | & | & \\ CH_2 & CH_2 & CH_3 \\ | & | & | \\ H-C-OH & C=O & C=O \;+\; CO_2 \\ | & | & | \\ CH_3 & CH_3 & CH_3 \end{array}$$

D-3-hydroxybutyrate   Acetoacetate   Acetone

Figure 7.12 The three ketone bodies. Both D-3-hydroxybutyrate and acetoacetate are produced in and released from the liver. Acetone is formed by the spontaneous decarboxylation of acetoacetate.

then subsequent formation of triglyceride or phospholipid, or entry into the mitochondrial matrix. During conditions favoring ketone body formation, entry of fatty acyl CoA into liver mitochondria is accelerated. Beta-oxidation of the fatty acyl CoA is greatly augmented, forming acetyl CoA at a rate that far exceeds the capacity of the liver mitochondria to oxidize it by the TCA cycle. Moreover, conditions favoring ketone body formation are characterized by low matrix concentrations of oxaloacetate in the liver. Accordingly, much of the acetyl CoA is directed to the formation of acetoacetate. Some acetoacetate is reduced to D-3-hydroxybutyrate. Figure 7.13 summarizes ketone body formation. The two major ketone bodies formed are acetoacetate and D-3-hydroxybutyrate. The ratio of 3-hydroxybutyrate to acetoacetate depends on the [NADH]/[NAD^+] ratio in the liver. Of course, acetone, the third member of the ketone body family, is formed primarily by spontaneous, nonenzymatic decarboxylation of acetoacetate.

The Fate of Ketone Bodies

In the liver, acetoacetate is a precursor for cholesterol synthesis. However, most ketone bodies are used in extrahepatic tissues (i.e., nonliver tissues). Ketone bodies, mainly acetoacetate and 3-hydroxybutyrate, are fuels for oxidation by mitochondria in a variety of tissues, chiefly skeletal muscle, heart, and the brain. Normally, the brain uses glucose as its primary fuel because FFA cannot pass the blood-brain barrier. However, as blood glucose concentration falls, and blood ketone body concentration rises, the brain can extract and use 3-hydroxybutyrate and acetoacetate. Ketone body oxidation in tissues (see figure 7.14) occurs as follows:

1. Ketone bodies are taken up by extrahepatic cells and transported into the mitochondrial matrix.
2. D-3-hydroxybutyrate is oxidized to acetoacetate using the enzyme 3-hydroxybutyrate dehydrogenase. This reaction generates acetoacetate and NADH.
3. Acetoacetate takes a CoA from succinyl CoA and forms acetoacetyl CoA.
4. Acetoacetyl CoA is split into two acetyl CoA using the last enzyme of β-oxidation of fatty acids, acyl CoA thiolase.
5. The acetyl CoA is oxidized in the TCA cycle.

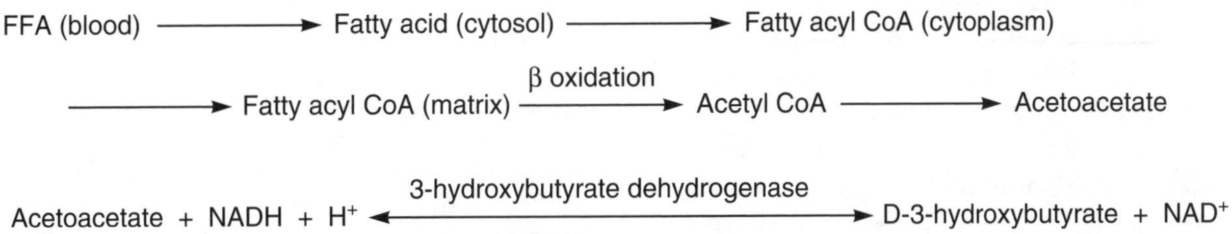

Figure 7.13 The steps in the formation of ketone bodies, starting with an increased blood free fatty acid concentration and leading to the formation of acetoacetate in the liver mitochondrial matrix. Acetoacetate can be reduced to D-3-hydroxybutyrate; thus the liver produces both acetoacetate and 3-hydroxybutyrate.

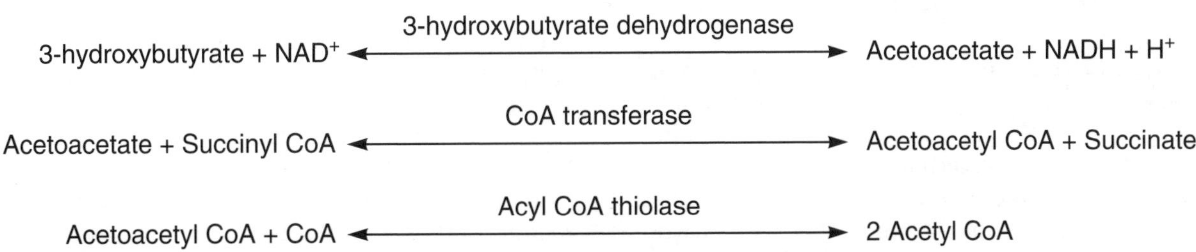

Figure 7.14 The fate of 3-hydroxybutyrate, taken up by a cell and catabolized to acetyl CoA in three major steps in the mitochondrial matrix.

Ketosis

Prolonged depletion of body carbohydrate stores or uncontrolled diabetes mellitus leads to a condition known as ketosis. This results from accelerated ketone body formation and is characterized by the following:

1. **Ketonemia**—an increase in ketone body concentration in the blood.
2. **Ketonuria**—loss of ketone bodies in the urine.
3. Acetone breath—loss of volatile acetone in expired air.
4. Elevated blood FFA concentration, resulting in accelerated formation of ketone bodies and low blood insulin.
5. **Acidosis** (or ketoacidosis)—a fall in blood pH due to the formation of H^+ when acetoacetate is formed and the loss of cations (e.g., Na^+) when ketone bodies are excreted. This can be fatal in uncontrolled diabetes mellitus.
6. Hypoglycemia (in normal persons) or hyperglycemia (in uncontrolled diabetes mellitus).

KEY POINT

Ketones can be a useful fuel during submaximal exercise, sparing the use of muscle glycogen and blood glucose. Use of ketone bodies to fuel exercise performance is enhanced with a week or more of a ketogenic diet. This increases the activity of enzymes needed to form acetyl CoA in mitochondria from ketone bodies. The net effect is to reduce the need to provide the TCA cycle with acetyl CoA from pyruvate.

Synthesis of Fatty Acids

Most fatty acids used by humans come from dietary fat. However, humans can synthesize fatty acids from acetyl CoA in liver, mammary gland, and adipose tissue, although this is not a major process in people with a relatively high-fat diet. The acetyl CoA comes from amino acids, carbohydrates, and alcohol, but carbohydrate will normally provide the bulk of the carbon atoms for fatty acid synthesis. The synthesis of fatty acids is often described as de novo lipogenesis, which means synthesis of fatty acids starting from the beginning. The 16-carbon palmitic acid is synthesized first and can be extended in length or desaturated to make some unsaturated fatty acids (but not the essential fatty acids needed in the diet). Synthesis of palmitic acid occurs in the cytosol, whereas oxidation occurs in the mitochondria. Synthesis and degradation of fatty acids also use different enzymes, permitting separate regulation of these two opposing processes.

Pathway to Palmitic Acid

Synthesis starts with the carboxylation of acetyl CoA to make a three-carbon molecule known as malonyl CoA (see figure 7.15). The cytosolic enzyme acetyl CoA carboxylase catalyzes this reaction, which involves the carboxylation of acetyl CoA. The actual substrate is the bicarbonate ion (HCO_3^-), which is transferred to acetyl CoA by the B vitamin biotin. This reaction is the committed step in fatty acid synthesis and is positively affected by insulin. We have also seen biotin involved in the pyruvate carboxylase reaction in gluconeogenesis. The malonyl CoA units are used to make fatty acids in the cytosol.

To make malonyl CoA in the cytosol, it is necessary to have a continuous supply of acetyl CoA. Recall that acetyl CoA is formed in the mitochondrial matrix primarily from pyruvate in the pyruvate dehydrogenase reaction and β-oxidation of fatty acids. Thus there must be a mechanism to get acetyl CoA from the mitochondrial matrix to the cytosol. Moreover, there is a need for NADPH in fatty acid synthesis, so there must be a way of producing this in the cytosol. NADPH can arise from the pentose phosphate pathway (see chapter 6). It can also be produced by malic enzyme:

$$\text{malate} + \text{NADP}^+ \xleftrightarrow{\text{malic enzyme}} \text{pyruvate} + CO_2 + \text{NADPH} + H^+$$

Figure 7.16 outlines how an excess of glucose can be used to make malonyl CoA, the precursor for fatty acid synthesis. Glycolysis converts glucose to pyruvate. It is translocated into the matrix and converted to acetyl CoA by pyruvate dehydrogenase. Acetyl CoA combines with oxaloacetate to produce citrate, using the first enzyme of the TCA cycle, citrate synthase. The citrate is transported into the cytosol by an antiport, with malate simultaneously going from

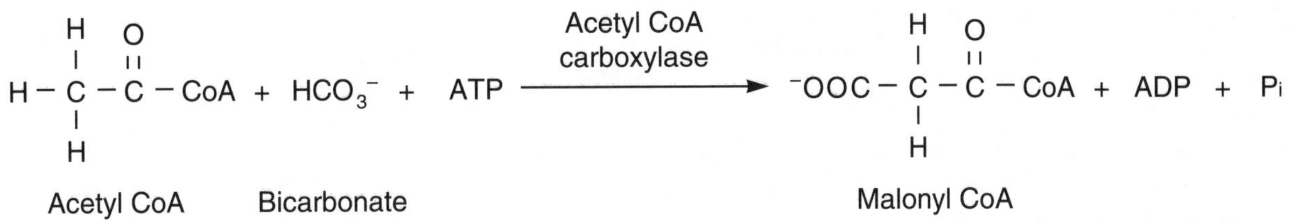

$$\mathrm{H{-}\overset{\overset{H}{|}}{\underset{\underset{H}{|}}{C}}{-}\overset{\overset{O}{\|}}{C}{-}CoA} + \mathrm{HCO_3^-} + \mathrm{ATP} \xrightarrow{\text{Acetyl CoA carboxylase}} \mathrm{{}^-OOC{-}\overset{\overset{H}{|}}{\underset{\underset{H}{|}}{C}}{-}\overset{\overset{O}{\|}}{C}{-}CoA} + \mathrm{ADP} + \mathrm{P_i}$$

Acetyl CoA Bicarbonate Malonyl CoA

Figure 7.15 Malonyl coenzyme A (CoA), the precursor for fatty acid synthesis, is formed by carboxylation of acetyl CoA.

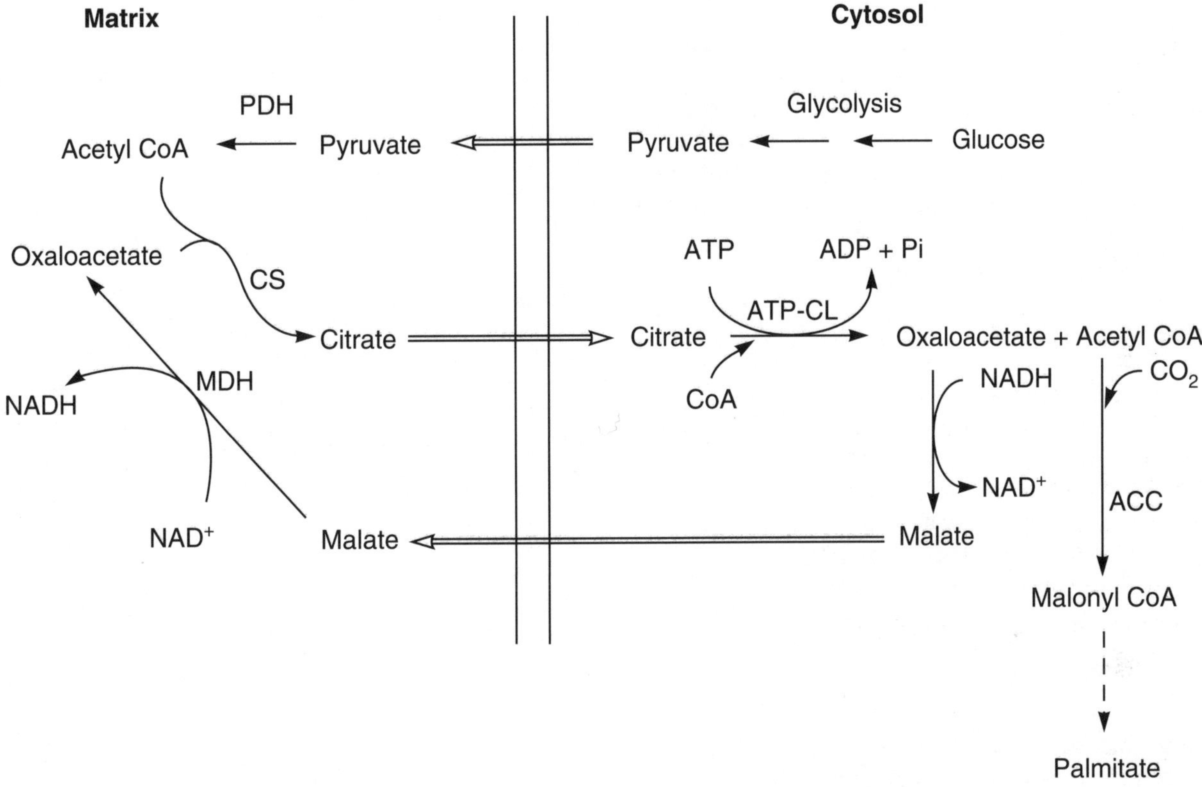

Figure 7.16 Conversion of glucose into fatty acids is a process that starts in the cytosol with glucose being converted to pyruvate in glycolysis. The pyruvate is translocated into the mitochondrial matrix and converted to acetyl CoA by pyruvate dehydrogenase (PDH). Since acetyl CoA cannot cross the inner mitochondrial membrane, it combines with oxaloacetate to make citrate, catalyzed by citrate synthase (CS). Citrate crosses the inner membrane to the cytosol using an antiport, with malate going the other way. In the cytosol, citrate is split into acetyl CoA and oxaloacetate by ATP-citrate lyase (ATP-CL). The acetyl CoA is carboxylated to malonyl CoA by acetyl CoA carboxylase (ACC). The oxaloacetate is reduced to malate, which returns to the matrix using the citrate-malate antiport.

the cytosol to the matrix. In the cytosol, citrate is cleaved into oxaloacetate and acetyl CoA by ATP-citrate lyase. This enzyme uses the free energy of hydrolysis of ATP to not only split citrate, but also to attach a CoA. The acetyl CoA in the cytosol is now a substrate for acetyl CoA carboxylase. The oxaloacetate is reduced to malate and transported back into the matrix by the citrate-malate antiport.

Palmitic acid synthesis occurs using a large enzyme, fatty acid synthase (FAS), which is composed of two very long, multifunctional polypeptide chains. The fatty acid synthase has seven distinct enzyme activities. Although we do not have the space to cover the details of palmitic acid synthesis, figure 7.17 shows the general process. To begin, an acetyl CoA enters, and the acetyl group binds to part of the fatty acid synthase. Next a malonyl CoA enters, and the malonyl group binds to another part of the fatty acid synthase. In a complicated mechanism, the acetyl and malonyl groups combine in a condensation reaction, the product of which binds to a domain of the FAS known as the acyl carrier protein (ACP). In this process, CO_2 is lost, resulting in the formation of a four-carbon molecule with two keto groups; the first of these is attached to the ACP. Following a reduction with NADPH, a dehydration, and a further reduction with NADPH, the second of the two keto groups is changed to a hydrocarbon. Addition of a new malonyl CoA followed by condensation to the four-carbon ACP attachment, and then decarboxylation, produces a six-carbon unit with two keto groups. One of these keto groups is again eliminated by a sequence of reduction with NADPH, dehydration, and reduction with NADPH. The process of condensation with a new malonyl CoA and elimination of the keto group continues until a 16-carbon atom palmitate is produced, still attached to ACP. Following hydrolysis from the ACP, palmitic acid is released.

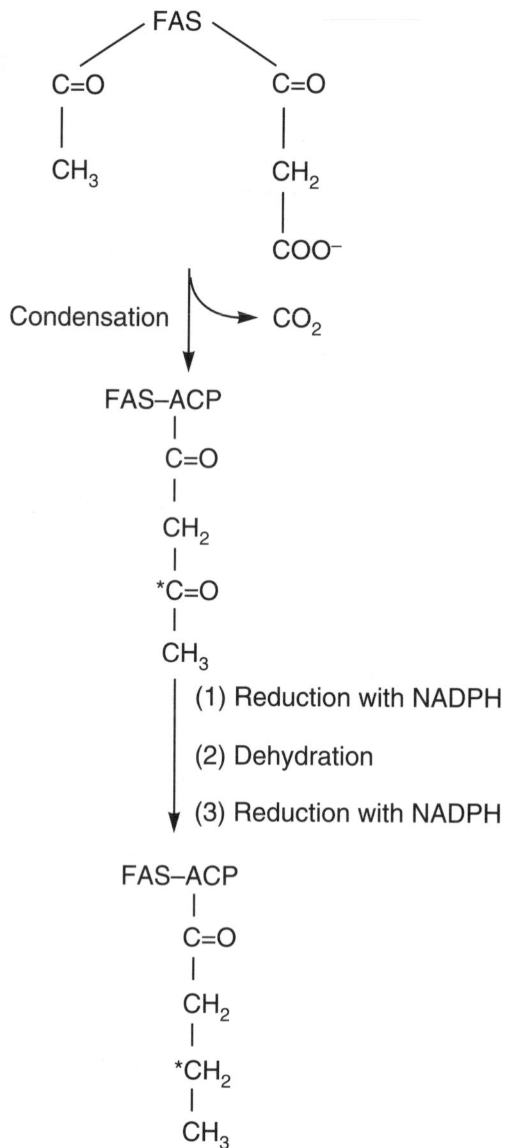

Figure 7.17 Fatty acids are synthesized on an enzyme known as fatty acid synthase (FAS). We start with an acetyl group contributed from acetyl CoA and a malonyl group contributed from malonyl CoA, each attached to a separate part of the FAS. The acetyl group condenses with the malonyl group and binds to a new domain on the FAS called the acyl carrier protein (ACP). In the condensation, CO_2 is eliminated. The beta keto group shown with the asterisk is eliminated in three steps, involving two reductions using NADPH before and following a dehydration. Condensation of another malonyl CoA, followed by reduction, dehydration, and reduction produces a six-carbon molecule attached to the ACP domain. This process continues until a 16-carbon palmityl group is formed.

Regulation of Fatty Acid Synthesis

Control of fatty acid synthesis in liver occurs through regulation of malonyl CoA formation by its enzyme acetyl CoA carboxylase (ACC). Citrate is a positive allosteric effector for acetyl CoA carboxylase, whereas fatty acyl CoA and malonyl CoA allosterically inhibit ACC. The potency of the inhibition by the fatty acyl CoA molecules is proportional to the carbon length of the fatty acyl units. An abundance of fatty acyl CoA and malonyl CoA signals a surfeit of products and therefore no need to make more fatty acids. On the other hand, an increase in citrate concentration indicates an overabundance of acetyl CoA, indicating a need to synthesize fatty acids. An abundance of carbohydrate means fatty acid synthesis from carbohydrate should take place. An accumulation of fatty acyl CoA reflects a lack of carbohydrate substrate and thus fatty acid oxidation should predominate.

Acetyl CoA carboxylase activity is inhibited by covalent phosphorylation, directed by a cyclic AMP-dependent protein kinase (protein kinase A–PKA) and an AMP-dependent protein kinase (AMPK). Recall that cAMP is really 3',5'-AMP, whereas AMP is 5'-AMP, where the numbers point out where the phosphate group is attached to the ribose (see figure 3.2). Dephosphorylation by protein phosphatases activates acetyl CoA carboxylase. Phosphorylation and dephosphorylation of ACC depend on the relative levels of insulin and glucagon. The latter results in inhibition of ACC. A relative rise in insulin concentration leads to a decrease in cAMP concentration by activation of cyclic AMP phosphodiesterase; insulin also activates phosphoprotein phosphatases, which dephosphorylate acetyl CoA carboxylase. At another level, a diet rich in carbohydrates leads to increased expression of the ACC gene and therefore more ACC. Prolonged fasting or a diet low in carbohydrate leads to decreased expression of the ACC gene and thus less of the ACC protein. In summary, when there is abundant carbohydrate, the insulin concentration will be elevated and glucagon concentration will be depressed. These relative changes favor de novo lipogenesis through activation of acetyl CoA carboxylase. On the other hand, with relative carbohydrate deprivation, glucagon will be elevated and insulin decreased, leading to a blockage in the conversion of carbohydrate to fat.

KEY POINT

As mentioned, lipogenesis from acetyl CoA generated by excess carbohydrates and amino acids is not a prominent feature of human metabolism. It can occur with a diet high in carbohydrate and protein and low in fat content. The major location where this occurs is the liver. As mentioned previously, a diet rich in carbohydrates or protein promotes enhanced oxidation of these fuels. Beyond this, lipogenesis from excess carbohydrate and amino acids can be viewed as a mechanism to save the energy from these fuels.

Malonyl CoA is also a potent inhibitor of the oxidation of fatty acids. In tissues where fatty acids can be oxidized to acetyl CoA in the mitochondria and where fatty acids can be simultaneously synthesized in the cytosol (primarily liver), malonyl CoA can inhibit the initial transfer of the fatty acyl group from CoA to carnitine at the cytosolic side of the inner membrane using the enzyme carnitine acyl transferase I (see figure 7.10). This process prevents futile cycling in which fatty acids are actively broken down in the mitochondria while at the same time being synthesized in the cytosol. Such regulation is active in the liver and adipose tissue, the main sites for fatty acid synthesis in humans. On another level, a relative increase in the intake of carbohydrate in the diet induces the transcription of genes coding for lipogenic enzymes. A high-fat diet, on the other hand, results in a decrease in the transcription of genes for lipogenic enzymes.

Fat as a Fuel for Exercise

Human fat stores are huge compared to carbohydrate stores. Moreover, considering the importance of glucose as fuel for the brain, it is important to use fat instead of carbohydrate during exercise, if this is possible. The lipid used to fuel muscular work comes from fatty acids released from adipose tissue, traveling to muscle as FFA. Intramuscular triglyceride is also a significant source of fatty acids. A third possibility is fatty acids released from plasma triglycerides by lipoprotein lipase, although the evidence suggests that this can provide at most 10% of fat oxidized during exercise lasting an hour or more.

Free Fatty Acids and Exercise

Blood glucose levels are well maintained during exercise lasting up to 60 minutes. This is due to the careful match between uptake of glucose from the blood and release of glucose to the blood from the liver. The relative constancy of blood glucose concentration during exercise is not observed with the concentration of FFA. With exercise, there is an increase in lipolysis, brought about mainly by the increase in epinephrine in the blood, increased sympathetic nervous activation of adipocytes through norepinephrine, and the decrease in plasma insulin concentration. Figure 7.18 illustrates the effect of 90 minutes of exercise at a

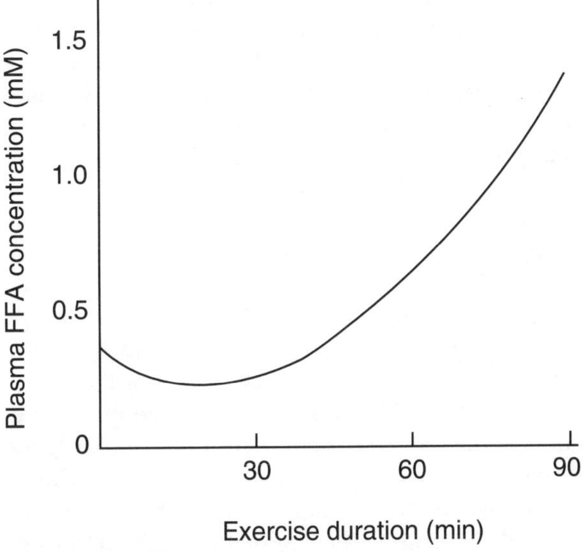

Figure 7.18 Plasma free fatty acid (FFA) concentration during 90 minutes of endurance exercise at a relative intensity of 60% of $\dot{V}O_2$max.

workload of 60% of $\dot{V}O_2$max on plasma FFA concentration for a subject in the postabsorptive state. The shape of the curve reflects the relative balance between release of fatty acids from adipose tissue and FFA uptake by exercising muscle. At the start of submaximal exercise, there is an immediate increase in the rate of FFA uptake by the exercising muscle. This exceeds the more slowly responding increase in adipose tissue lipolysis, such that the plasma FFA initially falls. With the gradual increase in lipolysis, the rate of release of FFA from adipose tissue meets then exceeds the rate of uptake of plasma FFA. As a result, the concentration of plasma FFA tends to rise over the course of the exercise period.

During more intense exercise, release of fatty acids from adipose tissue is reduced even though there will be a greater adrenergic stimulus for lipolysis. Two mechanisms have been proposed to explain this. At higher intensities of exercise there is a higher concentration of lactate in blood. Researchers have proposed that lactate inhibits lipolysis, but more recent data demonstrate that lactate probably increases reesterification of fatty acids into adipose tissue triglycerides. This means that rather than being released from the fat cell to enter the blood, the fatty acid is esterified to a new triglyceride molecule. Thus there may be a greater rate of lipolysis at 85% of $\dot{V}O_2$max compared to 65%, but more of the fatty acids generated by the lipolysis are simply reesterified back to form triglyceride molecules such that the rate of appearance of FFA in the blood is reduced. Another explanation is that adipose tissue blood flow is reduced at higher intensities of exercise, since blood flow regulation will attempt to direct flow to the active muscles by decreasing flow to the liver, kidney, and adipose tissues. This effect is related to the relative intensity of exercise. Therefore, fatty acids released by lipolysis are less likely to enter the general circulation due to a lower adipose tissue blood flow.

To focus only on changes in plasma FFA concentration to indicate fat oxidation during exercise can be very misleading as this does not indicate what is happening with these fatty acids. We could measure the respiratory exchange ratio (RER) to determine the relative use of fat, since we know that the lower the value, the more that fat is being oxidized. Again, this does not indicate what tissues are using the fat, although with a large increase in metabolic rate because of the exercise, it is safe to assume that exercising skeletal muscle is the major tissue contributing to the RER value. We could measure the concentration of FFA in arterial and venous blood across an exercising limb to determine the relative use of plasma FFA. However, accurate measures of FFA uptake into skeletal muscle require good values for blood flow, and this is not a simple technique. Moreover, subcutaneous adipose tissue can be an active source of FFA to the blood of the exercising limb, again leading to complications in the interpretation of the arterial and venous FFA concentration differences. A technique that is proving quite helpful is to infuse a labeled fatty acid into a vein at a precise rate. By sampling blood before and during exercise for changes in the concentration of the labeled fatty acid, predictions can be made about the rate of disappearance of fatty acids from the blood.

Use of Intramuscular Triglyceride

The difference between the fat provided by lipolysis in adipose tissue and the total fat oxidized can be accounted for primarily by the oxidation of fatty acids produced within exercising skeletal muscle fibers due to lipolysis of their stored triglyceride. There are two common methods to assess intramuscular triglyceride (IMTG) use during exercise. One technique is to chemically determine the amount of triglyceride in muscle samples obtained before and after exercise. This technique has been used in numerous studies, but marked decreases in IMTG concentration over an exercise bout may not be observed. We know that fatty acids taken up into skeletal muscle during exercise have two major fates; they can be immediately oxidized or they can be used to form IMTG, which will later support metabolism of the fiber. The fact that some studies have failed to note a significant decline in IMTG does not prove that this is not used. It may mean that some of the fatty acids taken up from the blood have been esterified into IMTG.

A second technique to determine the use of stored IMTG during exercise is to determine the release of glycerol from exercising skeletal muscle. IMTG lipolysis produces fatty acids and glycerol. We assume that the former is oxidized and the latter is released, due to an absence of the enzyme glycerol kinase in skeletal muscle. However, there are new data suggesting that skeletal muscle can metabolize glycerol. If these data are interpreted correctly, it means that the actual use of IMTG may be underestimated if it is based only on the release of glycerol. Indeed, there are a number of early studies showing that IMTG can support up to one half of the fat needs of exercising individuals. This is indeed a significant contribution, especially given the small amount of intramuscular triglyceride compared to adipose tissue triglyceride.

One of the confounding issues when attempting to understand the usage of lipid as a fuel is the fact that its use can be so dramatically influenced by the availability of carbohydrate stores in the form of muscle glycogen. Moreover, feeding before or during exercise, or the time from the last meal, can all play a major role in determining fuel utilization during exercise. The next section outlines some of the factors that contribute to our current understanding of exercise metabolism.

KEY POINT

While the body contains limited carbohydrate stores, there is an abundance of stored triglyceride. As we have mentioned previously, carbohydrate is a better fuel for exercising muscle for three reasons: (1) It can generate acetyl CoA for the TCA cycle at a much higher rate than can fatty acids from adipose tissue and intramuscular triglycerides; (2) there is more ATP per unit of oxygen consumed with carbohydrate; and (3) carbohydrate can generate ATP in the absence of oxygen via anaerobic glycolysis.

Metabolism During Exercise: Fat versus Carbohydrate

At rest, during the postabsorptive state, the RER is about 0.82, indicating that lipid is the predominant fuel being oxidized by the body. The RQ across rested muscle is even lower, demonstrating that inactive muscle uses fat as its primary fuel. As we noted in the previous chapter, it has been convincingly demonstrated that in well-nourished individuals during exercise, the RER increases in proportion to the increase in exercise intensity. This means that as exercise intensity increases the contribution of carbohydrate oxidation to ATP formation increases whereas that of lipid oxidation decreases. This is illustrated in figure 7.19, which shows the relative release of FFA and glucose into the blood and the utilization of glycogen during exercise at different intensities. Several points are immediately clear. The release of fatty acids into the blood from adipose tissue stores rises in parallel with exercise intensity to approximately 50% of $\dot{V}O_2$max, then gradually declines. On the other hand, release of glucose into the blood from the liver increases with exercise intensity. As we have noted, glycogen utilization increases exponentially with increasing exercise intensity.

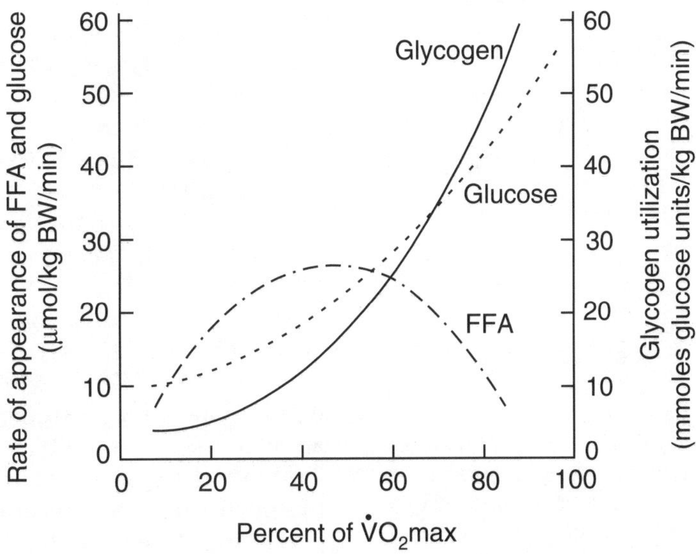

Figure 7.19 Use of glucose, glycogen, and free fatty acids (FFA) during exercise at various relative intensities, expressed as a percentage of maximum oxygen uptake ($\dot{V}O_2$max). The use of FFA and glucose are based on their rates of appearance in the blood. Glycogen use is determined by the rate of its loss from skeletal muscle. Figure obtained from G.A. Brooks, *Clinical and Experimental Pharmacology and Physiology* 24: 889–95, 1997, with permission.

George Brooks has proposed a "**crossover concept**" to explain the fuel utilization during exercise in terms of the balance between carbohydrate and fat. The crossover point is that relative exercise intensity where ATP formation from the use of carbohydrate exceeds that of lipid. Exercise at power outputs beyond this point will rely more and more on carbohydrate oxidation and less and less on fat. Brooks illustrated this with data obtained from a number of studies, and this is shown in figure 7.20. Data were obtained from a variety of experiments using four different species of animals, varying widely in terms of their $\dot{V}O_2$max expressed as milliliters per kilogram per minute. In all cases, the data generally fell along the lines shown in figure 7.20, indicating that the relative use of fuels is closely tied to relative intensity of exercise.

In the 1960s, Randle and his colleagues proposed a glucose-fatty acid cycle in which fatty acids could depress the oxidation of carbohydrate in heart and diaphragm muscle preparations. The mechanism to explain the glucose-fatty acid cycle was an inhibition of glucose utilization in the muscle cell, first at the pyruvate dehydrogenase reaction due to an increased concentration of acetyl CoA and NADH in the mitochondrial matrix (see Regulation of Pyruvate Oxidation in chapter 5). The elevated matrix acetyl CoA concentration would lead to an increase in citrate in the matrix and therefore an increase in cytosolic citrate. As discussed in chapter 6, phosphofructokinase, the prime regulatory enzyme of glycolysis, is inhibited by citrate, leading to a rise in the concentration of its substrate fructose 6-phosphate and thus glucose 6-phosphate. Since glucose 6-phosphate inhibits hexokinase, this would reduce the gradient for glucose transport into the muscle cell. Moreover, glycogen utilization would be depressed since glycogenolysis is depressed if glucose 6-phosphate concentration is elevated. Overall, muscle carbohydrate oxidation would be depressed and the difference would be made up by fatty acid oxidation.

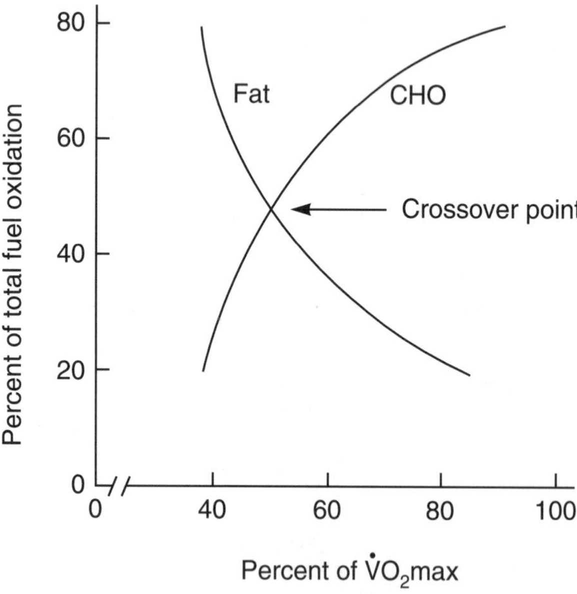

Figure 7.20 Relative contribution of lipid (fat) and carbohydrate (CHO) oxidation to energy needs during exercise at different intensities. The crossover point is the exercise intensity beyond which energy produced by carbohydrate oxidation exceeds that of lipid. The actual lines represent means from a variety of studies using four different species (goats, dogs, rats, and humans). Figure obtained from G.A. Brooks, *Clinical and Experimental Pharmacology and Physiology* 24: 889–95, 1997, with permission.

Many experiments have been performed using a variety of physical activity conditions to confirm the glucose-fatty acid cycle in humans. In general, a marked elevation in blood FFA concentration reduces overall carbohydrate utilization as well as skeletal muscle carbohydrate utilization in resting humans. Studies during exercise are less clear, although there are a number of recent reports that demonstrate that when blood FFA levels are acutely and sharply increased, muscle glycogen utilization is reduced during moderate intensity exercise. However, a FFA-induced depression in the use of carbohydrate during mild to moderate intensity exercise has not been observed in all exercise studies. During higher intensity exercise, an acute increase in FFA does not depress carbohydrate utilization since, as we have discussed earlier, there is a clear limit to the extent to which oxidative phosphorylation in muscle can be supported by acetyl CoA derived entirely by fatty acid oxidation.

The original mechanism proposed by Randle to account for the depression in glycogen utilization with elevated FFA does not appear adequate. Based on more recent studies, it would appear that an acute increase in FFA induced by infusion of lipid into the blood stream might depress carbohydrate oxidation at several sites. If the FFA concentration is elevated for several hours, it likely depresses glucose uptake through a decrease in the response of the GLUT-4 transporters to glucose because of the presence of elevated FFA. Alternatively, the increase in FFA may reduce the concentration of allosteric activators in muscle fiber cytosol (AMP and Pi) such that the activation of glycogen phosphorylase or phosphofructokinase is reduced. If flux through the glycolytic pathway to pyruvate is reduced, either by depression in glucose uptake, inhibition of phosphofructokinase, or decreased glycogenolysis, the availability of pyruvate for the matrix pyruvate dehydrogenase reaction will be lower. Overall, carbohydrate oxidation will be depressed, compensated by the increased availability of acetyl CoA from fatty acids.

Endurance exercise training increases the maximum capacity for oxidative phosphorylation that may exceed 100% in the trained muscle. This is due not only to an increase in the total amount and activity of TCA cycle and electron transport chain proteins, but also due to a parallel increase in the enzymes responsible for membrane transport and beta-oxidation of fatty acids. When the relative use of carbohydrate and fat is determined after versus before exercise at the same absolute work rate, the following is noted. The $\dot{V}O_2$ is the same after versus before training when exercising at the same absolute workload. The RER is lower during exercise after training, pointing to an increased oxidation of fatty acids. Therefore, use of both muscle glycogen and blood glucose is less after training. This point is particularly interesting because we know that endurance exercise training increases the total muscle content of the GLUT-4 transporter, that is, the sum of GLUT-4 in internal storage sites and within the muscle cell membrane. However, it is likely that the actual amount of GLUT-4 in muscle cell membranes is lower during exercise after training, compared to before training. People can exercise for a much longer period at the same absolute intensity of exercise after, compared to before, training. Besides the fact that the work is relatively easier because of the increase in fitness, much less carbohydrate is used.

If a person increases his or her $\dot{V}O_2$max by 20% following an endurance training program, he or she is obviously capable of exercising at a much higher intensity. If we compare the oxidation of carbohydrate and fat during exercise at the same relative intensity after training compared to before training, we see a different metabolic response. During exercise at say 60% of after-training $\dot{V}O_2$max the workload may be 180 watts, while 60% of pretraining $\dot{V}O_2$max may be 144 watts. Therefore, we can say with absolute certainty that after training the $\dot{V}O_2$ will be higher at the same relative workload, and more total fuel will be oxidized during exercise after training. The relative use of carbohydrate and fat during exercise with the greater energy demand is more controversial. Some studies show that the relative use of that fuel will be in about the same proportion as before training. Other studies show that the RER will be lower, suggesting that relatively more fat is being oxidized to support the higher $\dot{V}O_2$.

Carbohydrate Before or During Exercise

In the previous chapter, we pointed out that if a person does exactly the same amount and intensity of exercise with full muscle and liver glycogen stores, he or she will use more glycogen and produce a higher level of muscle and blood lactate compared to the same exercise performed with low glycogen stores. This tells us that if all things are equal, muscle will choose to use more muscle glycogen and blood glucose if it is available. At the time, we explained this on the basis of a mass action effect by which more carbohydrate in the form of higher blood glucose and muscle glycogen concentration forces the active muscle fiber to use more.

If a person doing prolonged, submaximal exercise does not ingest glucose, the RER during exercise will gradually decline, and after about one hour, blood glucose levels will decline. If the exercising subject takes in glucose during the exercise, blood glucose levels are better maintained, and the RER does not decline to the same extent. This tells us that the ingested glucose is being used as a fuel. Further, because the RER does not decline to the same extent, it also tells us that the ingested glucose is depressing the oxidation of fatty acids compared to the exercise situation where no glucose is ingested. There is no question that the ingested glucose is being oxidized, but what is it replacing? Since RER does not decline, and indeed may actually rise, this clearly reveals that fatty acid oxidation is decreased. The question that remains is whether the ingested glucose spares the use of muscle glycogen, depresses liver glycogenolysis and gluconeogenesis, or causes both liver glycogen and muscle glycogen to be spared. Many recent studies have examined this issue. Most studies demonstrate that ingested glucose suppresses the output of glucose from the liver. This means that the exogenous glucose decreases the rate of liver glycogenolysis and gluconeogenesis, if indeed the latter is active. However, the question whether ingested glucose suppresses the use of glycogen within the exercising muscle has not been unequivocally answered. Some studies show a reduction in muscle glycogen utilization with glucose ingestion, but many others show no effect.

Regulation of Fatty Acid Oxidation in Muscle

In an earlier section of this chapter, we discussed the formation of malonyl CoA by the ATP-dependent carboxylation of acetyl CoA, catalyzed by the enzyme acetyl CoA carboxylase (see figure 7.15). We also indicated that malonyl CoA can inhibit the activity of the enzyme carnitine acyl transferase I (CAT I), which is responsible for permitting the exchange of carnitine for CoA on the cytosolic side of the inner membrane (see figure 7.10). Inhibition of this enzyme blocks entry of fatty acids into the matrix, thereby preventing their oxidation. In the above section, it was mentioned that carbohydrate could depress the oxidation of fatty acids in liver. In this section, we will look at how this occurs in skeletal muscle.

Skeletal and cardiac muscle are important consumers of fatty acids, but they do not make fatty acids. In liver malonyl CoA is the precursor for de novo fatty acid synthesis, but it also regulates entry of fatty acids into the mitochondrion. Skeletal muscle is not known as a lipogenic tissue, which raises the question what is the reason for making malonyl CoA in skeletal muscle? The answer now seems clear; it is there to regulate fatty acid metabolism by muscle.

We know that the oxidation of fatty acids increases during fasting and light exercise. Under both conditions, the concentration of malonyl CoA in muscle decreases, which would help facilitate the transfer of fatty acids into the mitochondria. On the other hand, if glucose and insulin levels are rapidly and acutely raised, muscle malonyl CoA levels rise severalfold within 20 minutes. This would block entry of fatty acids into muscle mitochondria, sharply reducing the oxidation of fatty acids.

Malonyl CoA is synthesized in muscle by an isozyme of acetyl CoA carboxylase, ACC, known as ACC_β or ACC_2. This isozyme of acetyl CoA carboxylase is different from ACC_α, also known as ACC_1, which is found in the lipogenic tissues (liver and adipose tissue). Unlike the liver isozyme, the ACC_β content in muscle does not depend on the composition of the diet (e.g., high or low in carbohydrate), nor is it sensitive to insulin and glucagon concentrations. In this way it is distinct from the liver isoform. Moreover, ACC_β is not phosphorylated by the cyclic AMP-dependent protein kinase A (PKA). It is, however, phosphorylated by an AMP-activated protein kinase (AMPK).

Figure 7.21 outlines how malonyl CoA plays its central role in regulating fuel utilization in muscle, because its concentration is dictated both by the contractile state as well as the availability of fuels. Formation of malonyl CoA depends on the activity of ACC_β, which is subject to allosteric and phosphorylation control. Now let us see how this works under different conditions. At the onset of exercise, when muscle becomes active, the increased rate of ATP hydrolysis leads to an increase in the concentration of AMP (see chapter 4), which activates AMPK leading to phosphorylation of ACC_β. Phosphorylation of ACC_β decreases its activity, leading to a reduction in malonyl CoA, which allows entry of fatty acyl CoA into the mitochondrion for oxidation. In the recovery period after exercise, the AMP concentration will decrease along with the need for fatty acid oxidation to produce ATP. This will result in net dephosphorylation of ACC_β by the phosphoprotein phosphatases in muscle. Dephosphorylation activates ACC_β, leading to an increase in malonyl CoA to slow down the entry of fatty acyl CoA into the mitochondrion.

KEY POINT

The increase in AMP concentration in exercising muscle is related to exercise intensity. AMP activates phosphofructokinase (PFK), increasing the flux through glycolysis. AMP also leads to an increase in malonyl CoA concentration, which depresses the entry of long chain carboxylic acids into muscle mitochondria. Since fatty acid is not significant as a fuel for exercise at high intensities, the use of AMP level to manipulate glycolysis and fat oxidation within a muscle fiber is highly appropriate.

In the postabsorptive state, oxidation of fatty acids provides the bulk of the ATP needs for resting muscle fibers. In this state, ACC_β exists in an inactive state and malonyl CoA levels are low. With feeding, blood glucose concentration increases, leading to an increased blood insulin level. Muscle glucose uptake is increased, glycolysis is stimu-

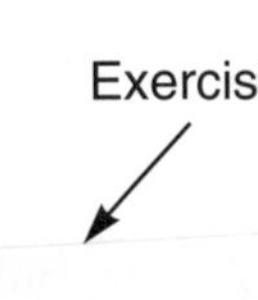

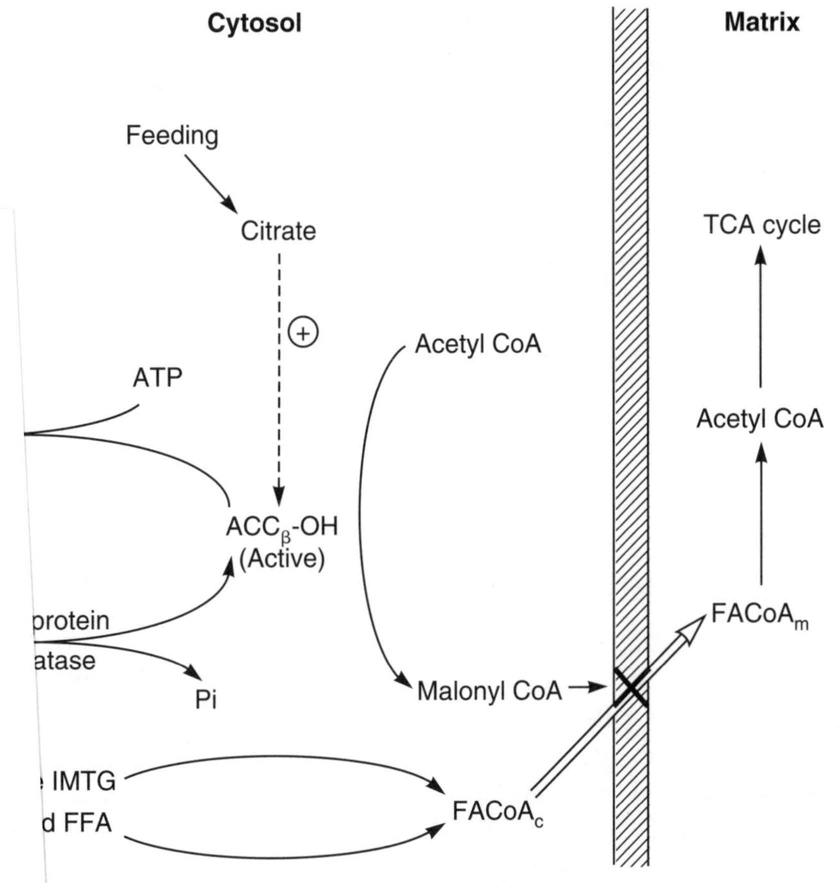

ration plays a central role in dictating fuel selection in muscle. Malonyl CoA blocks nitochondria where it will be converted into acetyl CoA by beta-oxidation and n by the large X across the transport of cytosolic fatty acyl CoA ($FACoA_c$) into the . Malonyl CoA is formed by the activated beta form of acetyl CoA carboxylase, $_\beta$-OH) and an inactive form (ACC_β-P). Conversion to the inactive form takes place the OH group of a serine residue catalyzed by an AMP-dependent protein kinase phosphate group, leading to the active form of the enzyme ACC_β-OH. Exercise n, activation of AMPK, and phosphorylation of ACC_β-OH to ACC_β-P and a decrease ormed and fatty acids can be oxidized. With feeding, there is a resultant rise in e both the phosphorylated and dephosphorylated forms of ACC_β. This raises the ibits fatty acid oxidation. The source of the fatty acyl CoA can be either intramus- borne free fatty acids (FFA).

effect, and pyruvate is formed at an increased rate. Entry of pyruvate into the mitochondrial matrix and conversion to acetyl CoA leads to an elevation of citrate concentration. An increase in matrix citrate leads to an increase in cytosolic citrate (see figure 7.16). Cytosolic citrate activates ACC_β, leading to increased formation of malonyl CoA and thus inhibition of fatty acyl CoA entry into the mitochondria. What happens from all this is that the minimal ATP needs of the resting muscle are handled to a greater extent by carbohydrate oxidation when there is abundant carbohydrate available.

During exercise when blood glucose levels are elevated, there will also be an increase in cytosolic citrate, based on the steps described in the previous paragraph. Even though the rate of ATP breakdown is greatly accelerated, the abundance of glucose will lead to depressed entry of fatty acids into the matrix such that carbohydrate will be more important as a fuel.

Medium chain triglycerides (MCT) containing fatty acids with 6, 8, or 10 carbon atoms have been used as a lipid supplement by some athletes. They are reputed to reduce carbohydrate oxidation because they are not metabolized in the same way typical ingested lipid is. Because of the short length of the fatty acid chain, they are not converted into triglycerides and then into chylomicrons in intestinal cells the way typical lipids are. Rather, they exit the gut in the blood and pass through the liver. From the general circulation, they enter cells and can pass into mitochondria without the need of the carnitine transport system. Therefore, unlike long chain fatty acids, their oxidation is not depressed by carbohydrate through a malonyl CoA mechanism. In a variety of exercise studies, MCT use has not been shown to have a significant impact on endurance performance, sparing neither muscle glycogen nor blood glucose use.

KEY POINT

A wonderful, theoretical rationale for MCT can be made, but the proof of this effect during exercise studies in the lab is not convincing. This is just one example out of many where hype of certain supplements and reality diverge. Although not prominently mentioned by proponents, MCT use can produce gastrointestinal distress that is really counter to exercise performance.

Cholesterol

Cholesterol is an important lipid, but we only seem to hear about its bad properties. Because of its hydrocarbon content (figure 7.22 illustrates the chemical structure), cholesterol is not soluble in water. In the blood, cholesterol either has a fatty acid attached to the cholesterol OH group, forming a cholesterol ester (about 70%), or is simple cholesterol (about 30%). Both cholesterol and cholesterol esters are in lipid-protein complexes called **lipoproteins**. The two main cholesterol-containing lipoproteins from a health perspective are the low density lipoproteins (LDL) and the high density lipoproteins (HDL).

Determining the concentrations of cholesterol and cholesterol-rich lipoproteins in the blood are important clinical tests. Total blood cholesterol concentration is reported three different ways: milligrams cholesterol/deciliter, milligrams cholesterol/liter, and millimolar. In the U.S. the first method predominates (mg/dl); many other parts

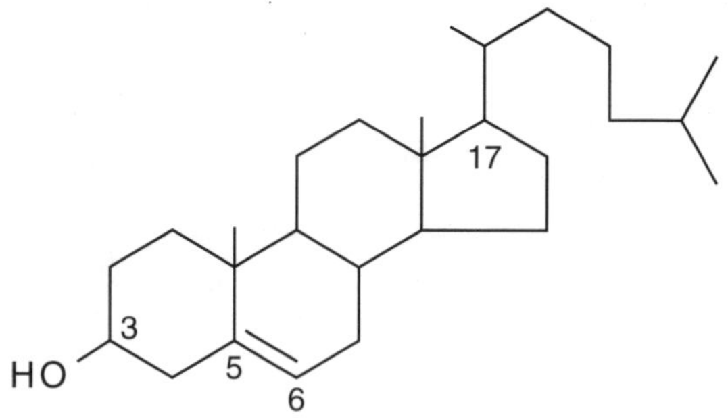

Figure 7.22 Shorthand illustration of cholesterol. The molecular formula is $C_{27}H_{47}O$. Can you find all the carbon atoms? The numbers illustrate the location of the hydroxyl (OH) group, the double bond, and the alkyl side chain.

of the world use the SI system (as mM units). If the molecular weight of cholesterol is 387, you should be able to convert a blood value of 150 mg/dl to the value of 3.88 mM. More important in determining a health profile is to determine the cholesterol concentration in the LDL and HDL fractions. From this perspective, it is best if the LDL cholesterol level is low while the HDL cholesterol is high.

Cholesterol is a component of membranes, a precursor for the synthesis of steroid hormones (e.g., cortisol, testosterone, estrogen, etc.), bile salts, and vitamin D, and a major component of myelin in nerves. Cholesterol is synthesized from acetyl CoA units. Although present in all cell types, cholesterol synthesis is most important in the liver, intestines, and adrenal and reproductive glands. For most people, about 60 to 70% of the body's cholesterol is synthesized; the remainder comes from the diet. Although no clear evidence exists that regular physical exercise decreases blood cholesterol levels, it may alter the type of lipoprotein carrying cholesterol in the blood, raising the level of HDL. New drugs have been shown to markedly decrease total blood cholesterol levels.

Summary

In the body, lipids exist primarily as fatty acids, triacylglycerols (triglycerides), phospholipids, and cholesterol. Triacylglycerols consist of three long chain fatty acids containing saturated or unsaturated carbon-to-carbon bonds, attached by ester bonds to the three-carbon alcohol glycerol. Triacylglycerols are mainly stored in specialized cells called fat cells or adipocytes. In these cells, triacylglycerols are made by joining fatty acids to glycerol. The reverse reaction, lipolysis or lipid mobilization to yield fatty acids and glycerol, is regulated by hormone-sensitive lipase. Triacylglycerol formation is favored and lipolysis inhibited by insulin, whereas epinephrine from the adrenal medulla and norepinephrine from the sympathetic nervous system play a major role in enhancing the rate of lipolysis.

Fatty acids released from the fat cell during lipolysis travel in the blood stream attached to the protein albumin. The fatty acids can be used by other cells as fuel. For this to happen, the fatty acid is transported across the cell membrane using fatty acid transport proteins, converted to a fatty acyl CoA, then transported into the mitochondrial matrix through attachment to carnitine. In a four-step process known as beta-oxidation, long chain fatty acids are broken down into two-carbon acetyl units attached to CoA. These acetyl CoA units can then feed into the TCA cycle. When lipolysis is increased because body carbohydrate stores are low, fatty acids can be a source of carbon by the liver to make ketone bodies. Ketone bodies are also a source of energy, but unlike fatty acids, they can be used by the brain as a fuel. Many of the problems associated with diabetes mellitus relate to uncontrolled formation of ketone bodies. Acetyl CoA units produced by the partial breakdown of excess carbohydrate or amino acids are used to make fatty acids.

We obtain most of our fatty acids from our diet, but the liver and adipose tissue have a limited ability to make new fatty acids. Unlike lipolysis, the synthesis of new fatty acids (de novo lipogenesis) occurs in the cell cytosol and involves an intermediate known as malonyl CoA. Malonyl CoA is formed by acetyl CoA carboxylase, whose activity is regulated by allosteric and phosphorylation mechanisms.

The oxidation of fat can play a major role in ATP formation by an active muscle. However, because of a limited rate at which fatty acids can be oxidized, their use is more important to a rested muscle or a muscle that is moderately active. The relative use of carbohydrate and fatty acid oxidation by muscle depends on a variety of factors, including exercise intensity, exercise training, and diet. Malonyl CoA plays an important role in determining fuel oxidation by muscle.

Key Terms

acidosis 128
adipose tissue 119
adrenergic receptors 120
autocrine effect 114
carnitine 123
chylomicrons 117
crossover concept 134
diacylglycerol 115
essential fatty acid 113
estrogens 122
free fatty acids (FFA) 119
growth hormone 120
hydrophilic 115
ketonemia 128
ketonuria 128
lipolysis 117
lipoproteins 138
long chain carboxylic acids 113
medium chain triglycerides 137
norepinephrine 120
pancreatic lipase 117
phosphatidic acid 115
phosphatidyl inositols 115
phospholipids 113
second messenger 120
triacylglycerols (triglycerides) 114
very low density lipoproteins 117

Review Questions

1. Draw the chemical structures for arachidonic acid [20:4 (5, 8, 11, 14)] and elaidic acid, which is the trans form of oleic acid.
2. In the synthesis of malonyl CoA, how many malonyl CoA units are needed? How many NADPH + H^+ are needed?
3. In the blood draining an adipose tissue site, the ratio of FFA to glycerol is 1.5 to 1. What proportion of triglyceride molecules being broken down are resynthesized?
4. It is reported that ingestion of caffeine can increase the concentration of FFA compared to an identical protocol where a placebo is given. If caffeine does not have a direct influence to increase lipolysis, by what other mechanisms might it act?
5. Assuming that the average molecular mass of stored triglyceride is 860 grams/mole, what percentage of this is glycerol?

6. During rhythmic leg kicking exercise, it is determined that 3 kilograms of quadriceps muscle is active over a one-hour period. The average $\dot{V}O_2$ for this exercise is 1.0 liter/minute. Before the exercise there are 12 millimoles/kilogram of stored intramuscular triglyceride and 10 millimoles/kilogram after 60 minutes of exercise. Assuming that the metabolic response during this exercise was three times that of rest, approximately how many kilocalories (kcal) of energy generated were due to the exercise itself? What is the approximate percentage of energy devoted to kicking that was derived from intramuscular triglyceride? What would be the other sources of energy?

CHAPTER 8

Amino Acid Metabolism

We first looked at the chemistry of amino acids in chapter 1. In the present chapter we view amino acids from a different perspective. Amino acids are obtained in dietary protein and are considered macronutrients. Protein provides approximately 17 kilojoules/gram (4 kcal/g) of physiological energy, compared with the same value for carbohydrates and 38 kilojoules/gram (9 kcal/g) for lipids. Humans typically ingest about 10 to 15% of their calories in the form of dietary protein. We eat protein to obtain the amino acids that are used to make body protein and other specialized substances. Amino acids are also used as a source of energy.

Overview of Amino Acid Metabolism

Figure 8.1 provides a summary of amino acid metabolism. Our principal source of amino acids is in the form of food protein. During digestion, protein is broken down to free amino acids that are absorbed into the blood. The amino acids in the blood and extracellular fluids represent an **amino acid pool**. Amino acids enter this pool from the gut. They also enter the pool following release from cells. Each cell makes its own specific proteins, taking up amino acids from the pool. The body of a normal 60-kilogram woman has approximately 10 kilograms of protein and about 170 grams of free amino acids. As we mentioned in chapter 1, there is a constant protein turnover. Some of the amino acids freed during protein degradation can be used to make new proteins. Others exit the cell, becoming part of the amino acid pool. In addition, the liver is capable of synthesizing some of the 20 amino acids needed to make our body proteins. These are described as nonessential amino acids, as opposed to the **essential amino acids** that cannot be synthesized and must be obtained in the diet from a variety of food sources. Amino acids that are synthesized in the liver are used within the liver or released to the blood. Thus the blood and extracellular fluids contain a pool of amino acids resulting from dietary intake, catabolism of cellular protein, and amino acids synthesized in and released from the liver. The amino acids in the pool are in a constant state of flux, with amino acids entering and leaving it.

We have no ability to store amino acids as we do for carbohydrate (glycogen) and fatty acids (triglyceride). Moreover, the protein content of our adult bodies is remarkably constant, and so we might expect that we would oxidize amino acids at a rate commensurate with our intake, that is, about 10 to 15% of our daily energy expenditure. In other words, those amino acids not used in protein synthesis or converted to other substances (e.g., heme groups; hormones such as serotonin, adrenaline, and noradrenaline; nucleotides; creatine; etc.) are simply used as fuels. As we mentioned in chapter 6, amino acids first lose their amino group(s) or other N atoms, and the resulting carbon skeleton can be oxidized directly, used to make glucose (gluconeogenesis), or converted into fat for storage.

> **KEY POINT**
>
> Although the carbon skeletons from excess amino acids can be degraded to acetyl CoA, converted to malonyl CoA, and used to make palmitate in the liver (see chapter 7), this is not very significant compared to fatty acid synthesis from excess carbohydrate. Immediate oxidation of amino acid carbon skeletons or conversion to glucose (gluconeogenesis) is the principal fate for excess amino acids.

Not shown in figure 8.1 is the transport of amino acids from the amino acid pool into cells. Because amino acids have charged groups, they need protein transporters to move them from the extracellular to the intracellular compartment or from the intracellular to the extracellular compartment. There are a number of amino acid transporters, which fall into two broad categories. The transporters can be sodium ion dependent or sodium independent. If the transporter is sodium dependent, the amino acid moves into a cell down a Na^+ concentration gradient. Therefore, this is a symport transport system. Amino acid transporters may have broad specificity, recognizing a number of amino acids, or narrow specificity, recognizing only one or two closely related amino acids. (For a review of the various categories of amino acids, refer to figure 1.4.) Amino acid transporters are also subject to regulation by hormones and growth factors.

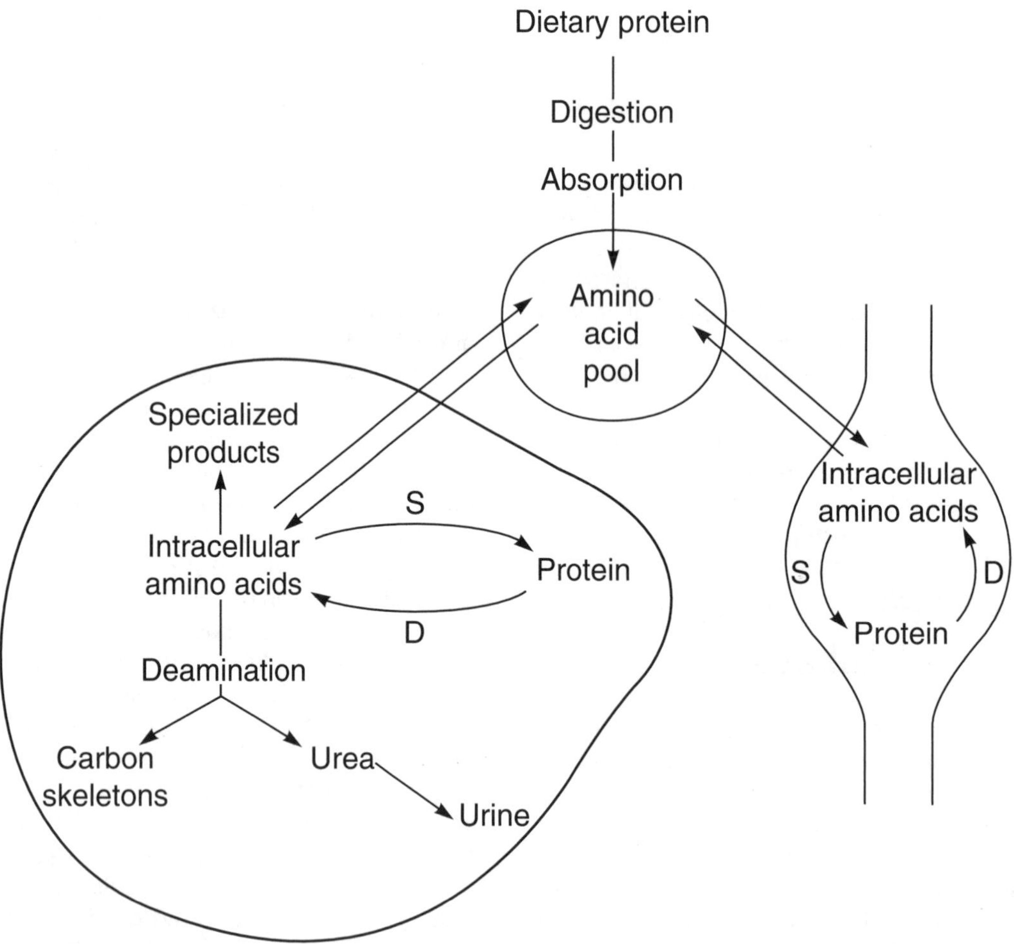

Figure 8.1 An overview of amino acid metabolism in the body showing two representative tissues—skeletal muscle and liver. Dietary protein is the principal source of the amino acids to meet our needs. Following digestion and absorption, these enter an amino acid pool, which is the blood and extracellular fluids. Cells take up amino acids from the pool and use them to synthesize proteins (S). Protein undergoes constant turnover, so degradation (D) releases amino acids that may be used again in protein synthesis or released to the amino acid pool. The liver is responsible for making a variety of specialized products from amino acids and also for the breakdown of most amino acids. Following removal of the amino groups by deamination, urea is formed and excreted from the body. Carbon skeletons can be a source of glucose, oxidized, or converted into lipid.

Insulin, in particular, is important to stimulate amino acid uptake, particularly into skeletal muscle fibers.

Degradation of Amino Acids

Amino acids undergo constant oxidative degradation during the following metabolic circumstances:

- During normal synthesis and degradation of proteins in the body, amino acids released during the constant breakdown process may not be immediately reused in synthesis. Since we cannot store amino acids, they will be degraded.
- When we ingest more amino acids than our bodies can use to make proteins or convert to other substances, these amino acids are degraded. In North America, most people eat more protein than they need.
- During starvation, fasting, dieting, or uncontrolled diabetes mellitus, when carbohydrates are either not available or not used due to an absence of insulin, amino acid catabolism accelerates.
- When we overtrain such that there is an imbalance in protein turnover favoring protein degradation, amino acids are degraded.

Catabolism of individual amino acids has two major stages. First, the amino acids must lose their nitrogen atoms because we cannot obtain usable energy from nitrogen groups. Second, the resulting carbon skeletons are fed into specific energy-yielding pathways to retrieve their chemical energy. The liver removes the amino groups from most amino acids, although skeletal muscle has a significant capacity for **deamination** of the branched chain amino acids (BCAAs). Exercise increases the catabolism of BCAAs by muscle.

Transamination

We have already seen transamination in the catabolism of BCAAs in chapter 6. In this reaction, **aminotransferase enzymes** (formerly known as transaminases), which contain vitamin B6, transfer the amino group of an amino acid to α-ketoglutarate, making glutamate and a new keto acid (see figure 8.2). The vitamin B6 is a precursor for the coenzyme PLP (see table 2.1). Notice that the ketone group in α-ketoglutarate is next to the carboxylate. Nearly all transaminase reactions involve amino transfer between an amino acid and α-ketoglutarate, forming a new α-keto acid and

$$\underset{\alpha\text{-ketoglutarate}}{{}^{-}OOC-CH_2-CH_2-C(=O)-COO^{-}} + \underset{\text{Amino acid}}{{}^{+}H_3N-CH(R)-COO^{-}} \xleftrightarrow{\text{Aminotransferase}} \underset{\text{Glutamate}}{{}^{-}OOC-CH_2-CH_2-CH({}^{+}NH_3)-COO^{-}} + \underset{\alpha\text{-keto acid}}{R-C(=O)-COO^{-}}$$

$$\text{BCAAs in muscle} + \alpha\text{-ketoglutarate} \xleftrightarrow{\text{BCAA aminotransferase}} \text{Glutamate} + \text{BCKAs}$$

$$\text{Most other amino acids in liver} + \alpha\text{-ketoglutarate} \xleftrightarrow{\text{Other aminotransferases}} \text{Glutamate} + \text{Other keto acids}$$

Figure 8.2 Aminotransferases, or transaminases, catalyze reactions in which amino groups are transferred from an amino acid to α-ketoglutarate, creating a new keto acid and glutamate. The top reaction shows a general transamination reaction using chemical structures. R is the side chain of the general amino acid. The middle equation shows transaminations that typically occur in muscle using the branched chain amino acids (BCAAs) and generating branched chain keto acids (BCKAs). The bottom equation shows the general transamination reaction of other amino acids that usually takes place in the liver.

glutamate. Additional aminotransferase enzymes are specific for amino acids other than glutamate. All transamination reactions are freely reversible with equilibrium constants of about one and standard free energy change values near zero. Their net direction depends on the relative concentrations of the four reactants. The overall effect is to transfer amino groups from a variety of amino acids to α-ketoglutarate (2-oxoglutarate), making glutamate.

Figure 8.2 summarizes the reactions for muscle, which preferentially transaminates BCAAs, and liver, which handles most of the other amino acids. In chapter 6 we also looked at the branched chain amino acids and their metabolism (see figures 6.13 and 6.14). Figures 6.13 and 6.14 show the transfer of amino groups on the BCAAs to pyruvate to form alanine using two aminotransferase enzymes, BCAA aminotransferase and alanine aminotransferase. In this way, the amino groups on BCAAs can be moved from skeletal muscle to liver for disposal.

Deamination

The body rids itself of amino groups by forming urea. Therefore, the amino group on glutamate (which comes from amino acids) must be transferred to the liver (if not already there) and into the urea molecule. Nitrogen from amino groups in the liver in the form of glutamate can be released as ammonia (principally the ammonium ion NH_4^+) in the glutamate dehydrogenase reaction (see figure 8.3). As shown, the coenzyme is NAD^+, but $NADP^+$ can also be used. The freely reversible glutamate dehydrogenase reaction takes place only in the mitochondrial matrix, whereas most aminotransferase enzymes exist in both the mitochondrial matrix and the cell cytosol. The glutamate dehydrogenase reaction, as the reaction proceeds to the right, is a deamination reaction since it removes the amino group. Because NADH is also formed, it is an oxidative deamination reaction. The glutamate dehydrogenase reaction removes nitrogen from glutamate, which can be used to make urea. As we will discuss later, exercising muscle releases ammonia (as the NH_4^+ ion). The glutamate dehydrogenase reaction in muscle is a source of this ammonia. As well, ammonia is produced by the deamination of AMP in the AMP deaminase (adenylate deaminase) reaction we discussed in chapter 4. This latter reaction is generally not significant unless muscle is working at a high intensity. Overall, the production of ammonium ion and its release from muscle is proportional to the intensity of the exercise.

KEY POINT

We have been using ammonium ion—the principal form we would find in the blood—rather than ammonia, although the former is simply protonated ammonia. Ammonia (ammonium ion) released from muscle will be taken up by liver (primarily) and kidney. In the liver, ammonia can be used to make urea.

Glutamine

The content of the amino acid glutamine is high in a variety of cells. It is particularly abundant in skeletal muscle, out of proportion to the amount of glutamine in muscle proteins. Indeed, it is by far the most abundant amino acid in skeletal muscle, accounting for about 60% of the total amino acid pool. It is also the most abundant amino acid in plasma. Glutamine is an important fuel for the gut and immune system. Glutamine can be synthesized from glutamate, using the enzyme glutamine synthetase according to the equation in figure 8.4. Notice that the additional nitrogen in glutamine is an amide of the side chain carboxylate. Skeletal muscle is an important site for glutamine synthesis.

$$\text{glutamate} + H_2O + NAD^+ \xrightleftharpoons{\text{glutamate dehydrogenase}} \alpha\text{-ketoglutarate} + NADH + H^+ + NH_4^+$$

Figure 8.3 Glutamate dehydrogenase can deaminate glutamate, forming ammonium ion. Because it is freely reversible, the amination of α-ketoglutarate can also form glutamate.

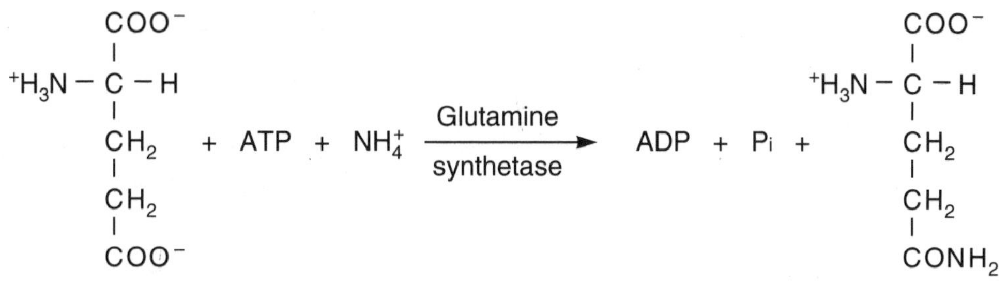

$$\begin{array}{c} COO^- \\ | \\ ^+H_3N-C-H \\ | \\ CH_2 \\ | \\ CH_2 \\ | \\ COO^- \end{array} + ATP + NH_4^+ \xrightarrow[\text{synthetase}]{\text{Glutamine}} ADP + P_i + \begin{array}{c} COO^- \\ | \\ ^+H_3N-C-H \\ | \\ CH_2 \\ | \\ CH_2 \\ | \\ CONH_2 \end{array}$$

Glutamate Glutamine

Figure 8.4 The glutamine synthetase reaction is very active in skeletal muscle, transferring ammonium ions to glutamate to make glutamine.

KEY POINT

Glutamine content in skeletal muscle and other tissues appears to have a regulatory role in whole body protein synthesis. During a variety of catabolic conditions, its content is decreased, and during anabolic states, intracellular glutamine content is elevated. Glutamine is also an important fuel for macrophages and lymphocytes, cells important to our health and well-being. Some studies have demonstrated that athletes who overtrain have an intracellular glutamine content that resembles catabolic states. These persons are also more prone to infections, particularly of the respiratory tract. Although much of the intracellular glutamine is formed by the glutamine synthetase reaction, supplemental protocols with glutamine have been attempted with mixed results. One of the problems is that much of the supplemental glutamine is oxidized by tissues of the intestinal tract or taken up by the liver and kidney.

The last reaction to consider is the removal of the amino group from the side chain of glutamine, which is the reverse of the glutamine synthetase reaction. Glutaminase, found mainly in liver mitochondria, catalyzes this deamination reaction:

$$\text{glutamine} + H_2O \xrightarrow{\text{glutaminase}} \text{glutamate} + NH_4^+$$

The Urea Cycle

Ammonia is quite toxic, especially to the brain. However, two safe forms of ammonia exist: the amino group of glutamate and the side chain amide nitrogen in glutamine. Although we can temporarily store ammonia in these innocuous forms, we must eliminate the nitrogen we cannot use. Animals convert the nitrogen to urea, whereas birds and reptiles eliminate amino groups by converting nitrogen to uric acid. Urea, whose structure follows, is a simple molecule formed in the liver and excreted from the kidney when urine is formed:

$$H_2N-\overset{\overset{\displaystyle O}{\|}}{C}-NH_2$$

The two amino groups in urea allow the body to rid itself of nitrogen. The carbonyl group comes from carbon dioxide. The nitrogen comes from the ammonium ion and from the amino group of aspartate. Figure 8.5 summarizes the path taken by nitrogen from amino acids in the body. Muscle releases alanine, which contains much of the muscle nitrogen. As we saw in figure 6.14, the source of the nitrogen on alanine is BCAAs. In the liver, the amino group on alanine is transferred to α-ketoglutarate to make glutamate, catalyzed by alanine aminotransferase. The glutamine released from muscle results from the amination of glutamate in the glutamine synthetase reaction. The liver is the major site for nitrogen removal from most of the amino acids (except the BCAAs). The amino groups end up on glutamate. As mentioned, muscle also releases ammonium ions, especially during exercise.

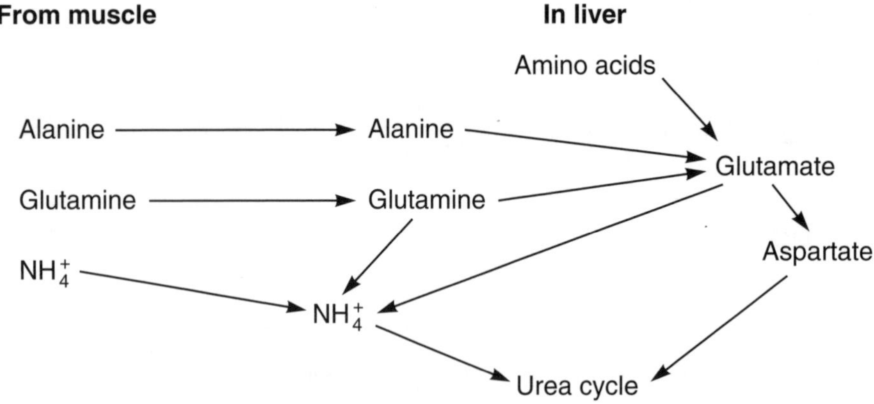

Figure 8.5 The nitrogen in urea comes from NH_4^+ and aspartate. Urea nitrogens come from two amino acids released in significant quantities from muscle and from most amino acids in liver. The release of NH_4^+ from muscle is insignificant during rest but quantitatively important during exercise.

Urea synthesis requires the ammonium ion that comes from ammonia taken up by liver from the blood, from glutamine via the glutaminase reaction, or from glutamate via the glutamate dehydrogenase reaction. Aspartate, which provides the other urea nitrogen, comes from a transamination reaction in which oxaloacetate is transaminated from glutamate to make aspartate and α-ketoglutarate. The enzyme aspartate aminotransferase catalyzes the following reaction.

$$\text{glutamate} + \text{oxalocetate} \xleftrightarrow{\text{aspartate aminotransferase}} \alpha\text{-KG} + \text{aspartate}$$

The urea cycle consists of four enzymatic steps that take place in the liver cells, both in the mitochondrial matrix and cytosol. Figure 8.6 illustrates the first reaction before the actual urea cycle begins. This first step occurs in the mitochondrial matrix and involves the formation of carbamoyl phosphate, catalyzed by carbamoyl phosphate synthetase. Carbamoyl phosphate contains nitrogen from the ammonium ion and a carbonyl group from the bicarbonate ion. The carbamoyl phosphate synthetase reaction is the committed step in urea synthesis and controls the overall rate of the urea cycle.

The urea cycle is shown in figure 8.7. The carbamoyl phosphate enters the urea cycle in the mitochondrial matrix, joining with ornithine to form citrulline in a reaction catalyzed by ornithine transcarbamoylase. The citrulline exits the mitochondrial matrix. The second nitrogen, in the form of the aspartate amino group, enters the cycle when aspartate combines with citrulline to form argininosuccinate. This step involves ATP hydrolysis and is catalyzed by argininosuccinate synthetase. The argininosuccinate is then cleaved to arginine and fumarate in a reaction catalyzed by argininosuccinate lyase. In the final step, arginine is cleaved by arginase to yield ornithine and urea. The ornithine is now regenerated and enters the mitochondrion to combine again with carbamoyl phosphate.

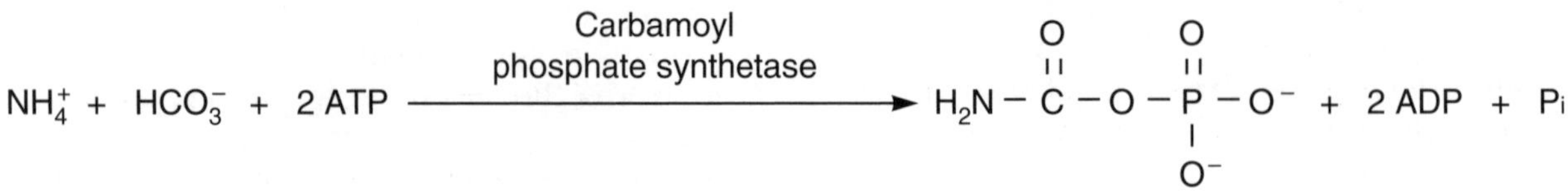

$$NH_4^+ + HCO_3^- + 2\,ATP \xrightarrow{\text{Carbamoyl phosphate synthetase}} H_2N-\overset{\overset{\large O}{\|}}{C}-O-\overset{\overset{\large O}{\|}}{\underset{\underset{\large O^-}{|}}{P}}-O^- + 2\,ADP + P_i$$

Figure 8.6 The reaction controlling the rate of urea formation is catalyzed by carbamoyl phosphate synthetase in the matrix of liver mitochondria.

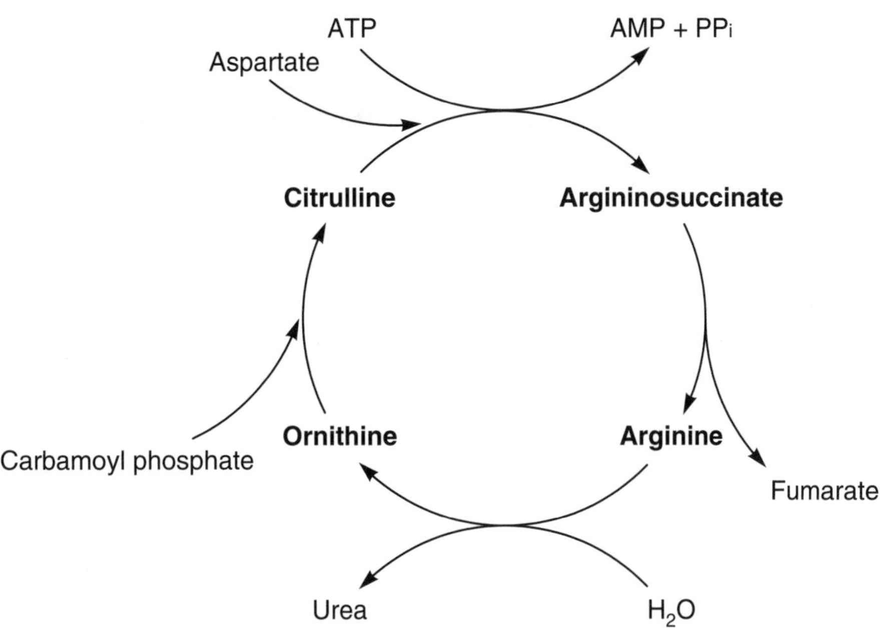

Figure 8.7 The urea cycle is a sequence of four enzyme-catalyzed reactions that form urea from nitrogen on carbamoyl phosphate and aspartate. There is no net consumption or formation of ornithine or other intermediates in the cycle, shown in bold letters.

Urea is extremely water soluble. Thus it leaves the liver via the blood and enters the kidneys, where it is filtered out. The urea cycle is metabolically expensive because the ATP cost to make one urea molecule is four. Two are used to make carbamoyl phosphate. Can you think where the other two ATP come from? (Hint: How many phosphate groups are removed from ATP during the condensation of citrulline and aspartate?)

Fate of Amino Acid Carbon Skeletons

After the amino groups are removed from the amino acids, carbon skeletons remain, in many cases in α-keto acids such as pyruvate, oxaloacetate, or α-ketoglutarate (see figure 8.8). These carbon skeletons have a variety of fates, such as gluconeogenesis, because 18 of the 20 amino acids can be a source of glucose. These are sometimes described as glucogenic amino acids. Note in figure 8.8 that leucine and lysine do not produce either pyruvate or TCA cycle intermediates. Leucine and lysine are described as ketogenic amino acids because their carbon skeletons produce only acetoacetyl CoA and acetyl CoA. The carbon skeletons of all amino acids can also be used for immediate oxidation because they form TCA cycle intermediates or acetyl CoA. Finally, all are potential sources of carbon to make new fatty acids because all are also potential sources of acetyl CoA. As mentioned, amino acid carbon skeletons are not a very significant source for de novo fatty acid synthesis.

Figure 8.8 illustrates what the carbon skeletons of the amino acids have in common with the TCA cycle or substances directly related to this cycle. The amino acids shown can generate the TCA intermediates or related molecules by simple removal of their amino groups via transamination (alanine, glutamate, and aspartate) or through a number of steps not shown. As we discovered in chapter 6, the formation of TCA cycle intermediates through removal of the amino groups from amino acids is known as anaplerosis. It is an important mechanism to maintain TCA cycle intermediate concentration, especially during exercise when demand on the TCA cycle is high.

Amino Acid Metabolism During Exercise

During exercise, carbohydrate and lipid supply most of the energy needs of the body as we have

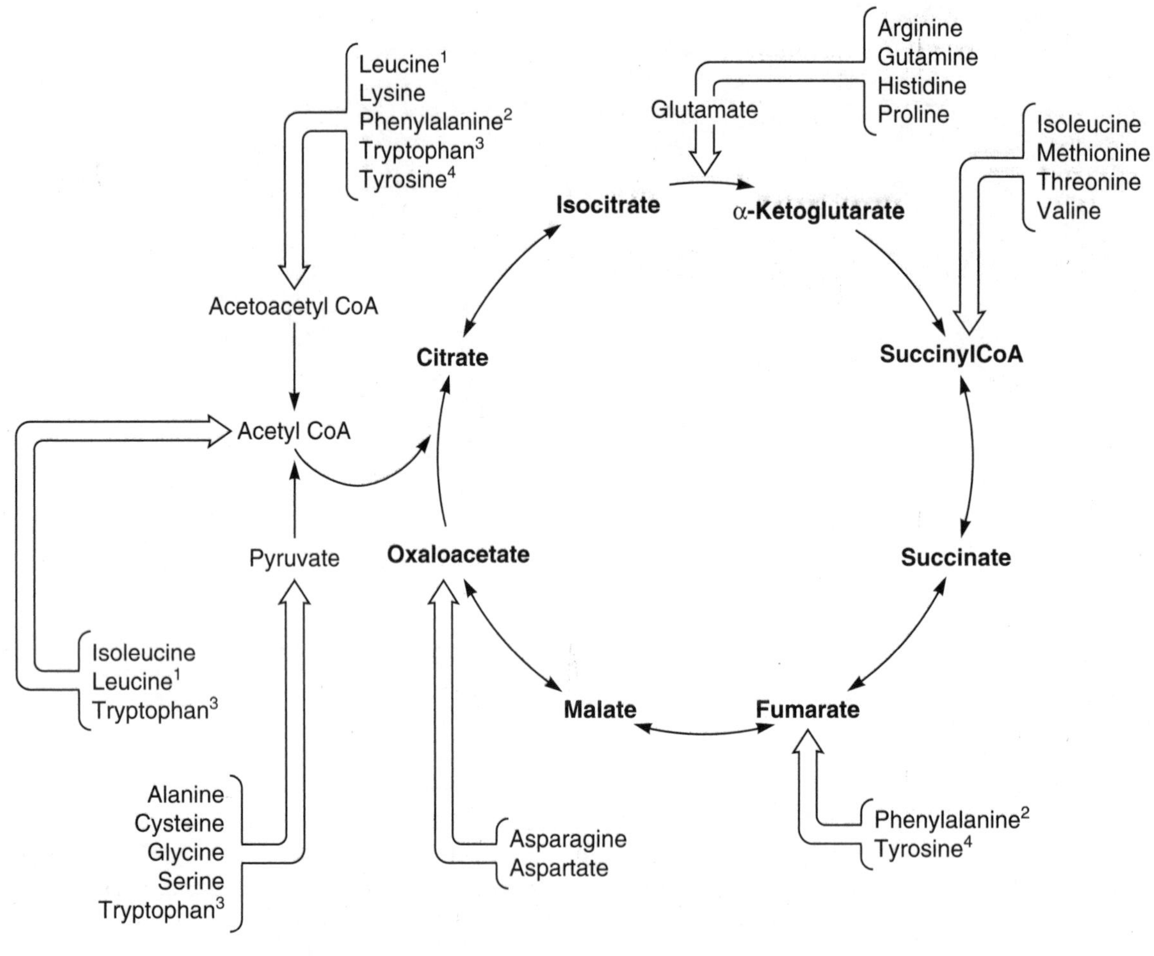

Figure 8.8 Following the removal of nitrogen groups by deamination or transamination, the amino acid carbon skeletons are converted into tricarboxylic acid cycle intermediates (shown in bold) or related molecules. Double arrows show the major paths taken by the carbon skeletons. The carbon atoms of some large amino acids, identified with superscript numbers, give rise to two or three different molecules.

already discussed. However, metabolism of amino acids also takes place at an accelerated rate in a manner designed both to transfer amino acid nitrogen (amino groups) out of muscle and to move their carbon skeletons from muscle to liver. To get a proper perspective on the effects of exercise, we will look at low to moderate intensity exercise and high intensity exercise, because the effects of the two are quite different. First, it is important to understand how we learn about muscle amino acid metabolism.

Metabolism of amino acids during exercise is a complex study, which in humans is understood by measuring the concentrations of amino acids in the arterial and venous blood across an exercising muscle. From concentration differences and muscle blood flow measures, one can determine what amino acids are being taken up and released from muscle and the rate at which this is taking place. One could then describe the release of an amino acid from muscle in terms of the concentration difference in arterial and venous blood times the blood flow, producing such numbers as micromoles of amino acid released per minute. In addition, muscle biopsies can be taken to determine the concentrations of amino acids and a variety of other metabolic intermediates in the muscle samples obtained before, during, and after exercise. For example, the lack of release of an amino acid or metabolite from muscle may mean that nothing is happening with it compared to the inactive condition or that its concentration is increasing within the muscle and that is why it is not being released. Thus both changes within muscle and changes in exchange across the muscle are needed.

Even when good measures of muscle metabolites, arterial and venous concentration differences, and blood flow can be obtained there may still be difficulties in getting an accurate picture of muscle amino acid metabolism during exercise. The reason for this is estimating with reasonable accuracy the mass of muscle involved in the actual exercise. In other words, we need to be able to describe changes taking place in muscle amino acids by describing releases or uptakes in terms of micromoles per minute per unit mass of muscle (i.e., micromoles per minute per kilogram of muscle). Although a detailed analysis of this is beyond the scope of this discussion, it should point out that much of the published data in current literature on amino acid metabolism in human muscle during exercise probably represents reasonable estimates and not precise values.

The Purine Nucleotide Cycle

In chapter 4 we looked at the deamination of AMP, catalyzed by the enzyme AMP deaminase (adenylate deaminase). This enzyme converts AMP to IMP (inosine monophosphate) and ammonia as shown.

$$\text{AMP} + \text{H}_2\text{O} \xrightarrow{\text{AMP deaminase}} \text{IMP} + \text{NH}_3$$

Protonation of the base NH_3 generates ammonium ion (NH_4^+). As we discussed in chapter 4, the AMP deaminase reaction results in an actual decrease in the total adenine nucleotide content (TAN) of the muscle because of the relationship of ATP, ADP, and AMP in the AMP kinase (adenylate kinase) reaction.

$$2\ \text{ADP} \xleftrightarrow{\text{AMP kinase}} \text{AMP} + \text{ATP}$$

AMP deaminase is more active in type II (fast twitch) than type I (slow twitch) muscle fibers. Therefore, it is unlikely to be activated significantly in low intensity exercise since the force production will be handled to a greater extent by motor units containing type I muscle fibers. Furthermore, AMP deaminase activity is low in rested or slowly contracting muscle, increasing only if the pH is decreased and the concentrations of AMP and ADP increase. Not only do type I fibers have less AMP deaminase, but they are unlikely to develop the rise in AMP and decrease in pH that is needed to activate the enzyme since these fibers have a higher capacity for oxidative metabolism and lower glycolytic activity. Therefore, we can anticipate that AMP deaminase is unlikely to be significant during low to moderate levels of exercise.

Since the AMP deaminase can reduce TAN levels by up to 50% in very severe exercise, there must be a way to regenerate the adenine nucleotides. The process for doing this is known as the purine nucleotide cycle and is shown in figure 8.9. In reaction one, which is catalyzed by AMP deaminase, IMP and NH_3 are produced. As mentioned, this reaction is active during intense muscle activity. AMP is regenerated by two additional reactions. In the first, aspartate combines with IMP at the cost of a GTP, forming adenylosuccinate. This is catalyzed by adenylosuccinate synthetase. In the final step, adenylosuccinate is split into AMP and fumarate, catalyzed by adenylosuccinate lyase. The net reaction shows that it costs energy to regenerate AMP in the form of a GTP. In the process, an aspartate is deaminated producing fumarate, a TCA cycle intermediate.

Moderate Intensity Exercise

Figure 8.10 summarizes some of the important changes in amino acid metabolism that take place in exercising skeletal muscle. During exercise in the postabsorptive state, skeletal muscle is in a net protein catabolic state, in which protein breakdown exceeds the rate of protein synthesis. Most of the amino acids produced by the net protein catabolism are released from the muscle. The major exception is glutamate, because there is a net uptake of glutamate at rest and even more so during exercise. Of the amino acids released, glutamine and alanine are the most dominant, far out of proportion to the other amino acids and far out of proportion to glutamine and alanine in skeletal muscle proteins. This suggests that both amino acids are being formed in skeletal muscle at an accelerated rate during exercise. In addition, skeletal muscle releases ammonium ion. Figure 8.10 illustrates where the ammonia, alanine, and glutamine are derived.

We have already discussed the fact that during glycolysis, approximately 1% of the pyruvate produced in glycolysis is converted to alanine. Transamination of pyruvate with glutamate by alanine aminotransferase produces alanine and α-ketoglutarate.

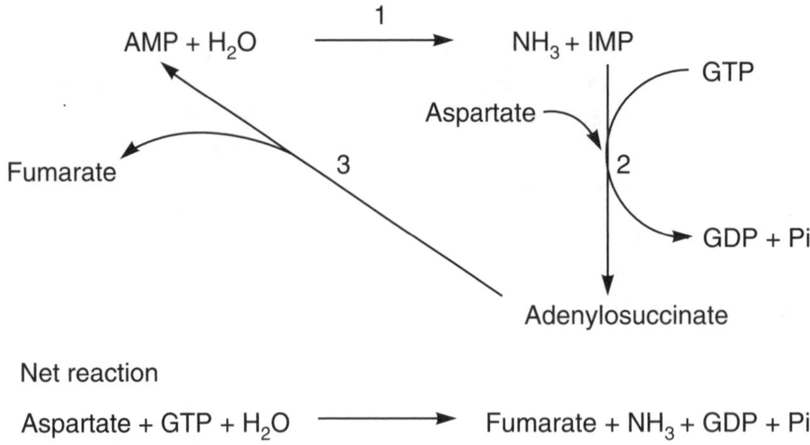

Figure 8.9 The purine nucleotide cycle is a simple pathway in which AMP is deaminated during intense muscle activity, catalyzed by AMP deaminase (1). The next two steps regenerate AMP. First, IMP combines with aspartate, catalyzed by adenylosuccinate synthetase (2). Secondly, adenylosuccinate is split into AMP and fumarate, catalyzed by adenylosuccinate lyase. The net reaction shows that the regeneration of AMP costs a GTP, and the amino acid aspartate is deaminated and a fumarate is produced.

As shown previously, the alanine leaves the muscle and travels in the blood to the liver (see figure 6.14). In the liver, alanine aminotransferase converts alanine to pyruvate and α-ketoglutarate. The pyruvate is a source of carbon to make glucose by gluconeogenesis. If the glucose produced in liver is released to the blood and catabolized to pyruvate in muscle, a cycle is produced that is known as the glucose-alanine cycle.

KEY POINT

If you consider the net effect of the glucose-alanine cycle, the carbon atoms of glucose are simply recycled from liver to muscle (as glucose) and muscle to liver (in alanine). However, this cycle allows the muscle to transfer nitrogen from branched chain amino acids (BCAAs) to liver for urea synthesis.

The principal source of the amino group that is released from muscle in alanine is from the BCAAs. As mentioned, these amino acids are preferentially transaminated in skeletal muscle by branched chain amino acid aminotransferase. This reaction converts α-ketoglutarate to glutamate and the deaminated products of the BCAAs, known as branched chain keto acids. These may undergo further oxidative metabolism within the muscle or be released to the blood for metabolism in the liver.

During low to moderate intensity exercise, muscle releases alanine at a rate far in excess of the rate at rest. As shown in figure 8.10, the source of the pyruvate for alanine formation is that produced during glycolysis from glucose and, more importantly, stored glycogen. We also know from chapter 6 that the contribution of glycogen to ATP production during prolonged exercise declines as the glycogen stores are gradually reduced. Although this is partially compensated by an increased use of blood glucose, it is typically noted that the RER during constant intensity exercise declines over time. With glucose and glycogen being the sources of pyruvate to make alanine, we would expect that the release of alanine declines with exercise duration, and this is what is observed.

As shown in figure 8.10, skeletal muscle also releases glutamine, especially during exercise. Amination of glutamate, using the glutamine synthetase reaction (shown in figure 8.4), is the major source of glutamine, because muscle proteins are not rich in the amino acid glutamine. The ammonium groups needed to form glutamine from glutamate can be derived from the glutamate dehydrogenase reaction and also from the deamination of AMP by the enzyme AMP deaminase. With its net uptake from the blood, glutamate is the source for glutamine synthesis and some of the ammonia produced. Indeed, so much glutamate is needed to make glutamine that the concentration of glutamate actually declines during exercise, unlike that of the other amino acids.

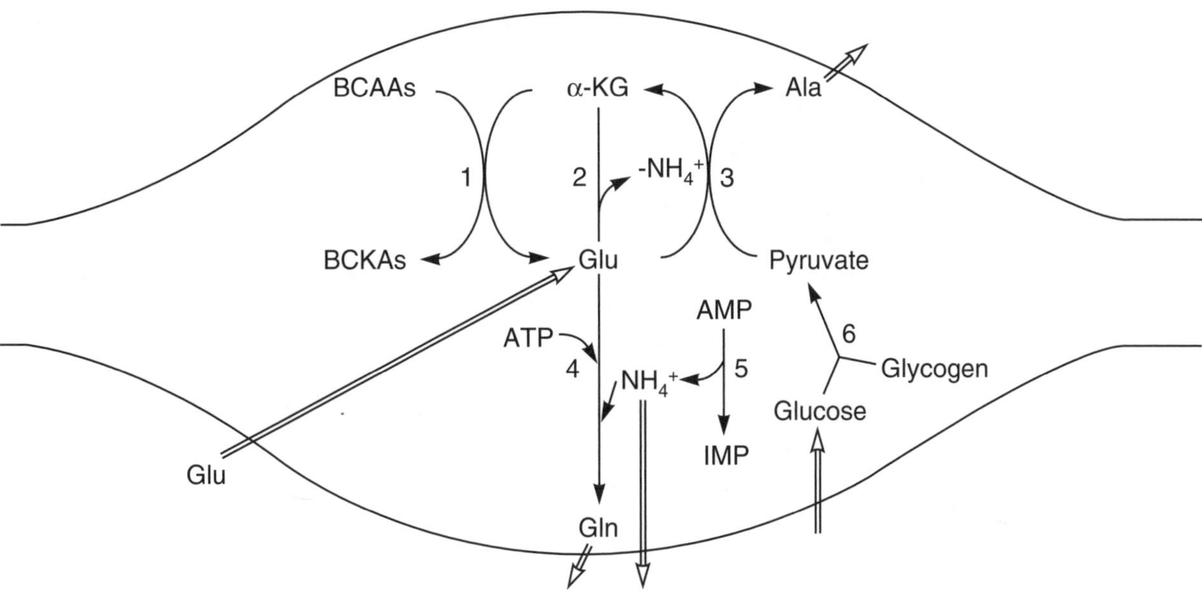

Figure 8.10 During exercise there is a net release of most amino acids from skeletal muscle. The exception is glutamate, for which there is a net uptake. Alanine (Ala) and glutamine (Gln), formed by reactions 3 and 4, respectively, are released from muscle at rates far beyond the other amino acids. Alanine is synthesized from pyruvate and an amino group donated by the branched chain other amino acids (BCAAs). The source of the pyruvate is glycolysis (6), using glucose or glycogen as a precursor. Glutamine is synthesized from glutamate (Glu) (4). Glutamate may also be deaminated by glutamate dehydrogenase (2), forming α-ketoglutarate (α-KG). Active in muscle during vigorous exercise is AMP deaminase (5), which produces ammonium ion (NH_4^+) and IMP. Exercising muscle releases NH_4^+. Reaction 1 is catalyzed by branched chain amino acid aminotransferase. Reaction 3 is catalyzed by alanine aminotransferase.

The changes just noted for moderate intensity exercise can be modulated depending on the nutritional state of the exercising person as well as the concentration of muscle glycogen. In a fasted, low glycogen state, there will be less pyruvate formed in glycolysis to produce alanine. In this state, protein breakdown is likely much greater, with a corresponding increase in the release of most amino acids from muscle. Early exercise studies have also suggested that the rate of catabolism of amino acids by the liver increases in a fasted, low muscle glycogen state as evidenced by increased formation of urea and larger concentrations of urea in sweat. Some studies have also noted an increase in muscle IMP and a decrease in TAN in untrained subjects during prolonged, moderate intensity exercise when muscle glycogen concentration is severely reduced. Apparently the steep decline in muscle glycogen leads to an increase in ADP concentration and, thereby, AMP concentration. This can result in a modest flux through AMP deaminase by a mass action effect, producing IMP even though conditions favoring complete activation of AMP deaminase are not present.

KEY POINT

Many athletes ingest carbohydrate-containing fluids during training. While they may be thinking that this is to provide fuel for their training, it will also help to maintain protein balance in the muscle by modulating the rate of protein breakdown.

High Intensity Exercise

When the intensity of exercise is high, there are likely to be only modest increases in the release of glutamine and alanine from muscle. Glutamate uptake by muscle also appears to be little influenced by increased exercise intensity. Glutamine synthesis consumes ATP and its release represents a loss of glutamate from muscle. Since glutamate is a source for the TCA cycle intermediate α-ketoglutarate, accelerated formation and release of glutamine may not be a productive reaction when ATP demand is high.

The key reaction that is accelerated by an increase in exercise intensity is the AMP deaminase reaction. In particular, the metabolic changes in an intensely contracting fast twitch muscle fiber (increase in AMP and H^+) favor activation of AMP deaminase, with the attendant formation of IMP and ammonia. As a phosphorylated molecule, IMP is trapped within the fiber, but ammonia is released at an accelerated rate. Studies generally show a positive relationship between lactate formation and release from skeletal muscle and release of ammonia.

Summary

The content of protein in the adult body remains remarkably constant. Because most adults take in about 10 to 15% of their dietary energy in the form of protein, an equivalent amount of amino acids must be lost each day. The body cannot store excess amino acids. Those not needed to make protein lose their nitrogen groups, and the carbon skeletons are used to make glucose (gluconeogenesis) or are converted to acetyl CoA, which can feed into the citric acid cycle or be used to make fatty acids. The first step in disposing of excess amino acids is to remove the amino groups, which are primarily transferred to α-ketoglutarate by aminotransferase enzymes to produce glutamate and keto acids. This process occurs in the liver for most amino acids, but the branched chain amino acids (leucine, isoleucine, and valine) lose their amino groups primarily in muscle through the action of branched chain amino acid aminotransferase. Skeletal muscle contains a high concentration of glutamine, synthesized by glutamine synthetase from glutamate and ammonium ions. During exercise, muscle releases alanine, ammonium ions, and glutamine at accelerated rates. Amino groups on amino acids are removed from the body in the form of urea, a molecule synthesized in liver using the urea cycle and excreted in the urine. The simple urea cycle eliminates amino groups that cannot be oxidized. The purine nucleotide cycle is active in skeletal muscle. During vigorous exercises, AMP deaminase is activated, changing AMP into IMP and ammonia. Regeneration of AMP and thus, the other adenine nucleotides, is accomplished by two reactions of this cycle.

Key Terms

amino acid pool 141
aminotransferase enzymes 143
deamination 143
essential amino acids 141

Review Questions

1. Write out the three reactions of the purine nucleotide cycle, starting with the deamination of AMP, then write a net reaction summarizing the overall cycle.
2. The following data (see table below) were obtained from subjects at rest and exercising on a cycle ergometer at 70% of $\dot{V}O_2$max for one hour. Blood flow was measured and exchange of amino acids across the leg was determined from blood samples obtained from the femoral artery and femoral vein. Leg blood flow averaged 0.3 liters/minute during rest and 8.0 liters/minute during exercise. The average concentration difference between arterial and venous blood for glutamate, glutamine, alanine, and leucine is shown below for rest and exercise conditions. The molecular weight (MW) for each amino acid is also shown.

Amino acid	MW	Arterial minus venous blood concentration difference per minute (μmol/min)	
		Rest	Exercise
Glutamate	146	12	40
Glutamine	146	–25	–100
Alanine	89	–30	–200
Leucine	133	–1	–20

a. What does a negative arterial minus venous concentration difference mean?

b. What was the total volume of blood flow through the leg during the 60 minutes of exercise?

c. What was the fold increase in blood flow from rest to exercise?

d. How many millimoles of glutamate were exchanged by the leg during the exercise period?

e. What are the two major sources of carbon for the alanine released from the muscle?

f. How many grams of alanine were released from the muscle?

g. If the alanine released from muscle during the 60 minutes of exercise was entirely converted to glucose in the liver, how many grams of glucose would this be? How many millimoles of glucose would this represent? MW of glucose is 180.

h. What is the likely main source for the glutamine released from the muscle during exercise?

i. What is the major source of leucine released from the liver during exercise?

j. If this study had been performed with muscles already partially depleted of glycogen, what would be the difference in amino acid exchange values from the numbers shown above?

PART III

Transcription, Protein Synthesis, and Degradation

We have already discussed the essential role of proteins in everything that takes place in an organism. In this section we learn how proteins are made and placed in their most appropriate locations within and outside of cells. As mentioned in the first part of the course, the information to make a protein is coded in the sequence of bases in DNA in cell nuclei. This DNA information is copied to make a sequence of bases in an RNA molecule known as messenger RNA in a process known as transcription. The information in the base sequence of messenger RNA is translated into a precise sequence of amino acids in a protein. In chapter 9, we focus on the process of transcription, learning how it is controlled in an appropriate way. Chapter 10 deals with the translation of messenger RNA information into the amino acid sequence of a protein. In addition, we will learn how newly synthesized proteins are modified to their functional forms and located to the precise location where they will carry out their role. Since proteins are continuously turned over, we discuss processes that break down old proteins so that newly synthesized proteins can take their place.

Before we begin, a review of some terms is in order. The phenotype of an organism represents its physical, observable characteristics, whereas the genotype represents the genetic factors responsible for creating the phenotype. The genome is the same in every cell nucleus in an organism with the exception of the sex cells, which have half the number of chromosomes. Although it represents all the genes in all the chromosomes in the cell nucleus, not all of the genome may be expressed. The word gene is familiar to most people, yet it can have a variety of definitions. From a classic biology perspective it may be described as the basic unit of inheritance. From a modern perspective it may be described as the segment of DNA that provides the information for the amino acid sequence of a polypeptide chain, because most genes provide this information. Lewin (1997) defines a gene as "the segment of DNA involved in producing a polypeptide chain." This includes not only the region that specifies the sequence of amino acids in a protein but also the regions in front and behind that regulate transcription. In humans, less than 5% of all DNA is actually expressed, about 60,000 genes. Put another way, of the more than 3 billion base pairs making up the 23 (for women) or 24 (for men) different DNA molecules in the nucleus, most are never used to make protein molecules.

CHAPTER 9

Transcription and its Control

In this chapter, we trace the path of gene expression from DNA base sequence to the making of a protein. Ribonucleic acid or RNA results when a section of DNA is copied in the nucleus during transcription. RNA plays an important intermediary role in the conversion of the DNA information into a sequence of amino acids in a protein. We learn about the different kinds of RNA, the genetic code, gene transcription, and how transcription is regulated. We pay particular attention to the regulation of transcription, an important area of current research in molecular biology.

Types of RNA

Messenger RNA (mRNA) is the actual template for protein synthesis in the cytosol. This means that the base sequence on mRNA specifies the sequence of amino acids in a polypeptide chain. Most genes will generate mRNA, the lifetime of which is short—usually several hours or more. Messenger RNA is also the least abundant of the three types of RNA, representing about 2% of total cellular RNA.

Transfer RNA (tRNA) is the smallest of the RNA molecules, usually between 73 to 93 nucleotides (Nt) in length. Transfer RNA attaches to specific amino acids and brings them to the complex of mRNA and ribosomes on which a polypeptide is formed. There is at least one tRNA molecule for each of the 20 amino acids involved in protein synthesis. All the tRNA molecules represent approximately 15% of the total cellular RNA.

Ribosomal RNA (rRNA) is the most abundant and represents more than 80% of all the RNA in a cell. A ribosome is a complex of protein and ribosomal RNA where proteins are synthesized. Figure 9.1 shows that the initial product of gene transcription must undergo modification to generate a specific kind of RNA molecule. There is one other type of RNA known as **small nuclear RNA** (snRNA). The snRNA is found in the nucleus associated with protein in particles described as small nuclear ribonucleoprotein particles or snRNPs. These are involved in the processing of the primary RNA transcripts during their conversion to mRNA molecules. We will mention them later.

The Genetic Code

The sequence of four bases in DNA is transcribed to form a sequence of four bases in mRNA that must specify the sequence of the 20 different amino acids used to make proteins. We may ask ourselves how 20 different amino acids can be uniquely described by only four different bases. The only way that four different bases in mRNA (identified as A, G, C, and U) can specify 20 different amino acids in a polypeptide chain is for the bases to be read in groups of three, known as

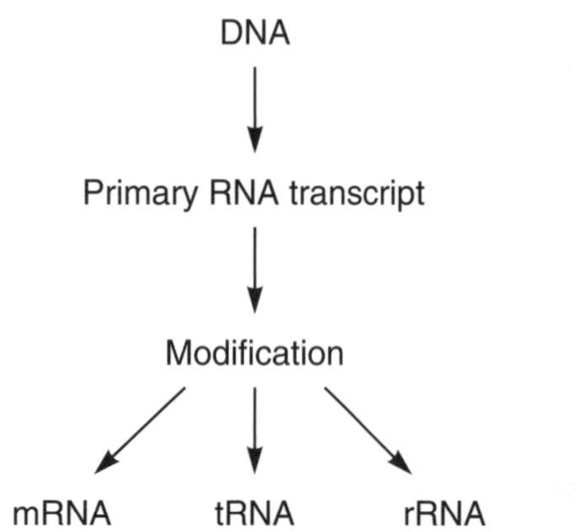

Figure 9.1 The three kinds of RNA generated by transcription of a gene and modification to produce the functional product.

codons. Four different bases read three at a time gives rise to 4^3 different possibilities, or 64 codons. This process is called triplet coding, that is, three bases read together to give a message.

KEY POINT

In chapter 3, we learned that a polynucleotide chain has two ends. At one end there is a ribose (for RNA) or deoxyribose (for DNA) sugar with a phosphate group attached to the 5' carbon. At the other end there is a ribose or deoxyribose sugar with a free 3' OH group. RNA is composed of a single polynucleotide chain, whereas DNA is composed of two complementary, antiparallel polynucleotide chains wound together in a double helix.

The genetic code is the relationship between the base sequence of DNA, transcribed to mRNA, and the sequence of amino acids in a polypeptide. It has a fixed starting point, and the bases are then read from this point in groups of three, nonoverlapping bases at a time. We usually consider the genetic code from the perspective of the codons of mRNA that spell out amino acids. In describing codons on mRNA, we consider the 5' end to be the starting point. Of the 64 possible codons, 61 specify amino acids, and the remaining three are stop signals. Because there are 61 codons for only 20 amino acids, most amino acids have two or more codons. We say, therefore, that the genetic code is **degenerate**. As shown in table 9.1, the codons for an amino acid with more than one codon are very similar. This similarity makes sense if minor errors are made. For example, the four codons for the amino acid glycine are GGA, GGG, GGC, and GGU. Notice that these all contain the same first two bases (letters). If amino acids are similar in structure, their codons are similar. For example, aspartic acid has the codons GAC and GAU, whereas glutamic acid, which is closely related in structure, has the codons GAA and GAG. If there is a base reading error, the same or a similar amino acid can still result.

The codon AUG is the initiation or **start codon** that signals the start of translation. This codon also represents the amino acid methionine; thus the first amino acid used in protein synthesis is always methionine. However, not all functional proteins have methionine as the first amino acid, for it can be removed after the polypeptide is completely formed. The codons UAG, UGA, and UAA are **stop codons**; they say that translation of the mRNA message is ended. The genetic code is universal for all organisms studied except in mitochondria, where some minor variations are noted. However, the amount of mitochondrial DNA is small—16,569 base pairs in humans—just enough to code for 13 polypeptides, 2 kinds of rRNA, and 22 tRNA molecules.

Transcription

During transcription, a small section of one strand of a huge, double-stranded DNA molecule is copied to yield an RNA molecule. The RNA formed is complementary to the DNA strand that is copied, with the exception that in RNA the base U is complementary to base A in DNA. Recall from chapter 3 that complementarity in DNA means that the bases A and T and the bases G and C are complementary in that they can form hydrogen bonds with each other. The precursors needed to make RNA are nucleoside triphosphates (NTP) such as CTP, GTP, ATP, and UTP. The strand of DNA copied is read in the 3' to 5' direction. The RNA will be formed in the 5' to 3' direction. These directions are the same as those involved in making a complementary copy of DNA when it is replicated before cell division, as was discussed in chapter 3.

The formation of RNA during transcription is catalyzed by a large oligomeric enzyme known as DNA-directed RNA polymerase, or simply **RNA polymerase**, of which there are three kinds. RNA

Table 9.1 The Genetic Code

5' base	Middle base				3' base
	U	C	A	G	
U	UUU Phe UUC Phe UUA Leu UUG Leu	UCU Ser UCC Ser UCA Ser UCG Ser	UAU Tyr UAC Tyr UAA Stop* UAG Stop*	UGU Cys UGC Cys UGA Stop* UGG Trp	U C A G
C	CUU Leu CUC Leu CUA Leu CUG Leu	CCU Pro CCC Pro CCA Pro CCG Pro	CAU His CAC His CAA Gln CAG Gln	CGU Arg CGC Arg CGA Arg CGG Arg	U C A G
A	AUU Ile AUC Ile AUA Ile AUG Met#	ACU Thr ACC Thr ACA Thr ACG Thr	AAU Asn AAC Asn AAA Lys AAG Lys	AGU Ser AGC Ser AGA Arg AGG Arg	U C A G
G	GUU Val GUC Val GUA Val GUG Val	GCU Ala GCC Ala GCA Ala GCG Ala	GAU Asp GAC Asp GAA Glu GAG Glu	GGU Gly GGC Gly GGA Gly GGG Gly	U C A G

#Codes for the amino acid methionine but also is the start or initiation codon.
*Stop codons do not have an amino acid assigned to them.

polymerase I (abbreviated RNAP I or sometimes Pol I) transcribes the more than 200 copies of the gene that generates most of the rRNA in humans. Pol I is the most abundant RNA polymerase. RNA polymerase II (RNAP II or Pol II) transcribes genes containing the information for the amino acid sequence in a polypeptide chain. This form of RNA polymerase gives rise to mRNA. RNA polymerase III (RNAP III or Pol III) is the least abundant of the RNA polymerase enzymes. Its products are tRNA and another small rRNA molecule that will form part of the ribosome. In the human genome, about 2,000 copies of this small rRNA gene are transcribed by RNAP III. The Pol III enzyme also generates the snRNA molecules, mentioned previously.

Figure 9.2 shows in simplified form what happens during transcription. Remember, there are two DNA polynucleotide chains. Part of one will be transcribed, producing a complementary copy in the form of the RNA molecule. Also remember, RNA contains uracil, not thymine. The DNA strand that is copied is the **template strand**. The other is the sense strand because it will have the same base sequence as the RNA, except U will replace T. The polarity of the sense strand and RNA are the same and opposite to that of the template strand. Directions are often described in terms of river flow. **Upstream** refers to the 5' direction, and, by convention, the frame of reference is the sense strand of DNA. Likewise, **downstream** refers to the 3' direction of flow.

	← Upstream direction	Downstream direction →
Sense strand	5′...A C G G T A A T G G C...3′	
Template strand	3′...T G C C A T T A C C G...5′	
RNA strand	5′...A C G G U A A U G G C...3′	

Figure 9.2 A section of double-stranded DNA showing the strand copied during transcription (template strand) and the untranscribed or sense strand. The RNA strand is complementary to the template and has the same base sequence as the sense strand, with uracil (U) in RNA replacing thymine (T) in DNA.

Figure 9.3 shows the three major phases of transcription. Initiation begins when general transcription factors (proteins, which we will discuss next) and RNA polymerase bind to the double-stranded DNA just upstream of the **start site**, forming a preinitiation complex. The start site is where transcription actually begins, and the first DNA base copied is given the number +1. Bases immediately upstream are in a region called the **promoter** and are given negative numbers. The promoter region can also include some bases just downstream from the transcription start site. When RNA polymerase and the general transcription factors bind to the promoter region, the DNA at the start site is unwound, exposing the template strand that will be copied. The first nucleotide triphosphate comes in, recognizing and binding to its complementary base on the template strand via hydrogen bonds. Then the second NTP comes in, recognizing its complementary base, number 2 on the template strand. A phosphoester bond is formed between the 3' OH of ribonucleotide one and the 5' phosphate of ribonucleotide two, and a pyrophosphate is released. During initiation, the RNA polymerase does not move along the DNA.

In the elongation phase, the RNA polymerase moves along the template strand of DNA, making a complementary RNA strand. As it moves, it un-

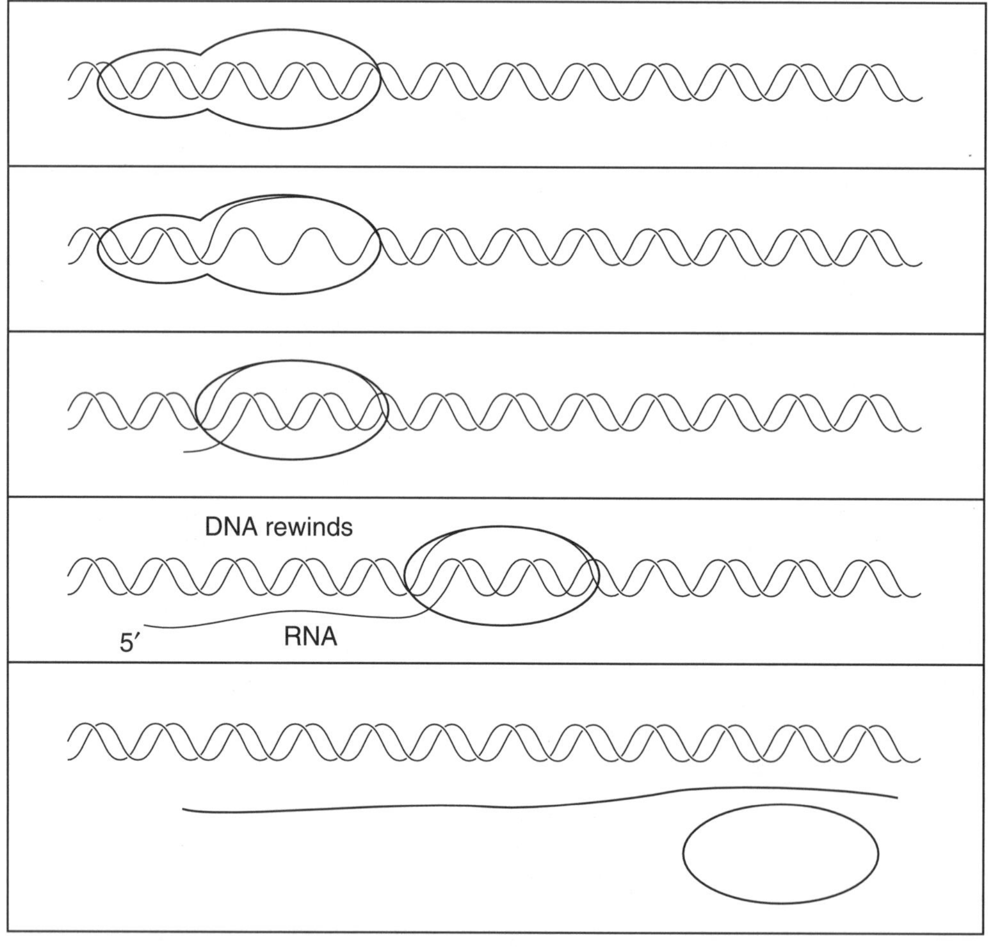

Figure 9.3 Transcription occurs in three distinct stages. **Initiation** involves RNA polymerase along with accessory proteins, which are shown as an extension to the polymerase enzyme, binding to double-stranded DNA (top panel). DNA is unwound, exposing the template strand to be copied (second panel). During **elongation,** the polymerase moves along the DNA, making a strand of RNA complementary to the DNA template by adding one base at a time (third panel). The DNA is unwound as the polymerase moves along, whereas behind the enzyme DNA rewinds (fourth panel). When a sequence of bases spelling out a **termination** message is reached, the RNA polymerase and the primary RNA transcript molecule dissociate from the DNA (bottom panel).

winds the DNA double helix, catalyzing the formation of the RNA strand. Behind it, the DNA double helix re-forms. The template strand is read in the 3' to 5' direction, and the RNA strand is formed in the 5' to 3' direction. When the bond forms between the terminal free 3' OH of the growing RNA polynucleotide and the incoming NTP, only one of the three phosphate groups is needed; the other two are released as inorganic pyrophosphate, PPi—the same as occurs in DNA replication. Figure 9.4 shows the details of the addition of a new ribonucleotide to the growing RNA chain. The termination phase begins when the RNA polymerase has moved along the template strand of DNA and reaches a sequence of bases that indicates the gene message is terminated. At this point, the RNA polymerase and the RNA strand dissociate from the DNA.

Regulation of Transcription

Each cell nucleus contains all the genes for that organism as base sequences in the DNA. However, cells only express the genes they need. For example, a liver cell will express genes also expressed by other cell types, but it will also express a number of liver-specific genes, different genes compared to a nerve or heart muscle cell. The human body contains more than 250 different kinds of cells. Also, as cells differentiate during development, different genes are expressed, giving rise to new proteins, and formerly active genes are no longer expressed. Furthermore, adult cells adapt to new circumstances by expressing previously inactive genes and not expressing previously active ones. Clearly, transcription must be regulated. This means that it is necessary to regulate which

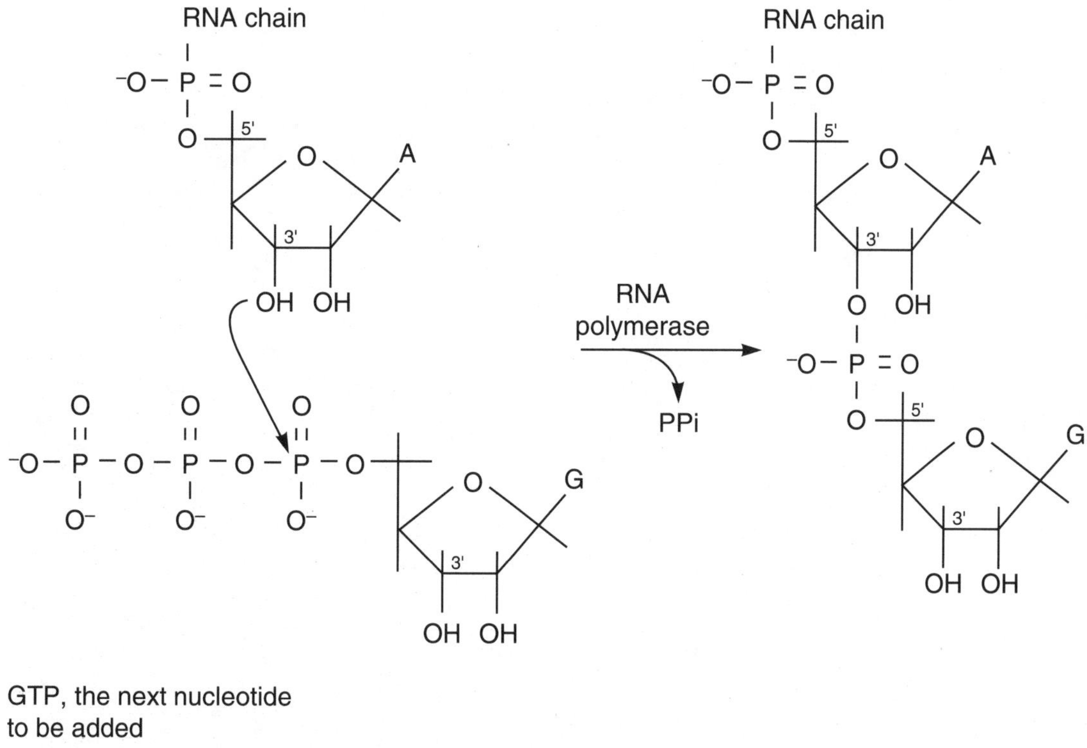

RNA chain

GTP, the next nucleotide to be added

RNA polymerase

PPi

RNA chain

Figure 9.4 The RNA chain grows during transcription by the addition of nucleoside triphosphates to the 3' end of the RNA chain. In the example shown, the new nucleoside triphosphate is GTP. On the left side of this figure, the OH group at the 3' end of the RNA chain attacks the alpha phosphate of GTP, creating a new phosphoester bond and releasing inorganic pyrophosphate (PPi). The chain on the right side is now lengthened by one nucleotide.

genes are expressed and at what rate. Although every human cell nucleus contains about 60,000 genes, only a fraction of these are transcribed at any one time, and those that are expressed depend on the cell type and its stage of development.

KEY POINT

In chapter one, we indicated the importance of proteins, which direct virtually all the events in a multicellular organism associated with a healthy life. The information for protein structure is based on the sequence of amino acids, dictated by the sequence of bases in a segment of DNA called a gene. Since multicellular organisms have highly specialized cells, we can assume that this specialization is due to the production and function of specific proteins. This means that specific genes must be properly transcribed in all cells.

Control of gene expression occurs primarily by directing the initiation stage of transcription. Two major types of control decisions must be described for a cell. Irreversible decisions turn specific genes on or off completely. For example, during embryo development, certain genes are initially expressed then turned off completely. Other genes are irreversibly turned on. But the second type of decision is adjustable in terms of transient increases or decreases in the rate of transcription of an already expressed gene in response to various environmental or metabolic conditions. Consider the analogy of a light switch with a dimmer. The irreversible decision is that the switch is either on or off. Once on, however, it can be adjusted to produce a low, moderate, or high level of light intensity.

Cis, Trans Base Sequences and Proteins

Proteins are responsible for most of what happens in an organism, including the regulation of gene transcription. Proteins that regulate transcription are known as **transcription factors**. They are products of their own genes, most likely on different DNA molecules from the gene or genes they regulate. They are also called **trans factors** or transactivating factors. If they promote transcription, they are **activators**, and if they have a negative effect on transcription, we call them **repressors**. We can define two major classes of trans factors: general factors needed for initiating transcription from most protein-coding genes and those that modulate general transcription factors and provide specificity for cell types and differentiation. We will discuss these shortly.

To regulate transcription, trans factors must bind to specific DNA sequences known as cis-regulatory elements, or cis elements, because they are on the same DNA molecule as the gene they regulate (usually very close to the base sequence being transcribed, although some are long distances away, many kilobases from where transcription actually starts). A cis-regulatory element is a base sequence described by any one of the following terms: a motif, a response element, or a box. They are normally short, usually 6 to 10 bp (base pairs), although some may be larger. These motifs, boxes, or elements have a negative charge because of the sugar phosphate backbone of the DNA strands. Figure 9.5 illustrates the region immediately up-

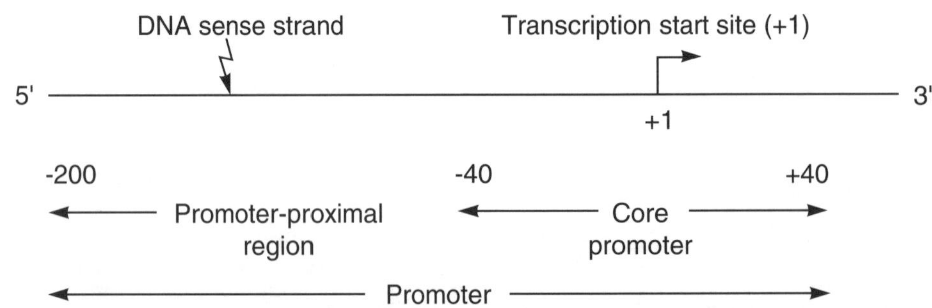

Figure 9.5 Control of transcription is always described in terms of the sense strand of DNA in the 5' to 3' direction. Transcription begins at base +1, identified as the transcription start site. Bases upstream of this have negative numbers and those downstream, in the 5' direction, have positive numbers. The core promoter ranges from roughly –40 to +40. The promoter-proximal region is upstream of this, extending to approximately –200. Together, the core promoter and promoter-proximal region can be called the promoter.

stream and downstream of the transcription start site known as the promoter or promoter region. Note that the description of transcription control always is in reference to the sense strand of DNA, which runs from left to right in the 5' to 3' direction. The transcription start site is base +1. The core promoter is the region that is –40 to +40 in terms of the start site. Immediately upstream of this region is the proximal-promoter region, ranging to –200 bases from the start site. Together, the core promoter and the proximal-promoter region can be called the promoter.

Upstream response elements, sometimes called upstream activating sequences or elements are cis elements upstream of the promoter, to which a variety of transcription factors can bind to influence transcription initiation from the promoter region. These upstream response elements may bind to steroid hormone receptors or other transcription factors. **Enhancers** are transcription control regions containing motifs that increase transcription from a gene promoter. What sets enhancers apart from the promoter or upstream response elements is that enhancers can influence transcription of their genes from variable locations. Unlike the promoter and upstream response elements, which must be in fixed locations, enhancers function from afar. Moreover, enhancers are not orientation specific, in that they could be not only moved to a different location but also turned to the opposite orientation, and they would still function. Enhancers can be found upstream, downstream, or even within the transcribed region. Enhancers create their influence through protein-protein interactions. The trans factors binding to enhancers interact with transcription factors within the promoter.

The trans factors (regulatory proteins) do not work in isolation. They must obviously bind to specific cis motifs (on DNA) and to other trans factors because more than one trans factor is always involved in gene transcription. Therefore, trans factors must have amino acid sequences that both recognize and bind to a specific cis motif on DNA, as well as amino acids that bind other trans factors to carry out their activating role. Accordingly, the amino acid region of the trans factors that binds to the cis motif must have a positive charge to interact with the negatively charged DNA phosphate groups. The region of the trans factors that recognizes and binds to other trans factors in activation must have a specific recognition conformation. A number of protein-DNA and protein-protein binding regions have been identified and given interesting names, such as the leucine zipper (every seventh amino acid is the hydrophobic amino acid leucine), zinc fingers (contains the metal ion zinc), and the helix turn helix (a short region of alpha-helix, followed by a short looping section, then another region of alpha-helix).

The Basal Transcription Apparatus

RNA polymerase II cannot initiate transcription by itself. Rather, it requires a number of general transcription factors interacting with it in the core promoter region. The proper combination of RNAP II and the general transcription factors in the core promoter ensure that the gene will be transcribed. However, the rate of transcription will be at a low or basal level because the influence of other trans factors bound to their cognate cis elements in upstream response elements and enhancers are necessary to maximize the rate of transcription of the gene. In this section, we focus only on the basal transcription of a gene.

Genes transcribed by RNAP II have a narrow region known as inr (initiator region) located between –3 and +5 where transcription is initiated. In addition, nearly all genes coding for proteins in eukaryotes have a **TATA box** in their promoter region. The TATA box is an adenine-thymine only 8-base sequence found about 25 to 35 bp upstream of the transcription start site, where a transcription preinitiation complex is formed involving general transcription factors and RNAP II (Pol II). Genes lacking a TATA box often have a stretch of GC-rich sequences approximately 100 bp upstream of the start site, which likely plays a role in transcription initiation.

The general transcription factors and RNAP II are assembled at the core promoter in a prescribed order to initiate transcription of a gene. The general transcription factors include TFIID, TFIIB, TFIIA, TFIIE, TFIIF, and TFIIH. These are oligomeric proteins with a variety of subunits. TFIID, in particular, contains a number of subunits, including a large TATA binding protein (TBP) subunit and several associated factors, known as TATA associated factors (e.g., $TAF_{II}250$, $TAF_{II}150$, and $TAF_{II}70$), which make specific contacts with the TATA box and other core promoter elements, including the inr. TFIID also contains other TATA-associated factors, which make contact with other elements of the general transcription apparatus and other upstream elements and enhancers.

The actual sequence of events involves recognition of the TATA box by subunits of TFIID (i.e., TBP), followed by formation of TFIID-IIA-IIB complex. Following the binding of the other members

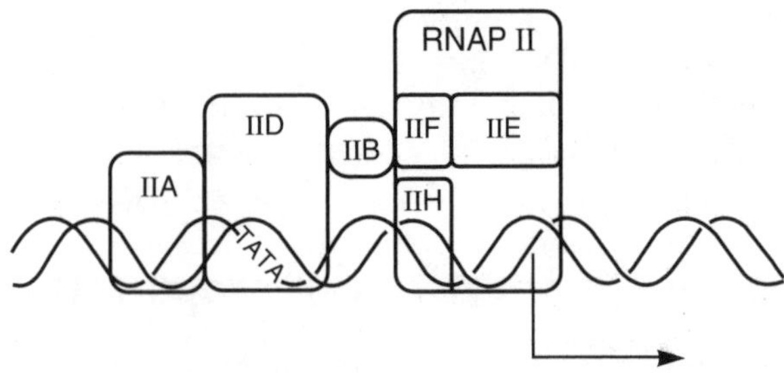

Figure 9.6 A schematic to show the final assembly of the general transcription complex necessary for proper transcription of genes and containing a thymine (T) and adenine (A) TATA box in the promoter. The individual transcription factors are shown as balloons, each identified by II and a letter. RNA polymerase II (RNAP II) is shown near the transcription start site, indicated by the bent arrow. Both strands of DNA are shown, and the only bases identified are those in the TATA box where transcription factor IID binds.

of the general transcription complex (TFIIE, IIF and IIH, and RNAP II), a fully activated preinitiation complex is formed. The preinitiation complex is sufficient to initiate transcription from TATA-containing promoters at a basal level. Figure 9.6 shows the binding of all the general transcription factors at the TATA box, creating an active preinitiation complex. There are a small number of genes with promoters lacking a TATA box, known as TATA-less promoters. For these genes, the same general transcription factors are involved in initiating transcription. It is likely that the inr is involved in positioning the general transcription factors.

Induced Levels of Transcription

In addition to the TATA box, which is so widely found in the promoter region of protein-coding genes, there are other base sequences recognized by specific proteins that can lead to higher levels of transcriptional control. These cis elements may be found in the promoter-proximal region beyond the core promoter (see figure 9.5), in upstream response elements and enhancers. From our description of these other cis elements, they may be close to or far removed from the TATA box along the DNA molecule. Each of these cis elements will bind its own specific transcription factors. These will interact with the general transcription factors to modulate transcription of specific genes, mostly by increasing the frequency of transcription initiation. This interaction is achieved in the preinitiation complex by bending or looping of DNA so distal cis elements with their bound trans factors can interact with the general transcription factors at the point where RNAP II binds just upstream of the transcription start point. Figure 9.7 illustrates in simple fashion how DNA looping brings distal transcription factors into contact with the general transcription factors at the start site of transcription to regulate the overall transcription of a gene.

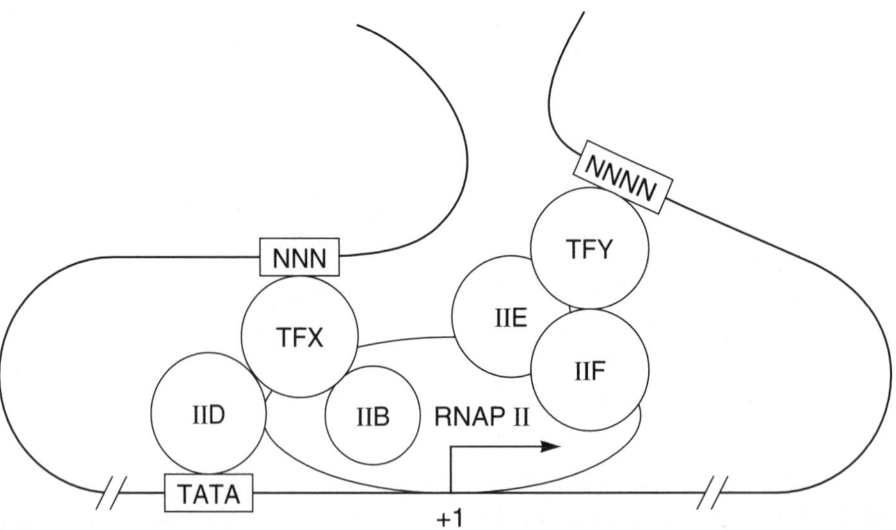

Figure 9.7 A model to illustrate how looping of DNA allows distal cis elements, identified by boxes and the letter N to indicate unspecified bases, to interact through their transcription factors (TFX and TFY) with the general transcription factors (IIB, IID, IIE, and IIF) at the site of the initiation. Only some of the general transcription factors are shown, along with RNA polymerase II (RNAP II). Only a single strand of DNA is shown. The transcription start site is indicated by the bent arrow and the +1.

There are also proteins that can act between a bound transcription factor and the proteins of the general transcription apparatus. These coactivators never actually contact a cis element but connect a distal transcription factor, brought close to the promoter by DNA looping, and one or more of the general transcription factors. Trans factors that augment transcription at the point of initiation may be ubiquitous, that is, found in and active in a variety of cell types and species, or they may be highly specific, expressed in a single cell type. If they are ubiquitous, the cis elements they bind to will be the same in a variety of species. We say that these are conserved sequences. For example, the CCAAT box is found between 80 and 100 bp upstream of the transcription start site of many mammalian genes. The CCAAT binding factor or CBF is the transcription factor interacting with the CCAAT motif. There is a cyclic AMP response element (CRE) in a wide variety of genes. This response element contains the nucleotide sequence (TGACGTCA) and binds a protein known as the cyclic AMP response element binding protein (CREB). This trans factor can be phosphorylated by protein kinase A, which, as we discussed, is responsive to cyclic AMP binding. This last example illustrates a classical signal transduction mechanism regulating expression of genes. An agonist in the blood (e.g., epinephrine) binds to its cell membrane receptor, raising the level of cAMP in the cell and promoting the transcriptional activation effect of CREB through phosphorylation and subsequent binding to CRE.

A variety of hormones and related molecules act within a variety of cell types by directly regulating gene expression. The **steroid hormones** (glucocorticoids, testosterone, estrogen, and progesterone), thyroid hormone, the active forms of vitamin D, and retinoids (from vitamin A) circulate in the blood and readily enter cells because their lipophilic nature permits them to diffuse across the cell membrane. Inside the cell, they bind to protein receptors, which are themselves products of specific genes. When the hormone (or ligand) is bound to its receptor, the receptor conformation is structurally altered.

The hormone-receptor complex can now bind to specific cis elements, known as hormone response elements (HRE), in the vicinity of target genes to regulate their expression. Gene regulation occurs when two hormone receptors with bound ligand bind at the HRE. This binding is a common feature of specific gene activation. If the two transcription factors (or, in this case, hormone receptor plus hormone) are the same, they are homodimers; if they are different, they are heterodimers. Figure 9.8 illustrates a common sequence for steroid hormone activation of a gene. Often, HREs are located upstream of the TATA box, but some, like enhancers, are also several thousand base pairs from the start of transcriptional initiation. The ligand-receptor complex, bound to its specific HRE, helps induce the formation of a transcription initiation complex and, along with RNAP II, promotes the beginning of transcription. Binding of the ligand-receptor complex to the HRE may promote a conformational change in DNA, opening the gene for transcription or repressing the gene by competitively blocking access of other transcription factors near the transcription start site. In figure 9.7, transcription factors X and Y (i.e., TFX and TFY) could represent hormone receptor complexes, bound at their specific HREs and interacting with the general transcription complex. Artificial analogues of testosterone, such as anabolic steroids, can stimulate skeletal muscle growth, likely by increasing the transcription of certain muscle protein genes.

KEY POINT

In previous chapters we have discussed metabolic regulation through the action of hormones, nerve signals, growth factors, and the like. Regulation by external signals is part of the overall process of generating appropriate cellular responses within an organism through a process called signal transduction. One effect of signal transduction is to influence the expression of specific genes.

Each cell type contains the same DNA and, therefore, the same cis regulatory elements. However, this does not mean that all cells can activate the same genes. For example, the manner in which the DNA is organized can influence the ability of a gene to be transcribed. We discuss this later. A particular cell type may only express a limited range of transcription factors—its own transcription factors, many common transcription factors found in other cells, plus the general transcription factors (i.e., TFIID, TFIIE, RNAP II, etc.). Even though a steroid hormone enters a cell, it cannot influence transcription unless that cell expresses the gene for the specific receptor for that hormone.

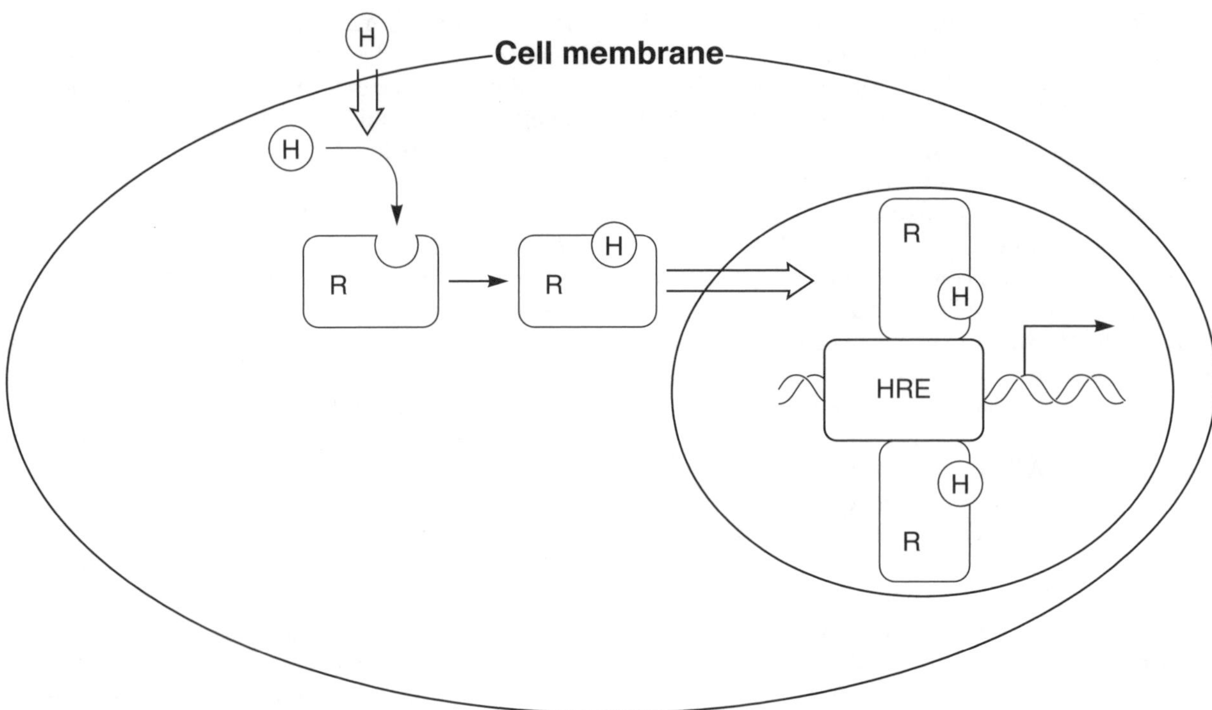

Figure 9.8 The likely sequence in which a steroid hormone (H) influences gene transcription. The hormone, being lipophilic, diffuses across the cell membrane and interacts with its specific receptor (R). The hormone receptor complex now crosses the nuclear membrane and interacts with its specific hormone response element (HRE) in the promoter region of a gene. Pairs of hormone receptor complexes bind as homodimers, along with the other members of the general transcription complex (not shown), to increase (or decrease) the rate of transcription initiation from the start site, shown by a bent arrow.

Thus, by controlling expression of transcription factor genes or genes for steroid hormone receptors, cells and tissues maintain their unique character. In addition, the presence of some trans factors (i.e., whether their genes are expressed) depends on signals originating outside the cell, such as nerve stimuli, growth factors, or hormones. Not all hormones enter the cell. Many hormones and all growth factors interact with specific receptors on the cell membrane and generate internal signals that influence gene expression as we saw with the case of epinephrine. Hormones, **growth factors**, and nerve signals allow the multicelled organism to control events within individual cells so that the organism operates in an integrated, cohesive manner.

Skeletal Muscle

The control of gene expression in skeletal muscle has been an area of active research for many years. Originally, the expression of genes in muscle in response to various models of increased or reduced activity was inferred from changes in the content of specific proteins in muscle, particularly mitochondrial proteins. With the ability to measure both steady state levels and rates of formation of specific messenger RNA molecules, we are able to understand more precisely how activity or inactivity influences skeletal muscle genes. This is particularly significant, because a change in muscle activity sufficient to influence gene expression will show up much earlier as a change in the mRNA for a protein. Indeed, the time before a change in activity results in a notable change in muscle fiber type may lag weeks behind a change in expression of myosin heavy chain genes. As mentioned previously, the myosin heavy chain composition of a fiber plays the dominant role in defining muscle fiber type, determining how rapidly ATP can be hydrolyzed.

Four highly specific trans factors, found only in skeletal muscle and essential for their development, are the **myogenic regulatory factors**: MyoD, myogenin, MRF4, and myf5. These proteins contain secondary and tertiary structural components known as the basic helix loop helix (bHLH) that recognize a six-base sequence (cis element) known as an E box. The myogenic regulatory factors form heterodimers with another class of proteins called E proteins at the E box. The E box is found in the regulatory region of some specific muscle genes, including those expressing the fast myosin light and heavy chains (see figure 1.12, showing heavy and light chains of myosin). Artificial expression of the MyoD gene family in other cell types can result in those cells expressing genes only transcribed in skeletal muscle. This emphasizes the significant role of the myogenic regulatory factors in the development of skeletal muscle.

Endurance exercise training can also increase the transcription of genes that code for enzymes involved in mitochondrial metabolism. These genes are primarily located in nuclear DNA, but genes on mitochondrial DNA are also transcribed more rapidly with endurance exercise. Evidence suggests that genes coding for enzymes involved in glycolysis are transcribed at a slower rate with aerobic training. Whether regularly performed exercise alters the expression of genes coding for contractile proteins is also being studied. For example, the gene for the IIX heavy chain of myosin may be transcribed less and the IIA myosin heavy chain gene expressed more during short-term endurance training and even after a few sessions of heavy resistance training. However, an actual switch in the transcription to the type I myosin heavy chain gene from the myosin IIA heavy chain has not been adequately demonstrated in human exercise training studies.

KEY POINT

Although many exercise scientists are interested in the role of various types of exercise in modulating gene expression in muscle, some exercise scientists look at more extreme forms of activity or inactivity to understand the limits of adaptation of muscle. For example, implanted electrodes around the nerve to a muscle can create artificial contraction situations where a slow muscle (e.g., the soleus) is forced to contract in bursts by high frequency stimulation. Alternatively, a fast muscle (extensor digitorum longus) is made to contract continuously with low frequency stimulation. Casting a limb to inhibit range of motion activity, suspending a hindlimb to unload muscles, or cutting the nerve to a muscle are other techniques used to probe the extremes of the continuum of activity and inactivity and the role played by the nerve in maintaining distinct muscle properties.

Calcium is the signal initiating the contractile event, as we discussed in chapter 4. Released from the sarcoplasmic reticulum in response to nerve activation of a fiber, the level of calcium in the muscle fiber reflects the frequency and duration of fiber activation. Besides binding to the thin filament protein, troponin, to initiate contraction, calcium can also bind to other cell proteins, including an ubiquitous calcium binding protein known as **calmodulin**. When calmodulin binds calcium, it can bind to and influence the activity of other proteins such as calcineurin. This is a protein phosphatase that dephosphorylates specific proteins containing one or more covalently bound phosphate groups (see chapter 2). One of the substrates for calcineurin is a transcription factor known as **NFAT** (the nuclear factor of activated T cells). NFAT is a transcription factor, first identified in T cells. However, it is also found in skeletal muscle, where it is believed to play an important role in skeletal muscle response to altered activity. As shown in figure 9.9, when activated by binding calcium-bound calmodulin, calcineurin can dephosphorylate NFAT. This allows it to translocate to the nucleus where it can bind to a specific NFAT response element (NRE) to regulate specific genes. Of course, reduced muscle activity due to injury or other forced inactivity means that the calcineurin pathway to upregulate the expression of specific genes is diminished. Therefore, a change in muscle activity, reflected by an altered level of calcium, can influence the expression of specific muscle genes.

Regulation by Organization of DNA

There is a DNA molecule in each of the 46 chromosomes in each cell nucleus. However, chromosomes

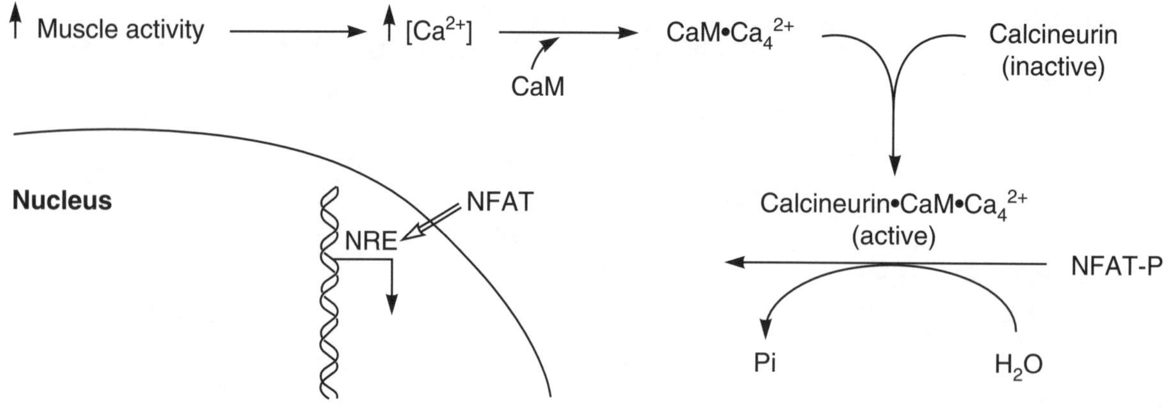

Figure 9.9 The calcineurin NFAT pathway can regulate the expression of genes in skeletal muscle. An increase in muscle activity increases the concentration of calcium [Ca^{2+}]. The Ca^{2+} binds to the calcium binding protein calmodulin (CaM), forming a complex CaM·Ca_4^{2+}. When inactive calcineurin binds CaM·Ca_4^{2+}, it forms a calcineurin·CaM·Ca_4^{2+} complex, which is active as a protein phosphatase. Dephosphorylation of NFAT-P produces NFAT, which translocates to the nucleus and activates genes containing a response element NRE.

are not visible as discrete structures except during cell division. As we mentioned in chapter 3, nuclear DNA is found in association with protein in a dense, tightly coiled mass known as chromatin.

There is another level of structure of DNA in nuclei. **Nucleosomes** are repeating subunit structures of chromatin. A nucleosome consists of 146 base pairs of DNA wrapped 1.65 turns around an octamer of eight histone proteins. The histone proteins are small proteins, rich in basic amino acids such as lysine and arginine, which carry a positive charge at pH 7 (see figure 1.4). Since each DNA strand has a negatively charged sugar phosphate backbone, the net positively charged histone proteins create a strong linkage. The histone octamer consists of two copies each of histones H2A, H2B, H3, and H4. Between adjacent nucleosomes there is a strand of DNA, known as linker DNA, containing roughly 50 base pairs. Figure 9.10 illustrates the general structure of a single nucleosome and three nucleosomes attached together by linker DNA.

The tight coiling of chromatin plus the nucleosome structure suggest that initiating transcription must be rather difficult, considering the intricate arrangement of transcription factors and DNA cis elements needed. Indeed, as they exist, nucleosomes block proper transcription initiation and make transcription elongation by the large RNAP II complex difficult at best. How then are genes transcribed? One of the fairly new discoveries about DNA is that certain lysine residues in core histone proteins can be acetylated, that is an acetyl group is added to the side chain ammonium group (see figure 1.4). Lysine acetylation of histones involves the changing of a small, positively charged ammonium group to a larger N-acetyl group without a charge. The tight binding of DNA to core histones is facilitated by attractions between the positive charge on the many lysine side chains to the negatively charged DNA backbone. Reduction in this binding facilitates the unraveling of DNA, exposing the promoter and other control regions and facilitating transcription elongation by RNAP II.

A class of enzymes found within the nucleus carries out this acetylation process. These are known as histone acetyltransferases or HATs. Acetylation is not a permanent state, so a related enzyme known as histone deacetylase removes the acetyl group, creating a positive charge and facilitating nucleosome formation. Consistent with the idea that histone acetylation facilitates transcription, it has been discovered that transcriptionally active regions of the genome are highly acetylated. Regions with low or zero rates of gene transcription have little histone acetylation. The accessibility of areas of DNA to hydrolysis by **nucleases** parallels the ability to be transcribed. This gives rise to the concept that areas of the genome

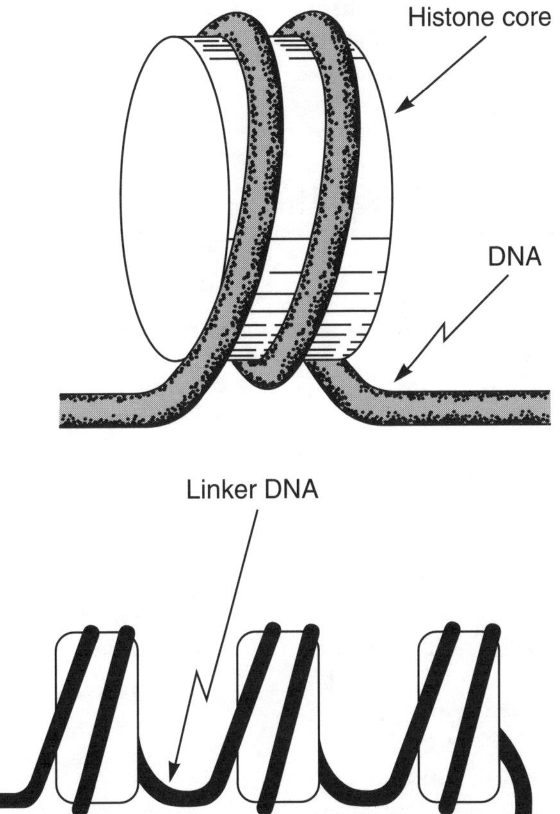

Figure 9.10 A single nucleosome, showing 1.65 turns of DNA wrapped around a histone core consisting of two of each of the histones H2A, H2B, H3, and H4 (top); three nucleosomes joined together by a strand of linker DNA (bottom).

are repressed, with low levels of gene transcription due to low levels of histone acetylation.

Found in the nucleus are ATP-dependent chromatin remodeling enzymes that change the structure of chromatin. These enzymes utilize the free energy of ATP hydrolysis to alter the conformation of chromatin, exposing repressed regions due to nucleosomes and tight higher levels of chromatin packaging. Serine residues in the histone H3 can be phosphorylated by a class of protein kinases that are activated by certain growth factors. The addition of the phosphate group to histone H3 is associated with increased gene expression. The nucleus also contains a class of enzymes that can add methyl groups to cytosine bases in DNA. This alters the ability of nearby genes to be transcribed. The complex protein-DNA and protein-protein interactions involved in transcription initiation, plus the ability of chromatin to be remodeled through histone acetylation and deacetylation, phosphorylation and dephosphorylation, and by ATP-dependent remodeling enzymes underscores the complexity of gene transcription.

Summary

Transcription of genes produces an RNA molecule called the primary RNA transcript. Depending on the gene transcribed, the transcript will be modified to produce a messenger RNA (mRNA), transfer RNA (tRNA), or ribosomal RNA (rRNA). The base sequence of the mRNA is read in groups of three called codons, which constitute the words of the genetic code. Sixty-one of the possible 64 codon combinations spell out the 20 amino acids, so the genetic code is said to be degenerate. Three of the codons (UAA, UGA, and UAG) stop the message. Transcription is carried out by a group of multisubunit enzymes known as RNA polymerases. These enzymes copy the genes that will become rRNA (RNAP I or Pol I), mRNA (RNAP II or Pol II), or tRNA (RNAP III or Pol III). During transcription, the polymerase unwinds the two DNA strands and copies part of one, reading it in the 3' to 5' direction. The nontranscribed DNA strand, called the sense strand, has the same base sequence as the primary transcript, except that thymine, in DNA, is substituted by uracil in RNA.

In the region of DNA before the transcription start site, specific protein molecules recognize base sequences to regulate the process of transcription. The base sequences, or promoters, are known as cis elements, and the proteins that bind there are called trans factors. Two types of trans factors are identified. General transcription (trans) factors are needed for the transcription of all genes. These bind upstream of the start site at a cis element called the TATA box and help position RNA polymerase and start the transcription process. Virtually all genes require other tissue or developmentally specific proteins to provide an additional level of transcription control. Circulating hormones and growth factors can influence transcription by entering the cell, binding to specific receptors, and modulating the process; members of the steroid hormone family work this way. Other hormones and growth factors bind to receptors on the cell membrane and generate second messenger molecules within the cell that alter the rate of transcription. The protein-DNA complex in the nucleus known as chromatin is organized into basic structures called nucleosomes. The tight DNA loops within nucleosomes and higher orders of DNA coiling in the nucleus repress transcription. Remodeling of chromatin, particularly by the addition of acetyl groups to histones or by other ATP-dependent chromatin remodeling enzymes, adds a further level of complexity to transcription control.

Key Terms

Review Questions

1. The ratio of A to G in human DNA is 1.56. It is 1.75 for T to C, and the ratio of purines to pyrimidines and A to T is 1.0. What is the mole fraction of A, C, G, and T in human DNA?
2. An RNA strand that is produced by the sense (as opposed to template) strand of DNA would be an antisense RNA, which should hybridize with the normal RNA strand because they are complementary. A potential strategy to block the action of a mRNA molecule producing a protein with a harmful effect would be to induce the sense strand of the gene to express a complementary antisense RNA to block the harmful mRNA. What would be the nucleotide sequence of the template strand of a section of duplex DNA that might be induced to generate antisense RNA against a mRNA with the partial sequence 5'-CCUACGGTCAAU-3'?
3. What is the sequence of amino acids in a section of polypeptide containing the codons 5'-UUACCGACAGUCUCC-3'?

CHAPTER 10

Protein Synthesis and Degradation

We have referred in chapters 1, 6, and 8 to the turnover of protein. This constant cycle of protein synthesis (translation) and protein degradation is so important that it represents the starting point for this book (see figure 1.1). A constant cycle of protein turnover serves three important functions. First, it allows the cell to destroy modified or wrongly formed proteins, reducing the potential for these to cause harmful changes in the cell. Second, body protein is also a potential source of oxidizable fuel for times when the organism is not ingesting sufficient food energy. Therefore, during starvation and fasting, rates of protein synthesis decline whereas rates of protein degradation increase, resulting in the increased appearance of endogenous amino acids for oxidation. Finally, changes in the rates of protein synthesis and degradation allow the overall organism and specific tissues to adapt to altered nutritional and environmental conditions. For example, a surfeit of carbohydrate in the diet induces increased formation of enzymes for de novo lipogenesis in liver. Exercise training can lead to changes in the content of specific proteins in the trained skeletal muscles.

In this chapter, we look at the mechanisms for creating and regulating polypeptide synthesis. In addition, we look at how newly synthesized polypeptides are combined with other polypeptides to make a functioning unit (often described as posttranslational processing). Finally, we look at the breakdown of proteins in cells, the various processes by which this occurs, and factors that control the rate at which this occurs. Proteins are synthesized on ribosomes in the cell cytosol. The messenger RNA (mRNA) message is translated into a sequence of amino acids. Before any of this can take place, however, the RNA molecules made by transcription must be modified to generate the active forms of transfer RNA (tRNA), ribosomal RNA (rRNA), and mRNA.

Posttranscriptional Modifications of RNA

In any cell, most of the expressed genes are single copy genes that provide information for the sequence of amino acids in a polypeptide. However, there are multiple copies of genes that provide the information for the formation of RNA molecules destined to become rRNA and tRNA.

Formation of mRNA Molecules

Following transcription of genes coding for polypeptides, the product is known as the **primary transcript** or pre-mRNA. Because in any nucleus, many hundreds of genes are being transcribed simultaneously, there will be a huge variety of pre-mRNA molecules. This may explain why some describe the recently transcribed messenger RNA

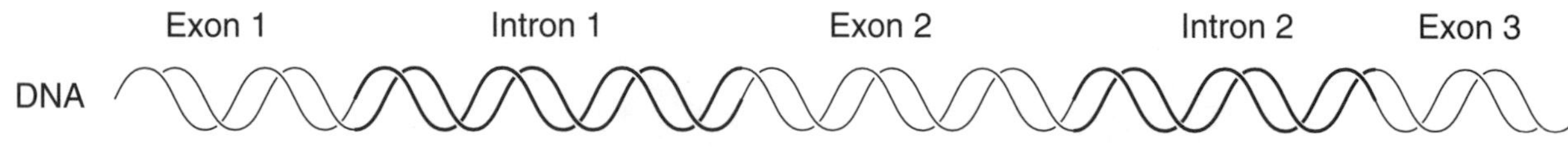

Figure 10.1 Most eukaryotic genes have alternating sections containing coding information (exons) and noncoding information (introns). The exons contain base sequences that will be spliced together to appear in mRNA, whereas the introns will be spliced out.

as heterogeneous nuclear RNA or hnRNA. This will undergo three kinds of changes before it will become a mature mRNA molecule. First, it will have a cap added to its 5' end. Next, it will have added a tail that consists of multiple copies of adenine nucleotides. Finally, it will have its introns removed. This last step is very important because the genes in eukaryotic organisms are interrupted. This means that these genes are in pieces. Parts of the gene, known as **exons**, contain information that will appear in mature mRNA; these can be called coding regions. Other parts contain base sequences, known as **introns** or intervening sequences, which will not appear in the mature mRNA (see figure 10.1). With a size of 2.4 megabases (millions of base pairs), the gene for Duchenne muscular dystrophy is the largest in the human genome. Its introns are as long as 100 to 200 kilobases, and it may take 24 hours to be transcribed.

As you will recall, transcription of genes coding for polypeptides (proteins) or for rRNA or tRNA proceeds in the 5' to 3' direction. Before the transcription process for a messenger RNA molecule is very far along, a cap is added to the 5' end. This involves adding a GTP to the free 5' end of the first nucleotide, creating a 5'-to-5' linkage with three intervening phosphate groups. Then, a poly A tail is added to the completed 3' end for most RNA molecules destined to become mature mRNA. The poly A tail consists of about 100 adenine nucleotides, added by the action of an enzyme known as poly A polymerase, using ATP as the substrate. In mammals, this enzyme begins adding the poly A tail 20 to 30 nucleotides after the base sequence AAUAAA on the RNA, at a point where a GU-rich sequence begins. The poly A tail is not part of the DNA transcription process but is added afterward.

The next step is to remove the introns from the capped, polyadenylated, primary gene transcript via a splicing process in which the junction between introns and exons is cleaved at both the 5' and 3' ends. The introns are then removed and contiguous exons joined up. The details of this process are complicated, involving small nuclear RNA molecules known as snRNAs in combination with the pre-mRNA molecule, forming a spliceosome. As a result of the 5' capping, the 3' polyadenylation, and intron removal by cutting and splicing, we now have what has been called a mature mRNA that leaves the nucleus for the cytosol where it acts as a template on which a polypeptide (protein) will be made, employing ribosomes.

For many genes, every intron is removed and every exon is incorporated into the mature mRNA. However, it is not uncommon for one gene to give rise to two or more mRNA molecules and thus two or more final proteins. This characteristic is common for some of the contractile proteins in skeletal, cardiac, and smooth muscle and can occur in these instances:

1. Transcription is initiated at different promoters, resulting in different 5' exons in the mRNA.
2. Transcription terminates differently because of more than one site of polyadenylation, resulting in different 3' exons.
3. Different internal exons in the gene are included or not included in the splicing process, giving rise to different mRNA molecules (thus two or more different polypeptides) in a process known as alternate mRNA splicing. This alternative splicing can occur in the same cell at the same time, in the same cell at different times during cell differentiation, or in different cells. Control of alternative splicing occurs through the action of specialized proteins.

A classic example of the use of different promoters and alternate splicing occurs with the myosin light chain ($MLC1_F/3_F$) gene. The gene is named this way because it gives rise to two light chain subunits found in the contractile protein myosin in type II muscle fibers. As shown in figure 10.2, the $MLC1_F/3_F$ gene contains 9 exons and two promoters. When transcription starts at first promoter (P_1), the primary transcript includes all nine exons. Alternative splicing of this primary gene product removes exons 2 and 3, producing the

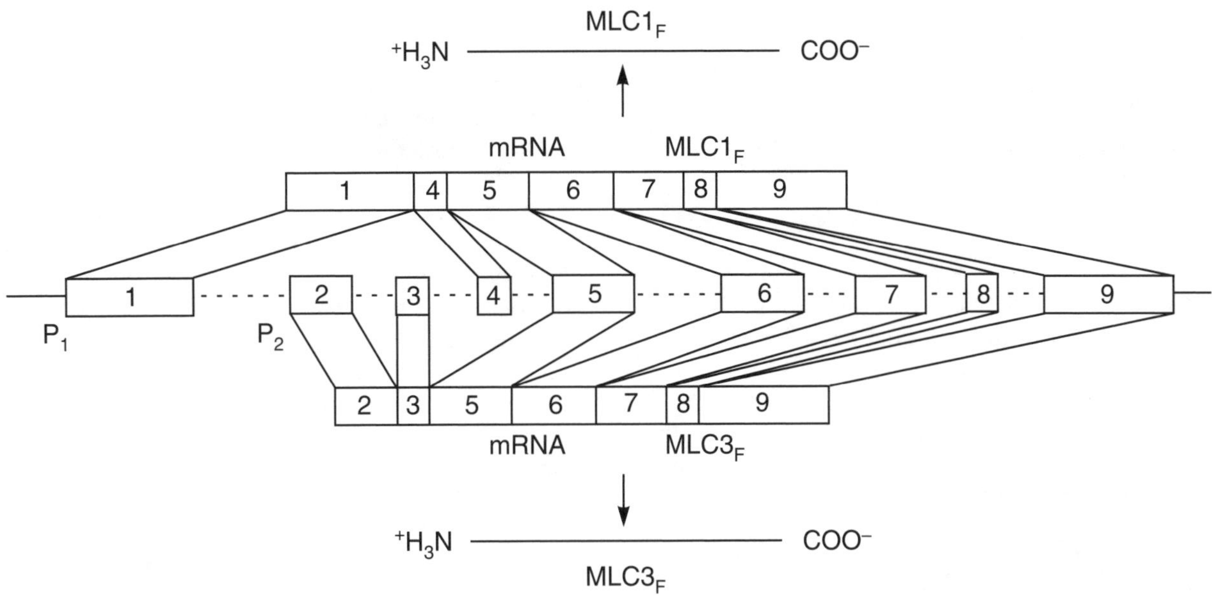

Figure 10.2 The myosin light chain $1_F/3_F$ gene contains 9 exons, shown as boxes with a number. Between the exons are the introns, shown as dashed lines. The solid line shows DNA upstream of exon 1 and downstream of exon 9. When transcription starts at promoter P_1, the primary transcript includes all nine exons. Alternative splicing of this removes exons 2 and 3, producing the mRNA molecule that, when translated, produces the polypeptide known as $MLC1_F$. Transcription that starts at promoter P_2 produces a primary transcript lacking exon 1. Alternative splicing of this primary gene product removes exon 4, producing the mRNA molecule that, when translated, produces the polypeptide known as $MLC3_F$.

mRNA molecule that, when translated, produces the polypeptide known as $MLC1_F$. Transcription that starts at second promoter (P_2) produces a primary transcript lacking exon 1. Alternative splicing of this primary gene product removes exon 4, producing the mRNA molecule that, when translated, produces the polypeptide known as $MLC3_F$. The ratio of $MLC1_F/MLC3_F$ depends on the frequency of initiation of transcription at the two promoters.

Formation of rRNA and tRNA Molecules

Modification of primary transcripts from genes for rRNA and tRNA also occurs. As mentioned, two different genes are needed to produce the rRNA molecules that make up much of the ribosomes. The products of these genes or primary transcripts are described by their velocity of sedimentation in an ultracentrifuge, expressed in Svedberg units (S).

Note: For very large molecules or molecular complexes, size is measured by how far they travel in an ultracentrifuge, expressed in Svedberg units rather than molecular weight. The larger the molecule, the faster it sediments in a centrifuge, and the larger the value of S. Unlike units such as kilodaltons (kD) or base pairs (bp), typically used to describe the size of protein or DNA molecules, respectively, Svedberg units are not linearly related to size. Thus a whole ribosome, with a mass of 80 S, is made up of two subunits of size 60 S and 40 S because the sedimentation behavior of particles in the centrifuge is not linearly related to size.

The two rRNA transcripts are described as 45 S and 5 S. The smaller transcript, 5 S (120 bases), remains unchanged, whereas the larger transcript, 45 S, undergoes some splitting and modification, resulting in three new rRNA molecules: 18 S, 28 S, and 5.8 S. The 18 S rRNA molecule combines with approximately 33 protein molecules to make the small ribosomal subunit described as the 40 S ribosomal subunit. The 5.8 S, the 28 S, and the 5 S rRNA molecules combine with about 49 protein molecules to make the large ribosomal subunit known as the 60 S ribosomal subunit. During protein synthesis, the 40 S and 60 S ribosomal subunits combine to make the complete ribosome, described as the 80 S ribosome.

KEY POINT

Although there are only two different genes for rRNA, there are multiple copies of these. We have already indicated that there is more rRNA by weight in the cell than all the mRNA and tRNA molecules combined.

Primary transcripts for genes for the various tRNAs undergo some modifications. For example, nucleotides are removed from the 5' and 3' ends, and, if present, an intron is removed. Then the 3' end has a nucleotide sequence CCA added. Finally, some of the bases are modified. Figure 10.3 shows the main features of a fully functional tRNA molecule. At one end there is a 5' phosphate group. The 3' end has the sequence CCA with a free 3' hydroxyl group to which the amino acid becomes attached. Notice the hydrogen bonding that helps maintain the rough cloverleaf structure, characteristic of the tRNA molecules.

Three bases at the bottom of tRNA represent the **anticodon** bases. These will correspond to the codon for the particular amino acid, except that they will be antiparallel and complementary. For example, if the codon is AAC (always described in the 5' to 3' direction), the anticodon will be UUG, in the 3' to 5' direction. Because there are 61 codons in mRNA specifying the 20 amino acids, we would expect that there would be 61 tRNA molecules each with one of 61 different anticodons. However, there is a bit of flexibility in the anticodon-codon binding because the first base in the anticodon (the one at the 5' end that pairs with the third base at the 3' end in the codon) can often recognize two bases. For example, the codons CUA and CUG could be recognized by the anticodon GAU. This flexibility in codon-anticodon recognition is known as **wobble**. As a result of wobble, there is a need for fewer than 61 different tRNA molecules to ensure that all 20 amino acids are able to participate in protein synthesis. Indeed, on a theoretical basis, looking at all the possibilities created by wobble, only 31different tRNA molecules would be needed. The actual number of different tRNA molecules found is a little more than 31.

Translation

In the process of translation, which takes place on ribosomes in the cytosol of the cell, the mRNA message is converted into a sequence of amino acids in a polypeptide chain.

Formation of Aminoacyl-tRNA

Each amino acid has at least one tRNA to which it will be attached. Each tRNA will have an anticodon that will match one of the codons (likely more due to wobble) for that amino acid, if it has more than one codon. Each amino acid will also have a specific enzyme to catalyze the joining of its alpha carboxyl group to the 3' OH group of the terminal adenosine of the tRNA using an ester bond. The enzyme that joins the amino acid to its tRNA is known as aminoacyl tRNA synthetase. It is the responsibility of each synthetase to match the correct amino acid and its tRNA, as shown in the following equation:

$$\text{amino acid} + \text{tRNA} + \text{ATP} \longrightarrow \text{aminoacyl-tRNA} + \text{AMP} + \text{PPi}$$

The energy needed for the formation of the ester bond between the amino acid and its tRNA is provided by the hydrolysis of ATP. The products of this hydrolysis reaction are AMP and inorganic pyrophosphate (PPi). Hydrolysis of PPi by inorganic pyrophosphatase ensures that the reaction above is driven to the right. Amino acids attached to their respective tRNA molecules are often described as charged amino acids. Some aminoacyl tRNA synthetases attach their amino acid directly to the 3' OH group of the terminal ribose on their tRNA, whereas others attach their amino acid to the 2' OH group of ribose. This latter attachment is only temporary as subsequently, the amino acid is moved to the 3' position. The role of the individual aminoacyl tRNA synthetase enzymes, to recognize the correct amino acid and the correct tRNA, places these enzymes in a critical position to ensure that correct proteins are produced by the cell.

KEY POINT

There are 20 different amino acids used to make proteins. There is at least one tRNA molecule for each amino acid. There are 20 different aminoacyl tRNA synthetases, one for each amino acid.

The Role of mRNA

Figure 10.4 illustrates the essential components of a functional mRNA molecule. The coding region, headed by the AUG start codon, contains the

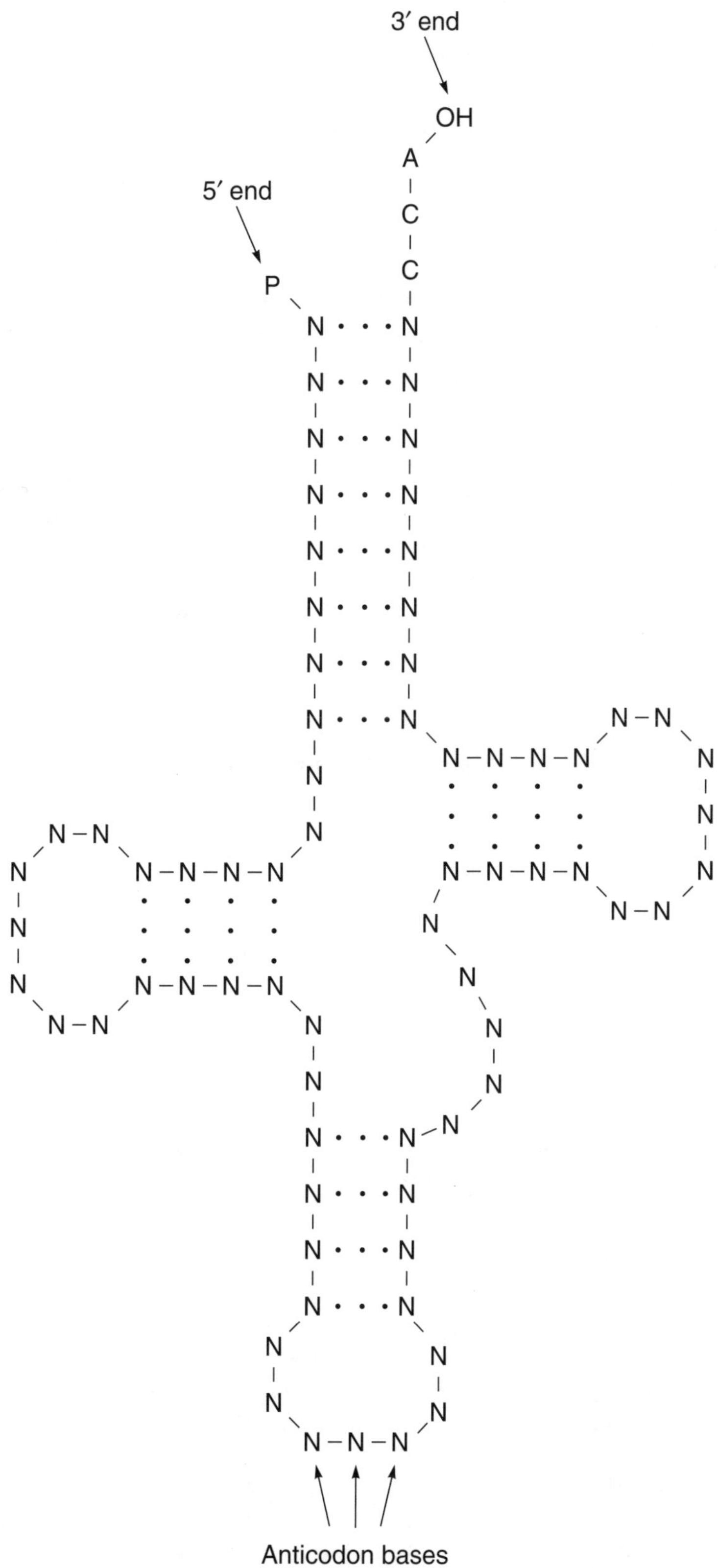

Figure 10.3 A schematic showing the rough cloverleaf shape of most tRNA molecules, which is maintained by hydrogen bonding (dotted lines), and the 5' end with its phosphate group. N represents unspecified nucleotides. The 3' end contains the bases cytosine and adenine (CCA), and the 3' OH (hydroxyl) group of the terminal adenosine is available to bind an amino acid. The three bases at the bottom represent the anticodon that recognizes the complementary bases on the mRNA codon.

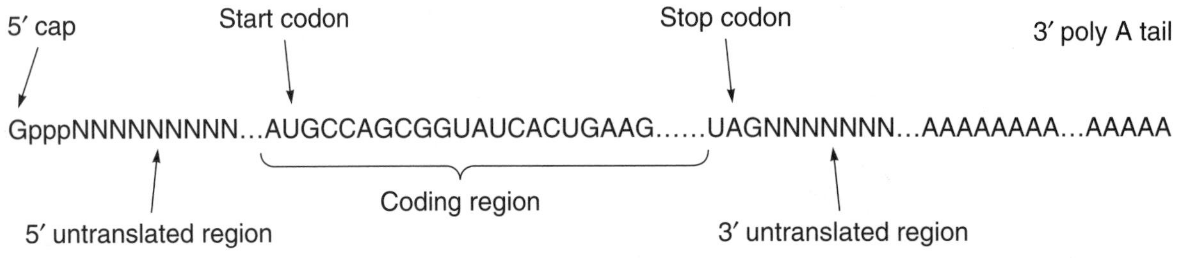

Figure 10.4 A model of mature mRNA that exits the nucleus and serves as a template for amino acid sequences in a polypeptide during translation in the cytosol. The coding region is flanked on the 5' and 3' ends by nucleotides that do not specify amino acid sequences. The message starts with the AUG codon and ends with either UAG (shown), UAA, or UGA stop codons. Unspecified nucleotides (bases) are shown as N.

information to indicate the amino acid sequence for the polypeptide chain, which is followed by a stop codon (UAG, UGA, or UAA). The term **open reading frame** (ORF) is often used to describe a region of triplet coding, bordered on one end by the start coding and at the other end by a stop codon. Flanking the coding region is a 5' noncoding region and a 3' region that is also noncoding. Because the base sequence in these regions is not translated into an amino acid sequence, they are also known as the 5' UTR (untranslated region) and 3' UTR, respectively. Most mRNA molecules contain the poly A tail. The functional mRNA molecule, once formed in the nucleus, is transported to the cytosol where protein synthesis takes place.

Initiation of Translation

Like transcription, the translation process has three stages: initiation, elongation, and termination. We will not go into great detail in this section but focus only on the overall process. The major players in the initiation of translation are the two ribosomal subunits, 40 S and 60 S, the mRNA molecule, the initial aminoacyl-tRNA (which will be methionyl-tRNA), a number of protein factors to control the initiation process (eukaryotic initiation factors, or eIFs, such as eIF1, eIF2, etc.), and a source of energy from the hydrolysis of ATP and GTP.

Figure 10.5 illustrates the initiation part of translation, which can be considered to take place as three discrete steps. In the first step, the methionyl-tRNA (met-tRNA), three initiation factors, the 40 S ribosome subunit, and GTP combine to form a 43 S preinitiation complex. In the second step, the preinitiation complex binds to the mRNA at the 5' cap site, then moves along the mRNA in a 5' to 3' direction searching for the initiation codon AUG. This process is facilitated by the intervention of additional initiation factors. The anticodon on methionyl-tRNA will recognize the AUG (initiation) codon on mRNA because it is complementary to it. In step three, the 60 S ribosomal subunit binds, generating the 80 S initiation complex. Before this can take place, the hydrolysis of GTP and the loss of the bound initiation factors are essential, because it is the presence of these, principally eIF3, that keeps the two ribosome subunits dissociated. Other initiation factors also participate in the formation of the 80 S initiation complex.

Elongation of Translation

The elongation phase involves the addition of amino acids, one at a time, to the carboxy terminal end of the existing polypeptide chain (see figure 10.6). Let us start with the 80 S initiation complex. The next aminoacyl-tRNA comes in, its anticodon recognizing the mRNA codon on the 3' side of the initiator codon. A peptide bond is then formed between the methionine carboxyl and the free amino group of the next amino acid, still attached to its tRNA. The formation of this peptide bond means the methionine is released from its tRNA. The methionine tRNA leaves, and we now have a dipeptide attached to the second tRNA. The 80 S ribosome complex then slides three bases along the mRNA molecule in the 3' direction. Then the next aminoacyl-tRNA comes in, recognizing the next mRNA codon. Another peptide bond is formed, and we now have a tripeptide. Again, the 80 S ribosomal complex slides along the mRNA. This process continues with the growing polypeptide chain still attached to the last incoming tRNA. The energy needed to make these peptide bonds comes from the hydrolysis of GTP to GDP and Pi. Two elongation factors, eEF1 and eEF2, participate in the elongation process.

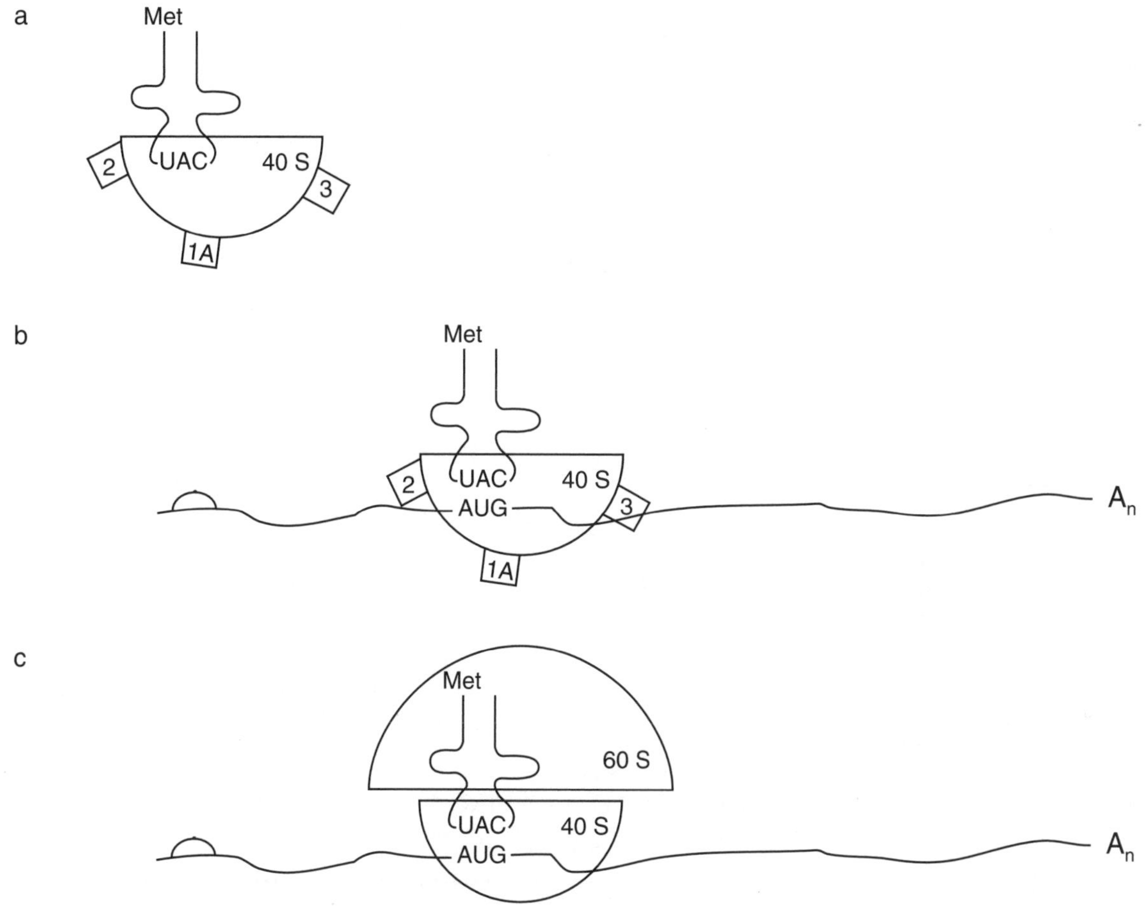

Figure 10.5 Three stages of translation initiation: (a) Formation of a 43 S preinitiation complex consisting of the 40 S ribosomal subunit, methionyl-tRNA (Met) and three initiation factors (eIFs), shown attached to the 40 S subunit and identified as 1A, 2, and 3. (b) The 43 S preinitiation complex binds to the mRNA at its 5' cap and moves along the mRNA until an AUG initiation codon is found. Base pairing between the anticodon on the methionyl-tRNA and the AUG codon ensures a correct match. Other eIFs also participate in this step, but are not shown. Note the 3' poly A tail shown as A_n. (c) Addition of the 60 S ribosomal subunit and removal of the initiation factors accompanied by GTP hydrolysis (not shown) mark the final step in translation initiation.

Termination of Translation

The process of elongation continues, with the ribosome and growing polypeptide chain moving along the mRNA three bases at a time, until a stop codon is reached (UAA, UGA, or UAG). At this point, a termination factor or release factor causes the release of the completed polypeptide chain from the last tRNA. The whole complex dissociates, and the 80 S ribosome dissociates into its 40 S and 60 S subunits. As with peptide chain elongation, energy for the termination of translation is provided by GTP hydrolysis.

Because mRNA is large, more than one ribosome is found on a single mRNA molecule—each containing a growing polypeptide chain. We call the mRNA and the two or more ribosomes, each with its growing polypeptide chain, a **polyribosome** or polysome (see figure 10.7). The overall process of translation occurs with remarkably few errors, roughly one error per 10,000 codons translated under normal conditions.

Regulation of Translation

When gene expression regulation is discussed, terms such as transcription control and posttranscription control are often used. In the previous chapter, we discussed transcription control primarily from the perspective of regulation of the initiation of gene transcription by RNAP II.

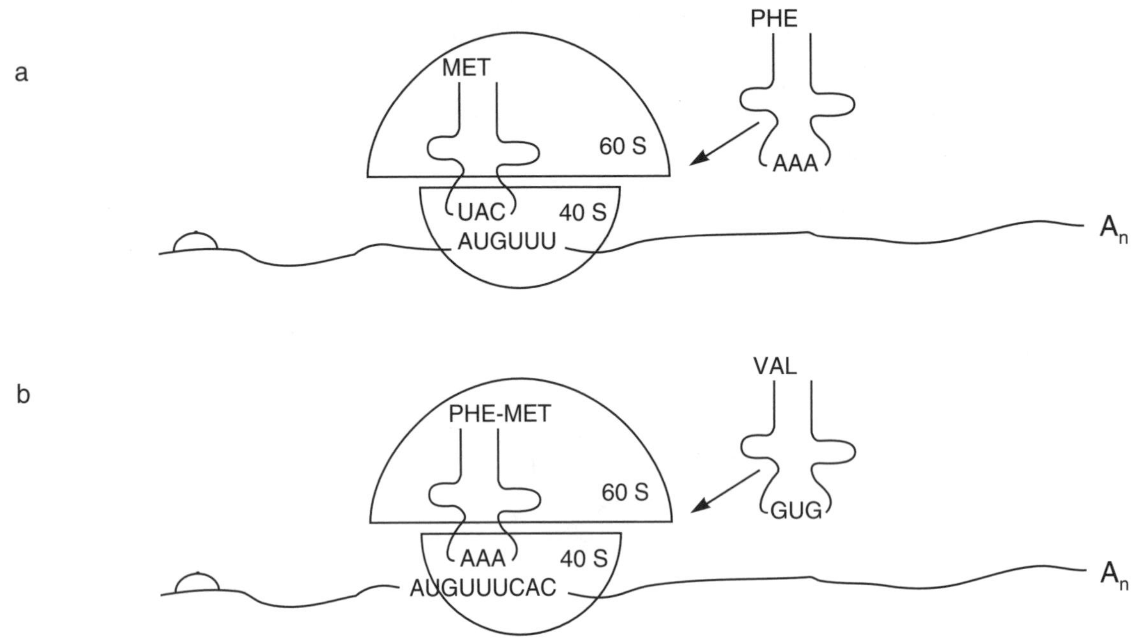

Figure 10.6 The elongation phase of translation: (a) The second aminoacyl-tRNA comes to the 80 S ribosomal complex, its anticodon AAA recognizing the second codon UUU, which codes for the amino acid phenylalanine (PHE). (b) A peptide bond is formed between the carboxyl group of methionine (MET) and the phenylalanine, still attached to its tRNA. The 80 S ribosome complex has moved three bases along the mRNA towards the 3' end where the poly A tail (A_n) is located. The 5' end contains the cap. Not shown are two elongation factors and the hydrolysis of GTP, which provides the energy for peptide bond formation and movement along the mRNA molecule.

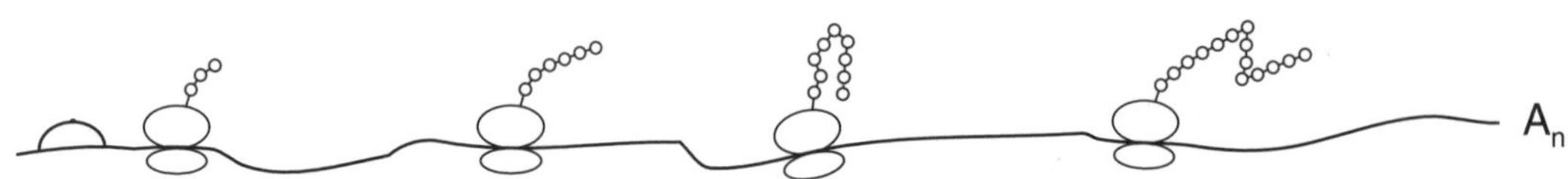

Figure 10.7 The length of most mRNA molecules, compared to that of the 80 S ribosome, allows for more than one ribosome, with its growing polypeptide, on each mRNA. Here, a single mRNA molecule has four ribosomes, each with a growing polypeptide chain shown as a string of beads. The overall complex is known as a polyribosome or a polysome.

Here, we look at posttranscriptional control because this refers principally to the regulation of translation of mRNA. Translation of muscle mRNA, or the formation of polypeptides in muscle, is exquisitely sensitive to a variety of nutritional and exercise states. Moreover, while protein synthesis is a continuous activity in cells, it does require energy expenditure since the equivalent of four ATP are needed for the synthesis of each peptide bond.

Measuring Protein Synthesis

Many of the early studies involved lab animals. Protein synthesis rate was measured as the incorporation of a radioactively labeled amino acid into muscle protein over a specific time period, expressed, for example, as nanomoles of amino acid incorporated per gram of muscle tissue per hour. Typically, phenylalanine is used for incorporation studies, as it is not metabolized by muscle, thus its only fate is to be used to make proteins. The rate of protein synthesis in specific fractions of muscle proteins can be obtained by isolating mitochondria, contractile proteins, or soluble cytosolic proteins and determining the incorporation rate of labeled phenylalanine in each fraction per hour. The efficiency of translation can be assessed by measuring the rate of protein synthesis and dividing this by the total content of RNA. In this way, an indication of the relative use of RNA can be determined, pointing out how effectively the mRNA, tRNA, and rRNA are being used in protein synthesis.

In studies with humans, ethical considerations preclude the use of radioactive amino acids for nonmedical studies. Therefore, scientists have employed isotopically labeled amino acids that are not radioactive, but which can be measured because of differences in their mass. These are known as **stable isotopes**. The amino acid enriched with a particular stable isotope can be infused at a constant rate into a vein. We know that it will be used to make muscle proteins. Therefore, muscle samples are obtained by biopsy at specific intervals and the amount of the amino acid with the stable isotope can be determined over time. Leucine, labeled at carbon 1 with an isotope of carbon (1-^{13}C-leucine) as opposed to the normal isotope of carbon (^{12}C), is often used. Recent experiments have also used phenylalanine labeled with deuterium.

Effects of Exercise and Diet

A variety of studies using both humans and animals has shown that the rate of protein synthesis is generally depressed during exercise. After exercise, protein synthesis increases for periods up to 48 hours before declining to baseline values. This has been shown following both endurance and resistance exercise, although the increase in protein synthesis with resistance exercise tends to be greater than what is observed following endurance exercise. When people who are new to resistance training are compared to those who routinely do resistance exercises, the former show a greater response in muscle protein synthesis following a training bout. Therefore, regular training appears to create a situation where the exercise stimulus has less influence on protein synthesis.

In the fasted state, muscle protein synthesis is increased following resistance exercise. This suggests that exercise by itself will stimulate synthesis. If the resistance or endurance training is followed by ingestion of carbohydrates or carbohydrates and protein, there is a dramatic improvement in muscle protein synthesis. This points to the important synergistic effect of food intake and exercise, particularly if the food intake closely follows the end of exercise. One of the explanations for the effect of food is that the blood insulin concentration is elevated. We know that insulin has an independent effect on stimulating muscle protein synthesis. Recent studies also demonstrate that ingestion of a mixture of amino acids after resistance exercise stimulates muscle protein synthesis, without leading to an increase in insulin. This suggests that simply increasing blood amino acid concentration can stimulate postexercise muscle protein synthesis, likely by a mass action effect.

How do exercise and diet increase muscle protein so quickly following the end of exercise? We might suggest that they increase the rate of transcription of a number of genes, leading to a general increase in the content of a variety of mRNA molecules. Alternatively, the exercise and diet could increase the transcription of genes for rRNA or tRNA, leading to increased capacity for protein synthesis. This explanation is highly unlikely given the short period of time in which the exercise or exercise and diet effect is noted—too short to transcribe genes, modify RNA, and get it into functional form in the cytosol where protein synthesis takes place. Moreover, these changes in protein synthesis take place without a measurable increase in total RNA or specific mRNA molecules. Therefore, the rate of protein synthesis increases without a change in total RNA, indicating an increased efficiency of translation. It has also been demonstrated that immobilizing a muscle leads to a dramatic decrease in muscle protein synthesis, without a change in total RNA content. Apparently, the overall translation process is very sensitive to muscle activity.

The search for the precise locus of control of protein synthesis, which is stimulated by exercise and nutrition, is complicated because there are so many steps and so many initiation factors. It seems reasonable that the initiation of translation should be a control point because it would be energetically wasteful to stop translation partway through, since the equivalent ATP cost to make one peptide bond is four. Moreover, stopping translation before it is completed would mean that the cell would need to dispose of partially formed polypeptides having no apparent function. Recent evidence points to the critical involvement of the initiation factors eIF4E and eIF4G in the final step of translation initiation as being responsible for the rate increase when amino acids were infused into rat blood. Additionally, many of the eIFs can exist in phosphorylated or dephosphorylated states, depending on the action of specific protein kinases and phosphoprotein phosphatases. The phosphorylation state of the initiation factors influences their role in translation initiation. This allows for more precise modulation of the rate of translation initiation of specific mRNA molecules.

There are control regions in some mRNA molecules in the 5' untranslated region (UTR), which can regulate the ability of the mRNA message to be initiated. For example, translation of the mRNA molecules for both transferrin, which binds and transports iron in the blood, and ferritin, which binds iron in a storage capacity, is regulated by protein binding in the 5' UTR. Control in the 5' UTR is similar to the kind of control exercised over transcription in that a protein recognizes and binds to certain base sequences on the mRNA, thereby modulating the ability to initiate translation.

The lifetime of mRNA molecules, measured as a half-life of mRNA, can vary from about five minutes to many hours and so must be considered as a level of control. As we have seen, mRNA is generated by transcription. How long the mRNA lasts in the cytosol is controlled by degradative enzymes known as ribonucleases. These enzymes cleave the sugar phosphate backbone of RNA, releasing individual nucleoside monophosphates. The length of time an individual mRNA molecule lasts in the cytosol depends on its being able to escape being degraded by the cytosolic ribonucleases. This is a very complicated area since there are proteins that stabilize and destabilize mRNA. Regulation of these proteins can therefore indirectly influence the lifetime of the mRNA as well as regulate the ability of the mRNA to be translated.

KEY POINT

Regulation of translation is important to many athletes attempting to increase muscle size. This means increasing the frequency of translation of each mRNA and increasing the lifetime of each mRNA. Simple strategies such as adopting an appropriate exercise training stimulus (which is not the same for everyone) and the timing and content of meals can play dominant roles. Such simple approaches may not be satisfying to the impatient athlete, and a host of alternative strategies such as supplements, drugs, and hormones have been attempted. Some of these are successful, but many have little or no influence or may be harmful.

Posttranslational Processing of Polypeptides

Translation of the mRNA message generates a one-dimensional amino acid sequence in a polypeptide. As we saw in chapter 1, there are higher levels of structure of polypeptide chains that must be created before the polypeptide can function. Even if a polypeptide can spontaneously assume its correct native shape after it is released from the translation apparatus, it may need to bind to other molecules or be in a different location in the cell to function properly. Moreover, in the crowded environment of the protein-synthesizing complex of ribosomes, mRNA molecules, other proteins involved in controlling translation, and a variety of newly formed polypeptides, some improper associations will certainly be made.

A group of proteins, known as molecular chaperones, binds to a fraction of completed polypeptide chains, helps them assume their final three-dimensional conformation, helps guide them to their site of action within the cytosol, or helps them cross a membrane if they are to work in a cell organelle such as a mitochondrion. Many of these proteins were initially called **heat shock proteins** (Hsp) because they were discovered in cells that had undergone sharp increases in temperature or other types of stress. Heat shock proteins are identified with a number, which identifies their molecular weight in kD (e.g., Hsp70, Hsp90). We now know that there are multiple kinds of heat shock proteins for each molecular weight class with different functions. **Chaperonins** are a subgroup of chaperone proteins that function primarily in protein folding and assembly of polypeptide subunits into molecular aggregates. Remember that some polypeptides must also bind with other polypeptides to generate the active oligomeric protein. For example, myosin is a hexamer, so six polypeptides must combine in the proper way to generate a functional myosin molecule. Moreover, once formed, myosin hexamers must be inserted into thick filaments in order to be able to hydrolyze ATP and generate a force (see chapter 4). Some polypeptides need to have a prosthetic group attached to them in order to be effective. For example, myoglobin, the protein in muscle that aids in storing and moving oxygen in the cytosol of cardiac and skeletal muscle, needs to have a heme group attached to its polypeptide. Other polypeptides must be packaged for export because they function in the blood. Others must be incorporated into membranes because much of the mass of a membrane is made up of a variety of proteins. It is clear that the process of translation only generates a polypeptide component and often many modifications must take place before that polypeptide can act in a concerted way in cellular function.

Protein Degradation

In chapters 1, 6, and 8, the concept of protein turnover was discussed. Turnover is a concept that considers the balance between the opposing processes of protein synthesis and protein degradation. The content of any protein in a cell is based on the relative rates of protein synthesis and protein degradation. This can be described by the equation

$$\Delta \text{ protein} = \text{synthesis rate} - \text{degradation rate.}$$

This tells us that the moment-to-moment change in the content of any protein is a reflection of the differences in rates of synthesis and degradation. We have already seen how proteins are synthesized. The opposing process is carried out by three major systems of protein degrading enzymes, known as proteinases.

For years, the content of a particular protein was thought to be regulated primarily by changes in the transcription of its gene and subsequent translation of this message. The degradation side of the above equation was considered to be relatively constant. We now know that protein degradation is extremely important, involved in regulating cellular processes such as growth and atrophy, transcription, rates of metabolic pathways, development, and some disease states such as cancer. There are three main protein-degrading systems operative in skeletal muscle: the ubiquitin-proteasome pathway, the lysosomal system, and the calpain system. As these are complex and incompletely understood mechanisms, we will briefly summarize how they function.

Ubiquitin-Proteasome Pathway

The ubiquitin-proteasome pathway plays a major role in targeting and degrading cellular proteins. This pathway must recognize and label a specific protein for degradation, then carry out the process. The recognition part is based on marking a protein by covalent attachment of a small ubiquitous protein known as **ubiquitin**. This is a protein containing 76 amino acids that is found in all eukaryotes, virtually unchanged. The proteasome is a large (700 kD), multicomponent, cylindrically shaped protein complex that, in concert with other regulators, breaks down the targeted protein. The proteasome is assembled from polypeptides derived from 14 distinct genes.

Proteins are marked for degradation by the addition of many ubiquitin units, described as polyubiquitination. Three conjugating enzymes are involved in this process, summarized in figure 10.8. In the first step (a), ubiquitin is covalently attached to conjugating enzyme E_1 using the free energy of hydrolysis of ATP. In step (b), ubiquitin is transferred to a second conjugating enzyme, E_2. Finally, (c) a third conjugating enzyme, shown as E_3, transfers a number of ubiquitin molecules to the target protein. Polyubiquitination of the protein marks it both for destruction and as a suitable substrate for the proteasome degrading complex.

The proteasome by itself is large, as mentioned, and it is described as the 20 S proteasome. In this form, it cannot degrade polyubiquitinated proteins but needs an additional regulatory protein. When this is bound to the 20 S proteasome, the resulting 26 S proteasome can degrade polyubiquitinated proteins when these pass through an opening in the cylindrical core of the proteasome exposing the target protein to protein degrading enzymes. The resulting products of small peptides are broken down to free amino acids within the cell. Regulation of the ubiquitin-proteasome pathway is complex and involves the synthesis and degradation of the polypeptide components of the pathway, including the three conjugating enzymes shown in figure 10.8 as well as regulatory proteins.

Lysosomal System

Lysosomes are organelles found in most cells that are responsible for degrading a variety of cell constituents. In particular, lysosomes contain a battery of protein-degrading enzymes known as cathepsins that break peptide bonds in the interior of the protein molecule. In addition, they also contain proteinases that cleave amino acids, one at a time, from the amino and carboxy ends of the polypeptide. To be degraded by lysosomes, proteins must enter via a process known as endocytosis.

Calpain System

Calpain is a calcium-activated proteinase in cell cytosol. It is found in several forms, each activated

a) $\text{Ubiquitin} + \text{ATP} + E_1 \longrightarrow E_1 - \text{ubiquitin} + \text{AMP} + \text{PPi}$

b) $E_1 - \text{ubiquitin} + E_2 \longrightarrow E_2 - \text{ubiquitin} + E_1$

c) $nE_2 - \text{ubiquitin} + \text{Target protein} \xrightarrow{E_3} \text{Target protein} - (\text{ubiquitin})_n$

Figure 10.8 Protein degradation by the ubiquitin-proteasome pathway first involves targeting the protein for destruction. This is carried out in a three-stage process with the aid of three conjugating enzymes shown as E_1, E_2, and E_3. The net effect is to add a number of molecules of ubiquitin to the target protein, producing a polyubiquitinated protein. This is now a suitable substrate for the proteasome complex.

by specific levels of intracellular calcium concentration. It is believed that calpains are activated by the rise in calcium with exercise and that they are responsible in part for initiating the increase in protein degradation that is associated with exercise. Moreover, severe exercise stress, known to produce muscle soreness and evidence of muscle damage at the microscopic level, is attributed to disturbances in internal calcium homeostasis, leading to activation of calpains. Because calpain acts on structural proteins in muscle, it is believed that its activation may be a first step in the response to exercise-induced muscle injury.

Summary

Protein synthesis occurs in the cell cytosol using mRNA, two ribosome subunits (described on the basis of their mobility in a centrifugal field as 40 S and 60 S), amino acids, energy sources, and a large number of protein factors. The modified rRNA molecules, created from two different kinds of RNA gene transcripts, are major components in the structure of ribosome subunits. A significant first step is the attachment of amino acids to specific tRNA molecules by a specific aminoacyl tRNA synthetase. Each tRNA contains a three-base sequence called an anticodon; the complementary base pairing between the anticodon on tRNA (containing its attached amino acid) with the codon on mRNA provides fidelity in amino acid sequence in a protein.

Translation initiation is the most complex step in protein synthesis. It involves the formation of a complete ribosome subunit (80 S) at the start (AUG) codon on mRNA with a methionyl-tRNA. Energy provided from GTP and the assistance of a number of eukaryotic initiation factors (eIFs) are also necessary. Elongation is the step where individual aminoacyl-tRNA molecules come to the ribosome complex and form a peptide bond between amino acids attached to their respective tRNAs. Individual ribosomes then move along the mRNA molecules three bases at a time. Each time the ribosome complex moves along the mRNA, a new amino acid is added to the growing polypeptide chain. Elongation continues the building of a polypeptide chain until a stop codon on mRNA is encountered. The ribosome complex then dissociates from the mRNA, and the completed polypeptide is released. Each mRNA molecule may have a number of ribosomes moving along independently, each with a growing polypeptide chain. Since the quantity of individual proteins in a cell can dictate rates of cell metabolism, translation, or protein synthesis, must be carefully regulated. Translation, mainly initiation, is controlled at a number of different sites. The lifetime of its mRNA molecules also determines the amount of a cell's protein.

The amount of a specific protein in a cell is related to its rate of breakdown as well as its synthesis. Protein degradation is an important process that plays a significant role in the function of a cell. Three major systems are involved in normal protein degradation in muscle. The ubiquitin-proteasome pathway involves targeting proteins for degradation by attaching multiple units of a small protein known as ubiquitin. Subsequent degradation to small peptides is carried out by a proteasome complex known as the 26 S proteasome. Lysosomes, internal organelles, degrade endocytosed proteins using a battery of proteinases. Calpains are calcium-activated proteinases that may play a significant role in the response of muscle to severe exercise stress.

Key Terms

anticodon 174
chaperonins 180
exons 172
heat shock proteins 180
introns 172
open reading frame 176
polyribosome 177
primary transcript 171
stable isotopes 179
ubiquitin 181
wobble 174

Review Questions

1. If a protein contains 200 amino acids, what is the minimum number of nucleotides in the mature mRNA molecule for this protein?
2. If we assume that there are 60,000 genes coding for polypeptide subunits in the human genome and that the average number of amino acids in each polypeptide is 400, estimate the percentage of the human genome (three billion base pairs) that is actually involved in coding for amino acid sequence in a polypeptide. Hint: Both strands of DNA can code for a polypeptide.

References

Alberts, B. 1998. The cell as a collection of protein machines: preparing the next generation of molecular biologists. *Cell* 92: 291-294.

Andel III, F., A.G. Ladurner, C. Inouye, R. Tjian, and E. Nagales. 1999. Three-dimensional structure of the human TFIID-IIA-IIB complex. *Science* 286: 2153-2156.

Angus, D.J., M. Hargreaves, J. Dancey, and M.A. Febbraio. 2000. Effect of carbohydrate or carbohydrate plus medium chain triglyceride ingestion on cycling time trial performance. *Journal of Applied Physiology* 88: 113-119.

Belcastro, A.N., L.D. Shewchuk, and D.A. Raj. 1998. Exercise-induced muscle injury: a calpain hypothesis. *Molecular and Cellular Biochemistry* 179: 135-145.

Bergman, B.C., G.E. Butterfield, E.E. Wolfel, G.A. Casazza, G.D. Lopaschuk, and G.A. Brooks. 1999. Evaluation of exercise and training on muscle lipid metabolism. *American Journal of Physiology* 276: E106-E117.

Bergman, B.C., E.E. Wolfel, G.E. Butterfield, G.D. Lopaschuk, G.A. Casazza, M.A. Horning, and G.A. Brooks. 1999. Active muscle and whole body lactate kinetics after endurance training in men. *Journal of Applied Physiology* 87: 1684-1696.

Blomstrand, E., and B. Saltin. 1999. Effect of muscle glycogen on glucose, lactate, and amino acid metabolism during exercise and recovery in human subjects. *Journal of Physiology (London)* 514: 293-302.

Bonen, A., S.K. Baker, and H. Hatta. 1997. Lactate transport and lactate transporters in skeletal muscle. *Canadian Journal of Applied Physiology* 22: 531-552.

Brooks, G.A. 1997. Importance of the "crossover" concept in exercise metabolism. *Clinical and Experimental Pharmacology and Physiology* 24: 889-895.

Brooks, G.A., M.A. Brown, C.E. Butz, J.P. Sicurello, and H. Dubouchaud. 1999. Cardiac and skeletal muscle mitochondria have a monocarboxylate transporter MCT1. *Journal of Applied Physiology* 87: 1713-1718.

Brooks, G.A., H. Dubouchaud, M. Brown, J.P. Sicurello, and C.E. Butz. 1999. Role of mitochondrial lactate dehydrogenase and lactate oxidation in the intercellular lactate shuttle. *Proceedings of the National Academy of Science USA* 96: 1129-1134.

Burelle, Y., F. Péronnet, S. Charpentier, C. Lavoie, C. Hillaire-Marcel, and D. Massicotte. 1999. Oxidation of an oral [^{13}C] glucose load at rest and prolonged exercise in trained and sedentary subjects. *Journal of Applied Physiology* 86: 52-60.

Cabrera, M.E., G.M. Saidel, and S.C. Kalhan. 1999. Lactate metabolism during exercise: analysis by an integrative systems model. *American Journal of Physiology* 277: R1522-R1536.

Chin, E.R., E.N. Olson, J.A. Richardson, Q. Yang, C. Humphries, J.M. Shelton, H. Uw, W. Zhu, R. Bassel-Duby, and R.S. Williams. 1998. A calcineurin-dependent transcriptional pathway controls skeletal muscle fiber type. *Genes and Development* 12: 2499-2509.

Combs, C.A., A.H. Aletras, and R.S. Balaban. 1999. Effect of muscle action and metabolic strain on oxidative metabolic responses in human skeletal muscle. *Journal of Applied Physiology* 87: 1768-1775.

Coppack, S.W., M. Persson, R.L. Judd, and J.M. Miles. 1999. Glycerol and nonesterified fatty acid metabolism in human muscle and adipose tissue in vivo. *American Journal of Physiology* 276: E233-E240.

Daniel, P.B., W.H. Walker, and J.F. Habener. 1998. Cyclic AMP signaling and gene regulation. *Annual Reviews of Nutrition* 18: 353-383.

Demartino, G.N., and G.A. Ordway. 1998. Ubiquitin-proteasome pathway of intracellular protein degradation: implications for muscle atrophy during unloading. In *Exercise and Sport Sciences Reviews,* volume 26, edited by J.O. Holloszy. Baltimore: Williams & Wilkins.

Demartino, G.N., and C.A. Slaughter. 1999. The proteasome: a novel protease regulated by multiple mechanisms. *The Journal of Biological Chemistry* 274: 22123-22126.

Derave, W., S. Lund, G.D. Holman, J. Wojtaszewski, O. Pedersen, and E.R. Richter. 1999. Contraction-stimulated muscle glucose transport and GLUT-4 surface content are dependent on glycogen content. *American Journal of Physiology* 277: E1103-E1110.

Donovan, C.M., and K.D. Sumida. 1997. Training-enhanced hepatic gluconeogenesis: the importance for glucose homeostasis during exercise. *Medicine and Science in Sports and Exercise* 29: 628-634.

Duchen, M.R. 1999. Contributions of mitochondria to animal physiology: from homeostatic sensor to calcium signaling and cell death. *Journal of Physiology (London)* 516: 1-17.

Eisenber, D. 1999. How chaperones protect virgin proteins. *Science* 285: 1021-1022.

Friedlander, A.L., G.A. Casazza, M.A. Horning, A. Usaj, and G.A. Brooks. 1999. Endurance training increases fatty acid turnover, but not fat oxidation, in young men. *Journal of Applied Physiology* 86: 2097-2105.

Garrett, R.H., and C.M. Grisham. 1999. *Biochemistry*. Fort Worth, TX: Saunders.

Gibala, M.J., M. Lozej, M.A. Tarnopolsky, C. McLean, and T.E. Graham. 1999. Low glycogen and branched chain amino acid ingestion do not impair anaplerosis during exercise in humans. *Journal of Applied Physiology* 87: 1662-1667.

Gibala, M.J., D.A. MacLean, T.E. Graham, and B. Saltin. 1999. Tricarboxylic acid cycle intermediate pool size and estimated cycle flux in human muscle during exercise. *American Journal of Physiology* 275: E235-E242.

Graham, T.E., and K.B. Adamo. 1999. Dietary carbohydrate and its effects on metabolism and substrate stores in sedentary and active individuals. *Canadian Journal of Applied Physiology* 24: 393-415.

Graham, T.E., and D.A. MacLean. 1998. Ammonia and amino acid metabolism in skeletal muscle: human, rodent, and canine models. *Medicine and Science in Sports and Exercise* 30: 34-46.

Gravholt, C.J., O. Schmitz, L. Simonsen, J. Bülow, J.S. Christiansen, and N. Møller. 1999. Effects of a physiological GH pulse on interstitial glycerol in abdominal and femoral adipose tissue. *American Journal of Physiology* 277: E848-E854.

Hayashi, T., J.F.P. Wojtaszewski, and L.J. Goodyear. 1997. Exercise regulation of glucose transport in skeletal muscle. *American Journal of Physiology* 273: E1039-E1051.

Hochachka, P.W., and G.O. Matheson. 1992. Regulating ATP turnover rates over broad dynamic work ranges in skeletal muscle. *Journal of Applied Physiology* 73: 1697-1703.

Horowitz, J.F., R. Mora-Rodriguez, L.O. Byerley, and E.F. Coyle. 1999. Substrate metabolism when subjects are fed carbohydrate during exercise. *American Journal of Physiology* 276: E828-E835.

Houston, M.E. 1999. Gaining weight: the scientific basis of increasing skeletal muscle mass. *Canadian Journal of Applied Physiology* 24: 305-316.

Howlett, R.A., G.J.F. Heigenhauser, E. Hultman, M.G. Hollidge-Horvat, and L.L. Spriet. 1999. Effects of dichloroacetate infusion on human skeletal muscle metabolism at the onset of exercise. *American Journal of Physiology* 277: E18-E25.

Jeukendrup, A.E., A. Raben, A. Gijsen, J.H.C.H. Stegen, F. Brouns, W.H.M. Saris, and A.J.M. Wagenmakers. 1999. Glucose kinetics during prolonged exercise in highly trained human subjects: effect of glucose ingestion. *Journal of Physiology (London)* 515: 579-589.

Jeukendrup, A.E., W.H.M. Saris, and A.J.M. Wagenmakers. 1998. Fat metabolism during exercise: a review. Part I: Fatty acid mobilization and muscle metabolism. *International Journal of Sports Medicine* 19: 231-244.

Jeukendrup, A.E., W.H.M. Saris, and A.J.M. Wagenmakers. 1998. Fat metabolism during exercise: a review. Part II: Regulation of metabolism and the effects of training. *International Journal of Sports Medicine* 19: 293-302.

Lee, C.P., Q. Gu, Y. Xiong, R.A. Mitchell, and L. Ernster. 1996. P/O ratios reassessed: mitochondrial P/O ratios consistently exceed 1.5 with succinate and 2.5 with NAD-linked substrates. *FASEB Journal* 10: 345-350.

Lewin, B. 1997. *Genes* VI. Oxford, NY: Oxford University Press.

McCormack, J.G., and R.M. Denton. 1994. Signal transduction by intramitochondrial Ca^{2+} in mammalian energy metabolism. *News in Physiological Sciences* 9: 71-76.

Monemdjou, S., L.P. Lozak, and M-E. Harper. 1999. Mitochondrial proton leak in brown adipose tissue mitochondria of UCP 1-deficient mice is GDP insensitive. *American Journal of Physiology* 276: E1073-E1082.

Odland, L.M., G.J.F. Heigenhauser, D. Wong, M.G. Hollidge-Horvat, and L.L. Spriet. 1998. Effects of increased fat availability on fat-carbohydrate interaction during prolonged exercise in men. *American Journal of Physiology* 274: R894-R902.

Pilegaard, H., K. Domino, T. Noland, C. Juel, Y. Hellsten, A.P. Halestrap, and J. Bangsbo. 1999. Effect of high-intensity exercise training on lactate/H^+ transport capacity in skeletal muscle. *American Journal of Physiology* 276: E255-E261.

Randle, P.J., P.B. Garland, C.N Hales, and E.A. Newsholme. 1963. The glucose fatty-acid cycle: its role in insulin sensitivity and the metabolic disturbances of diabetes mellitus. *Lancet* 1: 785-789.

Randle, P.J., E.A. Newsholme, and P.B. Garland.1964. Regulation of glucose uptake by muscle. *Biochemistry Journal* 93: 652-665.

Romijn, J.A., E.F. Coyle, L.S. Sidossis, J. Rosenblatt, and R.R. Wolfe. 2000. Substrate metabolism during different exercise intensities in endurance-trained women. *Journal of Applied Physiology* 88: 1707-1714.

Ruderman, N.B., A.K. Saha, D. Vavvas, and L.A. Witters. 1999. Malonyl CoA, fuel sensing, and insulin resistance. *American Journal of Physiology* 276: E1-E18.

Saraste, M. 1999. Oxidative phosphorylation at the fin de siècle. *Science* 283: 1488-1493.

Shearer, J., I. Marchand, P. Sathasivam, M.A. Tarnopolsky, and T.E. Graham. 2000. Glycogenin activity in human skeletal muscle is proportional to muscle glycogen concentration. *American Journal of Physiology* 278: E177-E180.

Talmadge, R.J. 2000. Myosin heavy chain isoform expression following reduced neuromuscular activity:

potential regulatory mechanisms. *Muscle and Nerve* 23: 661-679.

Thomason, D.B. 1998. Translation control of gene expression in muscle. In *Exercise and Sport Sciences Reviews,* volume 26, edited by J.O. Holloszy. Baltimore: Williams & Wilkins.

Tschakovsky, M.E., and R.L. Hughson. 1999. Interaction of factors determining oxygen uptake at the onset of exercise. *Journal of Applied Physiology* 86: 1101-1113.

Turcotte, L.P. 2000. Muscle fatty acid uptake during exercise: possible mechanisms. *Exercise and Sport Sciences Reviews* 28: 4-9.

Vary, T.C., L.S. Jefferson, and S.R. Kimball. 1999. Amino acid-induced stimulation of translation initiation in rat skeletal muscle. *American Journal of Physiology* 277: E1077-E1086.

Wagenmakers, A.J.M. 1998. Muscle amino acid metabolism at rest and during exercise: role in human physiology and metabolism. In *Exercise and Sport Sciences Reviews,* volume 26, edited by J.O. Holloszy. Baltimore: Williams & Wilkins.

Williams, M.H. 1999. *Nutrition for health, fitness, and sport.* 5th edition. Boston: WCB McGraw-Hill.

Wilson, D.F. 1994. Factors affecting the rate and energetics of mitochondrial oxidative phosphorylation. *Medicine and Science in Sports and Exercise* 26: 37-43.

Winder, W.W. 1998. Malonyl CoA: regulator of fatty acid oxidation in muscle during exercise. In *Exercise and Sport Sciences Reviews,* volume 26, edited by J.O. Holloszy. Baltimore: Williams & Wilkins.

Winder, W.W., and D.G. Hardie. 1999. AMP-activated protein kinase, a metabolic master switch: possible roles in type 2 diabetes. *American Journal of Physiology* 277: E1-E10.

Workman, J.L., and R.E. Kingston. 1998. Alteration of nucleosome structure as a mechanism of transcriptional regulation. *Annual Reviews of Biochemistry* 67: 545-579.

Zierler, K. 1999. Whole body glucose metabolism. *American Journal of Physiology* 276: E409-E426.

Glossary

acceptor control—Regulation of the rate of oxidative phosphorylation by the availability of ADP.

acid—A compound that can donate a proton.

acidosis—A metabolic condition in which the production of protons is elevated and the capacity to buffer these is diminished. The blood pH may be reduced.

actin—The principal protein of the thin filament in muscle. It can bind with myosin and enhance its ability to hydrolyze ATP.

activator—A DNA binding protein that enhances the expression of a gene or an allosteric effector that increases the activity of an enzyme.

active site—The region of an enzyme protein that binds substrate and converts it to product.

active transport—Transport across a membrane against a concentration gradient so that energy is needed for this to occur.

adenosine—The molecule formed when the base adenine binds to the sugar ribose.

adenosine 3', 5' cyclic monophosphate—*See* cyclic AMP.

adenosine diphosphate—*See* ADP.

adenosine triphosphate—*See* ATP.

adenylyl (adenylate) cyclase—A membrane-bound enzyme that converts ATP into cyclic AMP.

adipocyte—A single fat cell.

adipose tissue—A collection of adipocytes embedded in a connective tissue network.

ADP—Adenosine 5'-diphosphate, an adenine nucleotide produced when ATP is hydrolyzed and used as a substrate in reactions producing ATP.

adrenal gland—An endocrine gland located above the kidneys that secretes hormones from its cortex and medulla compartments.

adrenergic receptors—Membrane bound receptors categorized as alpha or beta that can bind epinephrine and norepinephrine.

aerobic—Requiring or taking place in the presence of oxygen.

allosteric enzyme—An enzyme that plays a regulatory role in metabolism because in addition to binding its substrate or substrates at its active site, it also binds small molecules at allosteric sites, which modulates the activity of the enzyme.

allosteric site—The site on an allosteric enzyme where effector molecules bind.

alpha-helix—A type of secondary structure of a protein in which the polypeptide backbone forms coils with maximum hydrogen bonding between amino acids.

amino acid pool—A collection of free amino acids in a compartment such as the blood stream.

aminotransferases—A class of enzymes that remove an amino group from one amino acid and transfer it to another, usually alpha-ketoglutarate.

AMP—Adenosine 5'-monophosphate, an adenine nucleotide.

amphipathic—A molecule that has both hydrophilic and hydrophobic properties.

amphoteric—Capable of accepting or donating protons, and thus able to act as a base or acid.

anabolism—The part of metabolism in which larger molecules are made from smaller molecules.

anaerobic—Occurring without air or oxygen.

anaplerosis—Reactions that can generate TCA cycle intermediates.

anhydride bond—A type of energy-rich bond formed when two acid groups combine with the elimination of water.

anticodon—A specific sequence of three bases in tRNA that is complementary to a mRNA codon.

antiparallel—Two polynucleotide chains that are opposite in polarity.

antiport—A membrane transport protein that simultaneously moves two molecules in opposite directions across the membrane.

apoptosis—Programmed cell death in which a regulated process leads through a series of steps to the ultimate death of a cell.

ATP—Adenosine 5'-triphosphate, an adenine nucleotide used as the energy currency in metabolism. The free energy released when ATP is hydrolyzed is used to drive reactions in cells.

ATPase—An enzyme that can hydrolyze ATP, producing ADP and inorganic phosphate.

ATP synthase—An enzyme complex in the inner mitochondrial membrane that uses the free energy released in oxidative phosphorylation to combine Pi with ADP to make ATP.

autocrine—The effect produced on a cell by a substance produced by that cell.

base pair—Two nucleotides in polynucleotide chains that can be paired through hydrogen bonding between their bases; for example base A with base T.

beta oxidation—The process through which fatty acids attached to coenzyme A are broken down in a sequence of four steps to produce acetyl CoA units.

beta-sheet—A type of secondary structure of a polypeptide chain in which hydrogen bonding between components of the chain creates a zigzag structure.

biotin—A B vitamin involved as a coenzyme in some carboxylation reactions.

branched chain amino acids—Three essential, hydrophobic amino acids: leucine, isoleucine and valine.

brown adipose tissue (BAT)—A form of adipose tissue in which much of the reduction of oxygen to make water is not coupled to ADP phosphorylation to make ATP. Therefore most of the energy release appears as heat.

buffer—A system that can resist changes in the pH of a solution.

calmodulin—A ubiquitous calcium binding protein in cells, capable of binding four calcium ions.

calpain—A calcium-dependent cytoplasmic proteinase.

carboxy terminus—The end of a polypeptide chain with a free alpha carboxylate group.

carnitine—A small molecule that aids in transport of long chain fatty acids across the inner mitochondrial membrane.

catabolism—That part of metabolism in which larger molecules are broken down to smaller components and energy is released.

chaperonins—A class of proteins that is involved in posttranslational processing of polypeptide chains.

chemiosmotic hypothesis—The proposal currently accepted to describe how the free energy released during electron transport from reduced coenzymes to oxygen is coupled to ATP formation in oxidative phosphorylation.

chromatin—A complex of DNA, histone, and nonhistone proteins from which chromosomes emerge during cell division.

chromosome—A discrete structure observed as such during cell division, consisting of a long DNA molecule with associated proteins.

chylomicrons—Lipoproteins formed in intestinal cells and released into the blood stream from the lymphatic system. These contain lipids obtained from the diet and some proteins to maintain their particulate form.

citric acid cycle—*See* tricarboxylic acid cycle.

codon—Three adjacent nucleotides that specify a particular amino acid. We generally consider these to be in mRNA.

coenzyme—A molecule generally derived from a B vitamin that is essential to a particular enzyme in its catalytic role.

coenzyme A (CoA)—A particular coenzyme formed in part from the B vitamin pantothenic acid, used to carry acyl groups such as acetyl groups or fatty acyl groups from fatty acids.

coenzyme Q—Also known as ubiquinone—a small molecule involved in the electron transport chain, leading to ATP formation.

competitive inhibitor—An enzyme inhibitor that resembles the normal substrate and blocks the enzyme by reversibly binding and releasing from the active site. Its effects can be overcome by an excess of substrate.

complementary—Two bases or two nucleotide chains containing purine and pyrimidine bases that recognize each other by their unique structures and can form stable hydrogen bonds with each other.

complementary DNA (cDNA)—A DNA molecule formed when an mRNA is copied by reverse transcriptase to make a complementary DNA strand. Creation of a new complementary DNA strand produces double-stranded cDNA.

cooperativity—Property of some proteins with multiple subunits in which binding of a substrate or ligand to one subunit alters the affinity of the other subunits for the substrate or ligand.

Cori cycle—A cycle in which glucose is broken down to lactate in a muscle fiber. The lactate is released to the blood, taken up by the liver and converted back to glucose, then released again to be used to form lactate by muscle.

cristae—Infoldings of the inner mitochondrial membrane.

cross bridge—A part of the myosin molecule with ATPase and actin binding activity. During ATP hydrolysis, the myosin cross bridge goes through a cycle of attachment to actin, force generation, and then detachment.

crossover concept—The point during graded exercise when the contribution of fat to ATP production is matched by that of carbohydrate. At intensities beyond this, carbohydrate becomes increasingly more important.

cyclic AMP (cAMP)—Adenosine 3', 5'-monophosphate in which the phosphate group forms an ester bond with OH groups attached to both the 3' and 5' carbon atoms of the sugar ribose.

cytochromes—Heme-containing proteins in the electron transport chain that can be alternately in an oxidized or reduced state.

cytoplasmic energy state—The concentration ratio of ATP/ADP × Pi in the cytosol of a cell.

deamination—An enzymatic process in which an amino acid loses an amino group.

degenerate code—Having more than one codon specifying each amino acid.

degradation—Breakdown, such as changing a protein to its component amino acids.

denaturation—Disruption of secondary and tertiary structure of a biological polymer such as a protein. With this is a loss of normal activity.

deoxyribose triphosphates—Nucleoside triphosphates in which the ribose has a hydrogen atom as opposed to an OH group at carbon 2'. These are the precursors for making DNA.

dephosphorylation—Removal of a phosphate group from a molecule to which it is covalently attached.

diabetes mellitus—Type I or insulin dependent diabetes caused by destruction of the insulin producing and secreting beta cells of the pancreas.

diacylglycerol—Also known as a diglyceride—a glycerol molecule with two fatty acids attached via ester bonds.

disaccharide—A molecule composed of two monosaccharides joined together by a covalent bond.

disulfide bond—A covalent bond joining two sulfhydryl (SH) groups on cysteine side chains in the same or different polypeptide chains.

DNA polymerase—The enzyme involved in replicating nuclear DNA prior to cell division (mitosis).

domain—A distinct section of a protein molecule with a distinct three-dimensional structure and specific function.

downstream—For a gene, it represents the direction RNA polymerase moves during transcription. It is considered to be in the 3' direction.

electrochemical gradient—The energy required to separate a charge and concentration difference across a membrane.

electron transport chain—A group of four protein-lipid complexes in the inner mitochondrial membrane that transfer electrons from reduced coenzymes to oxygen.

endergonic reaction—A reaction or process in which free energy is required to make the reaction occur.

enhancers—Regulatory base sequences in DNA to which specific protein transcription factors may bind to influence the expression of a gene. Generally these lie a considerable distance from the gene they regulate.

enthalpy change—The total energy liberated in a reaction or process.

entropy—The degree of randomness or disorder in a system.

epinephrine—Hormone released from the adrenal medulla in response to physical stress.

equilibrium—The state of a system when there is no net change in energy.

equilibrium constant (K_{eq})—The ratio of the product of the concentrations of products of a reaction to the product of the concentrations of reactants when the system is in equilibrium.

essential amino acids—Amino acids that must be obtained in the diet because the body cannot synthesize them or cannot synthesize them in adequate amounts. Examples are leucine, isoleucine, valine, threonine, phenylalanine, histidine, tryptophan, lysine, and methionine.

essential fatty acids—Fatty acids needed by the body but which cannot be synthesized by the body in amounts sufficient to meet physiological needs. Examples are linoleic acid and linolenic acid.

estrogens—Female sex hormones that stimulate and maintain female secondary sex characteristics.

euglycemia—A normal range of blood glucose concentration.

exergonic reaction—A reaction that can proceed spontaneously with the release of free energy.

exon—Part of the primary transcript that is found in cytoplasm in mRNA, tRNA, and rRNA.

exothermic reaction—A reaction that liberates energy when it proceeds from reactants to products.

facilitated diffusion—Diffusion in which a substance crosses a membrane, down its concentration gradient, with the aid of a specific carrier.

FAD (flavin adenine dinucleotide)—An oxidation-reduction coenzyme that oxidizes a class of substrate molecules, transferring their electrons to coenzyme Q.

faraday constant (F)—The number 96.5 kilojoules per mole per volt, which allows one to convert a change in redox potential, expressed in volts, into units of energy.

feedback inhibition—Inhibition of an allosteric enzyme by a product of its reaction or a product of a pathway in which the enzyme participates.

FFA (free fatty acid)—A fatty acid bound noncovalently to albumin, circulating in the blood.

free energy change—The fraction of the total energy released in a reaction or process that can be used to do useful work.

G protein—A GTP binding membrane protein that participates in the process of signal transduction, linking the binding of an external ligand to its membrane receptor into action within the cell.

gene—The segment of DNA that provides the information for the sequence of amino acids in a polypeptide chain.

genome—The total genetic information carried by a cell.

glucagon—Polypeptide hormone secreted from the alpha cells of the pancreas in response to a decrease in blood glucose concentration.

glucogenic amino acids—Those amino acids whose carbon skeletons can be used to make glucose in the liver.

glucose transporters—A family of membrane proteins that transport glucose across cell membranes, down its concentration gradient.

glycogen—A large branched polymer formed entirely of glucose, which is the principle way of storing carbohydrate in liver and muscle.

glycogenesis—The formation of glycogen from glucose.

glycogenin—A self-glycosylating polypeptide that is the precursor for glycogen synthesis.

glycogenolysis—The process in which glycogen is broken down in liver and muscle to yield glucose 6-phosphate.

glycolysis—A catabolic pathway that breaks down glucose 6-phosphate derived from glucose or glycogen into lactate and in the process generates ATP.

growth factors—A family of extracellular polypeptides that can bind to specific receptors on cell membranes, influencing the growth and differentiation of that cell.

growth hormone—A polypeptide hormone synthesized in and secreted from the anterior portion of the pituitary gland.

half-life—The time taken for the disappearance or decay of one half of the population of chemical substances.

heat shock proteins—A family of proteins originally shown to be synthesized in response to cellular or organism stress. Among the members of this expanded family are the chaperonins, which participate in posttranslational processing of new polypeptides.

heme—A prosthetic group containing a central iron ion.

Henderson-Hasselbalch equation—The equation relating the solution pH, the pK_a, and the ratio of undissociated to dissociated acid concentrations.

histones—A family of nuclear proteins carrying net positive charges because of a preponderance of basic amino acids. These are associated with the polynucleotide backbone of DNA in structures known as nucleosomes.

holoenzyme—A catalytically active enzyme containing the polypeptide subunit and any necessary cofactors.

hormone receptor—A protein located in a cell membrane or within a cell that recognizes and binds a specific hormone.

hybridization—Association of two, complementary polynucleotide chains because of base pairing through hydrogen bonding between complementary bases.

hydride ion—A hydrogen atom with an additional electron so that it carries a negative charge.

hydrogen bond—A weak, noncovalent bond between a hydrogen atom covalently attached to an electronegative atom and another electronegative atom in the same or different molecule.

hydrophilic—Polar molecules or charged ions that can interact with the polar water molecule.

hydrophobic—Nonpolar groups or molecules that cannot interact with water.

hyperglycemia—Elevated blood glucose concentration.

hypoglycemia—When blood glucose concentration is very low.

inorganic phosphate—Ionized phosphate group that contains a net negative charge at neutral pH.

inorganic pyrophosphate—An anhydride containing two phosphate groups joined together that can be hydrolyzed by the enzyme inorganic pyrophosphatase.

insulin—A polypeptide hormone produced in and secreted from the beta cells of the pancreas in response to an elevation in blood glucose concentration.

intermembrane space—The space between the outer and inner mitochondrial membranes.

International Unit—The amount of enzyme that converts one micromole of substrate to product in one minute.

intron—Part of the primary transcript removed by splicing during RNA processing in the nucleus.

isoelectric point (pI)—The pH where a molecule or ion has no net charge but contains at least one positive and one negative charge.

isozymes (isoenzymes)—Different molecular forms of the same enzyme, catalyzing the same reaction with the same mechanism, but having different kinetic parameters.

ketogenic amino acids—The carbon skeletons of these amino acids cannot be converted into glucose in the liver. Two examples are lysine and leucine.

ketone bodies—Acetoacetate, 3-hydroxybutyrate, and acetone together are known as ketone bodies.

ketonemia—A condition of elevated blood ketone body concentration.

ketonuria—Loss of ketone bodies in the urine.

kinetic parameters—V_{max} and K_m are considered the defining parameters of an enzyme.

Krebs cycle—*See* tricarboxylic acid cycle.

ligand—A small molecule that binds to a specific larger one, such as a receptor.

Lineweaver-Burk plot—A transformation of the Michaelis-Menten equation in which reciprocals of substrate concentration and velocity are plotted, yielding straight lines, the intercepts of which can produce values for the kinetic parameters.

lipoic acid—An organic coenzyme made in the body that works with two mitochondrial dehydrogenase enzymes.

lipolysis—The enzymatic breakdown of triglyceride molecules, catalyzed by a class of enzymes known as lipases.

lipoproteins—A family of protein-lipid particles found in the blood.

long chain carboxylic acids—Fatty acids with 16–20 carbon atoms, which may be saturated or unsaturated.

MAP kinases—A family of protein kinases that respond to cell stimulation by mitogens.

mass action ratio—The ratio of the product of the concentration of the products of a reaction to the product of the concentration of reactants at any time during the course of a chemical reaction.

matrix—That part of the mitochondrion that is bounded by the inner membrane.

maximum velocity (V_{max})—The maximum velocity of an enzyme-catalyzed reaction.

medium chain triglycerides—Triglyceride molecules containing fatty acids with a carbon chain length of 6–10.

messenger RNA (mRNA)—A class of RNA molecules, complementary to a portion of a strand of DNA that provides the information for the synthesis of a polypeptide chain.

Michaelis constant (K_m)—The concentration of substrate needed to produce one half the maximum velocity for an enzyme-catalyzed reaction.

monocarboxylate transporter—A membrane protein that transports small anions down their concentration gradient. Usually refers to transport of lactate and pyruvate.

monosaccharides—Simple sugars that cannot be digested or broken down further without destroying their structure.

mutation—A change in the nucleotide sequence of DNA that can be passed on.

myogenic regulatory factor—A class of transcription factors that, when bound to their cognate cis elements, regulate the expression of genes specific for development and maintenance of muscle properties.

myosin—The principal protein of the thick filament of muscle fibers. It has ATPase and actin binding activity.

myosin heavy chain—The largest subunit in the oligomeric protein myosin.

NAD^+ (nicotinamide adenine dinucleotide)—The oxidized form of the more common coenzyme involved in redox reactions in the cell.

NADH—Reduced form of NAD^+. Transfers its electrons to oxygen in the electron transport chain.

$NADP^+$ (nicotinamide adenine dinucleotide phosphate)—The coenzyme involved in redox reactions in which larger molecules are built out of smaller components. The synthesis of fatty acids uses the reduced form of $NADP^+$.

NFAT (NF-AT) (nuclear factor of activated T cells)—A transcription factor first identified in T lymphocytes but now known to operate in other cells.

nitrogen balance—The relationship between nitrogen intake into the body in the form of food protein and nitrogen loss from the body in urine, feces, and sweat.

noncompetitive inhibitor—Irreversible inhibition of an enzyme that causes loss of enzyme function. This kind of inhibition decreases V_{max} but has no effect on K_m.

norepinephrine—Also known as noradrenaline, it is the chief neurotransmitter in the sympathetic nervous system.

nucleases—Enzymes that hydrolyze the phosphodiester backbone of nucleic acids.

nucleoside—Product produced when a purine or pyrimidine base is attached to a ribose sugar.

nucleoside triphosphate—A sugar plus base (nucleoside) with three attached phosphate groups.

nucleosome—A basic structural component of chromatin consisting of a DNA strand wound around a core of histone proteins.

nucleotide—A nucleoside with one or more attached phosphate groups.

oligomer—A small polymer, usually of nucleotides (oligonucleotide), amino acids (oligopeptide), or monosaccharides (oligosaccharide).

open reading frame—A sequence of nucleotide codons in mRNA that code for amino acids, but there are no stop codons present.

osmotic pressure—The pressure that must be applied to a solution to prevent solvent (water) from flowing into it.

oxidation—A chemical reaction in which a substrate loses electrons.

oxidative phosphorylation—The formation of ATP from ADP and Pi in association with the transfer of electrons from substrate to coenzymes to oxygen.

oxygen consumption ($\dot{V}O_2$)—The utilization of oxygen to accept electrons in oxidative phosphorylation; volume consumed per minute.

P/O ratio—The ratio of the number of ATP molecules produced per pair of electrons transferred from a substrate to oxygen to reduce the oxygen.

pancreatic lipase—A lipase produced in and secreted by the pancreas. Its function is to break down dietary triglyceride.

pantothenic acid—A B vitamin that forms a major part of coenzyme A.

pentose phosphate pathway—A pathway to interconvert hexoses and pentoses that is a source of NADPH for synthetic reactions. It is also called the phosphogluconate pathway.

peptide bond—A specialized form of amide bond that joins two amino acids together.

phosphagen—A molecule such as phosphocreatine that can transfer its phosphate group to ADP to make ATP.

phosphate ester—Bond formed between an alcohol OH group and phosphoric acid.

phosphatidic acid—A diglyceride with a phosphate group attached in ester linkage to the third OH of glycerol.

phosphatidyl inositols—A family of phospholipids in which inositol or phosphorylated derivatives are attached to phosphatidic acid.

phosphocreatine—A phosphagen capable of rapidly transferring its phosphate group to ADP to quickly regenerate ATP.

phospholipids—A lipid containing one or more phosphate groups.

phosphoprotein phosphatase—An enzyme that removes a phosphate group attached in ester linkage to a protein.

phosphorylation—Covalent attachment of a phosphate group to a molecule.

phosphorylation potential—*See* cytoplasmic energy state.

polypeptide—A molecule composed of 20 or more amino acids joined together by peptide bonds.

polyribosome—A mRNA molecule with two or more ribosomes carrying out protein synthesis on it.

primary structure—The sequence of amino acids in a peptide or protein, starting from the N-terminus.

primary transcript—The initial product of transcription.

promoter—The region upstream of the start site of a gene to which transcription factors can bind to influence the initiation of transcription.

prosthetic group—A cofactor (metal ion or organic compound) that is covalently attached to a protein and is necessary for that protein to function.

proteasome—A cylindrical-shaped protein complex responsible for much of the degradation of cellular proteins.

protein balance—The relationship between protein synthesis and protein degradation, which affects the net content of protein in an organism. This is often inferred from nitrogen balance.

protein kinase—An enzyme that attaches a phosphate group from ATP to the side chain of a serine, threonine, or tyrosine residue in the protein.

purine—A nitrogen-containing heterocyclic base found in nucleic acids. Principal examples are adenine and guanine.

pyrimidine—A nitrogen-containing heterocyclic base found in nucleic acids. Principal examples are thymine, cytosine, and uracil.

quaternary structure—The three-dimensional organization of a protein that is composed of subunits, dealing with the way the subunits interact with each other.

rate-limiting—That step in a reaction or pathway that limits the overall rate of the reaction or pathway.

redox potentials—The tendency of a substance to be reduced, measured in volts.

redox reaction—A reaction in which electrons are transferred from one substrate (oxidation) to another substrate (reduction).

reducing equivalents—A pair of electrons that can be used to reduce something. Generally transferred as a hydride ion or pair of hydrogen atoms.

reduction—What occurs when a substrate gains one or more electrons.

repressor—A protein that binds to a cis regulatory element associated with a gene and blocks its transcription.

respiratory exchange ratio (RER)—The ratio of carbon dioxide produced to oxygen consumed as measured by collecting expired air at the mouth of an individual.

respiratory quotient (RQ)—The ratio of carbon dioxide produced to oxygen consumed as measured in an isolated system or across a tissue or organ.

retrovirus—A virus whose genome is RNA rather than DNA. These viruses have a reverse transcriptase enzyme that allows them to replicate their genome as DNA in a host.

reverse transcriptase—The enzyme found in retroviruses that allows them to make a DNA copy from RNA.

ribonuclease—A class of enzymes that can hydrolyze the sugar-phosphate backbone of RNA.

RNA polymerase—The enzyme that transcribes a section of DNA producing an RNA molecule complementary to the copied strand of DNA.

sarcolemma—The cell membrane of a muscle cell or fiber.

sarcoplasmic reticulum—A specialized form of endoplasmic reticulum found in striated muscle that holds, releases, and takes up calcium.

second messenger—A molecule or ion produced in or released from the interior of a cell in response to the binding of a ligand (first messenger) to a cell membrane receptor.

secondary structure—A regular, repeating structure taken by the polypeptide backbone of a polypeptide. Examples are the alpha-helix and the beta-sheet.

semiconservative—Process that describes the replication of DNA in which each strand acts as a template upon which a complementary strand is produced.

signal transduction—The process through which an extracellular signal is amplified and converted to a response within the cell.

small nuclear RNA (snRNA)—Small RNA molecules in the nucleus that play a role in RNA processing, such as splicing out introns.

sodium-calcium antiport—A membrane protein that transports calcium against its gradient using the energy produced when sodium flows down its gradient.

stable isotopes—Isotopes of common elements that are not radioactive.

standard free energy change ($\Delta G^{\circ\prime}$)—The free energy change for a reaction under standard conditions at pH 7.0 and with all reactants and products at one molar concentration.

standard redox potential ($\Delta E^{\circ\prime}$)—The tendency for a substance to be reduced, measured in volts and based on a standard value.

start codon—The codon AUG, which signals the start of translation; also the codon for methionine.

start site—The nucleotide on the sense strand of DNA that indicates the beginning of transcription.

state four—The state of oxidative phosphorylation that is limited by the availability of ADP.

state three—The state of oxidative phosphorylation when all substrates and components are present in adequate amounts.

steady state—A nonequilibrium condition of a system in which there is a constant flux, yet the intermediates within the system are at constant concentration.

stereoisomers—Isomers that have the same composition of components, yet have different molecular arrangements.

steroid hormones—A class of hormones based on the structure of their precursor, cholesterol.

stop (termination) codons—Three codons, UAG, UAA, and UGA, which signal the end to translation.

substrate—The specific compound that is chemically modified by an enzyme.

substrate-level phosphorylation—Process whereby ADP is converted into ATP by any mechanism other than oxidative phosphorylation.

TATA box—A cis element located about 25–35 base pairs upstream of the start of many eukaryotic genes that helps to locate RNA polymerase.

template strand—The strand of DNA that actually is the strand copied in a complementary way by RNA polymerase.

thiamine pyrophosphate (TPP)—A coenzyme involved in several mitochondrial oxidative decarboxylation reactions.

thymine—A pyrimidine base found in DNA.

total adenine nucleotide (TAN)—The sum of the concentrations of ATP, ADP, and AMP.

trans (transcription) factor—A protein transcription factor that binds to a cis element near a gene and increases the frequency of initiation of transcription.

transamination—Transferring an amino group from an amino acid to a keto acid to make a new amino acid and a new keto acid.

transcription—The enzymatic reaction whereby a segment of a DNA strand is copied by making RNA.

translation—The process whereby the sequence of codons on an mRNA molecule is used to direct the sequence of amino acids in a polypeptide chain.

triacylglycerols (triglycerides)—A triester consisting of the alcohol glycerol and three fatty acid molecules.

tricarboxylic acid (TCA or citric acid or Krebs) cycle—A sequence of reactions taking place in mitochondria where acetyl units attached to CoA are degraded to carbon dioxide, and the electrons produced are transferred to the coenzymes NAD^+ and FAD.

troponin—A calcium-binding protein in the thin filament of striated muscle.

T-tubules—Invaginations in the sarcolemma of striated muscle.

ubiquinol (reduced coenzyme Q or QH_2)—The reduced form of an electron carrier in the mitochondrial inner membrane.

ubiquinone (oxidized coenzyme Q or Q)—The oxidized form of an electron carrier in the mitochondrial inner membrane.

ubiquitin—Small protein found in all cells that becomes covalently attached to proteins destined for degradation by the proteasome complex.

uncoupled electron transport—Electron transport from reduced coenzymes to oxygen that is not associated with ADP phosphorylation to make ATP.

uncoupling protein—Inner membrane protein that allows protons to flow down their electrochemical gradient without ADP phosphorylation.

uniporter—A membrane protein that allows a polar molecule or ion to flow across the membrane down a concentration gradient.

upstream—Term to describe a direction with respect to a gene, that is, in the 5' direction, based on the sense strand of DNA.

uracil—A pyrimidine base found in RNA.

urea—Small molecule produced in the liver in the urea cycle and the major form in which the body expels nitrogen groups from amino acids.

urea cycle—A sequence of reactions in the liver in which urea is synthesized.

very low density lipoproteins (VLDL)—A triglyceride-rich lipoprotein fraction, synthesized in liver.

wobble—The relatively loose base pairing between the third (3') base in a codon and the first base (5') of an anticodon.

zwitterion—A molecule with one positive and one negative charge.

Index

Note: Page numbers followed by t or f refer to the table or figure on that page.

F

G

H

About the Author

Michael Houston, PhD, completed his doctorate in 1969 before joining the department of kinesiology at the University of Waterloo, where he taught biochemistry to exercise science students for more than 25 years. In addition to teaching biochemistry, he has taught a graduate course in exercise metabolism. One of his major teaching interests is to prepare exercise physiology students for the molecular developments at the forefront of the exercise physiology field.

Dr. Houston currently is professor and head of the department of human nutrition, foods and exercise at Virginia Polytechnic Institute and State University in Blacksburg, Virginia. He is former president of the Canadian Society for Exercise Physiology and a member of the American College of Sports Medicine. He has published more than 100 articles in prominent scholarly journals on the integration of biochemical and physiological mechanisms in the exercising muscle.